Fortschritte der Chemie organischer Naturstoffe

Progress in the Chemistry of Organic Natural Products

51

Founded by L. Zechmeister
Edited by W. Herz, H. Grisebach, G.W. Kirby,
and Ch. Tamm

Authors:
M. Gill, W. Steglich

Springer-Verlag
Wien New York 1987

Dr. W. Herz, Professor of Chemistry, Department of Chemistry,
The Florida State University, Tallahassee, Florida, U.S.A.

Prof. Dr. H. Grisebach, Biologisches Institut II, Lehrstuhl für Biochemie der Pflanzen,
Albert-Ludwigs-Universität, Freiburg i. Br., Federal Republic of Germany

G.W. Kirby, Sc. D., Regius Professor of Chemistry, Chemistry Department,
The University, Glasgow, Scotland

Prof. Dr. Ch. Tamm, Institut für Organische Chemie der Universität Basel,
Basel, Switzerland

© 1987 by Springer-Verlag/Wien
Softcover reprint of the hardcover 1st edition 1987

Library of Congress Catalog Card Number AC 39-1015

ISSN 0071-7886

ISBN-13:978-3-7091-7456-2   e-ISBN-13:978-3-7091-6971-1
DOI: 10.1007/978-3-7091-6971-1

Contents

List of Contributors

GILL, Dr. M., Department of Organic Chemistry, University of Melbourne, Parkville, 3052, Victoria, Australia.

STEGLICH, Professor Dr. W., Institut für Organische Chemie und Biochemie der Universität Bonn, Gerhard-Domagk-Strasse 1, D-5300 Bonn 1, Federal Republic of Germany.

Pigments of Fungi (Macromycetes)

By M. Gill, Department of Organic Chemistry,
University of Melbourne, Parkville, Victoria, Australia

and

W. Steglich, Institut für Organische Chemie und Biochemie
der Universität Bonn, Bonn, Federal Republic of Germany

Contents

1. Introduction

The Macromycetes exhibit a great variety of colours and colour changes which have traditionally been used by mycologists as taxonomic characters. The scientific bases of these phenomena have attracted the attention of organic chemists for over a century, the beginnings of chemical investigations of the pigments of mushrooms and toadstools being readily identified with the works of BACHMANN (57, 58),

STAHLSCHMIDT (*591, 592*), THÖRNER (*653, 654*), ZELLNER (*726*), and
ZOPF (*732*). Much of the chemistry which emerged from these pioneering studies has been reviewed in the older literature (*112, 174, 363,
424, 519*). The elucidation of the structures of polyporic acid and atromentin by KÖGL (*423, 425, 426, 431, 433*) represented a significant
advance in this field but later work by this author, principally on 'boletol' (*427, 428*), 'muscarufin' (*429*) and the structure of thelephoric
acid (*430*), has unfortunately not withstood the test of time and mean-
while has caused considerable confusion.

The pigments of Higher Fungi have been reviewed more recently
by EUGSTER (*222*), and by STEGLICH (*597, 598*), and comprehensive
lists of fungal metabolites, including pigments, have from time to time
appeared (*485, 578, 667, 668*). Quinonoid pigments of fungal origin
form an integral part of THOMSON's excellent treatise on natural quinonoid colouring matters (*655*).

An area of interdisciplinary research which is still continuing to
increase in importance is the use of specific pigments or groups of
pigments as taxonomic markers in fungi. GABRIEL (*265*) was the first
to employ chromatography as an aid to the chemotaxonomy of macromycetes and the subsequent application and expansion of her ideas,
in particular by the Lyon school and the groups led by BRESINSKY
and by MOSER, among others, has clearly demonstrated the significance
of pigments in fungal systematics (*40, 45, 75, 122, 123, 444, 445, 483,
499, 669*). The value of chemotaxonomic data to the mycologist has
increased as the chemical structures of the individual pigments have
become known, while reciprocal benefit has been derived by organic
chemists who have been guided by taxonomic studies to new and fruit-
ful areas for structural investigations.

The study of the chemistry of mushroom pigments has gained considerable impetus during the past fifteen years from the rapid development of more efficient methods for purification and the availability
of increasingly sophisticated techniques for structural elucidation.
Thus, we are now able to address and overcome problems such as
the instability of many pigments and their frequent occurrence in complex mixtures which frustrated and confused earlier workers. Consequently, many colouring matters have been isolated from macromycetes
and their chemistry delineated in recent years.

In reviewing the chemical, biological and mycological literature to
the early part of 1986 we have endeavoured to restrict our attention
to those fungi which, to the observer in the field, are recognised as
forming conspicuous fruit bodies. We have included in this group the
slime moulds (Myxomycetes) since these intriguing organisms are com-
monly bracketed with the true fungi in the popular literature. Pigments

elaborated by macromycetes growing in mycelial culture are included in those cases which provide additional chemical interest or taxonomic information. Included are also some colourless secondary metabolites where these compounds bear an obvious and important relationship to the fungus pigments themselves. We have not dealt with the extensive chemistry of pigments from microfungi although some information in this regard may be gleaned by reference to the various compendia and reviews of fungal metabolites which are cited in the bibliography.

Many of the pigments of fungi belonging to Macromycetes possess structures which betray their likely biosynthetic origins. The pigments discussed here have been arranged according to their probable biogenesis following Turner's classification (667, 668), but it should be realised that in only a few cases have biosynthetic pathways been firmly established.

A concluding section deals briefly with those reactions between fungal sporophores and chemical reagents which produce colour changes in the fruit body.

The naming of fungi follows in the main the works of Moser (498), Jülich (396) and Singer (582) for Basidiomycetes, Dennis (183) for Ascomycetes, and Martin and Alexopoulos (473) for Myxomycetes.

2. Pigments from the Shikimate-Chorismate Pathway

Many fungi have developed pathways by way of which the aromatic products of shikimate metabolism may be further elaborated to pigments (659, 660, 661, 662, 702). The initial steps in the biogenesis of these pigments are the well known reactions of primary metabolism which lead from shikimate to chorismate and thence to arylpyruvic acids, aromatic amino acids (e.g. phenylalanine and tyrosine), hydroxycinnamic acids and p-hydroxybenzoic acid (243, 347). Pigments arising from each of these four precursor groups are now discussed in turn. Compounds of shikimate origin which are derived via anthranilic acid and p-aminobenzoic acid are considered in Sections 5 and 6, respectively.

2.1. Compounds Derived from Arylpyruvic Acids

2.1.1. Grevillins

The grevillins are a unique group of pyrandione pigments present in fruit bodies of boletes belonging to the genus Suillus (Table 1). They may be easily detected in chromatograms of the extracts of these mush-

rooms by their colour reactions with concentrated sulphuric acid: grevillin-A (1) (red-violet), grevillins-B (2), -C (3) and -D (4) (violet), and
anhydrogrevillin-D (5) (carmine) (461, 463, 603). The structures were
deduced from spectroscopic data (101, 461, 463, 603).

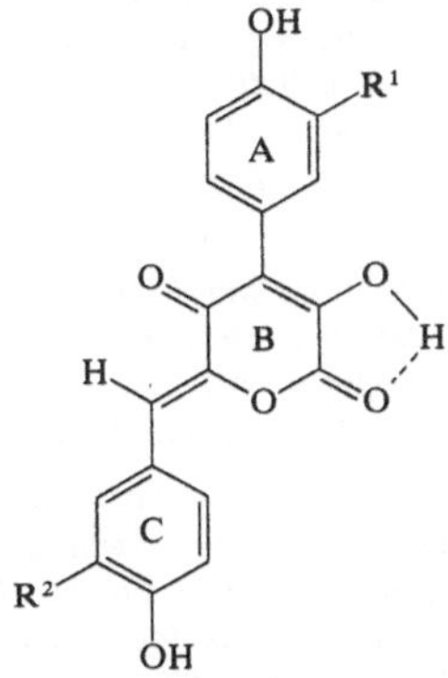

	R¹	R²
(1)	H	H
(2)	H	OH
(3)	OH	OH

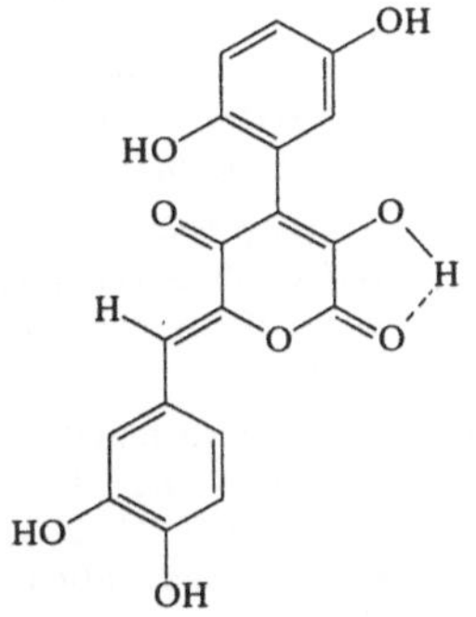

(4)

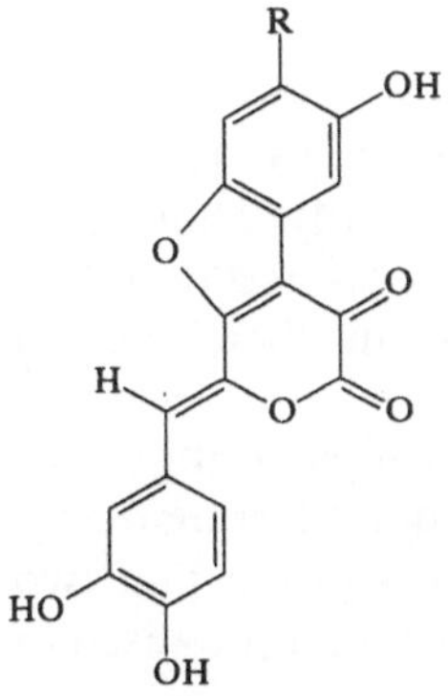

(5) R = H
(6) R = OH

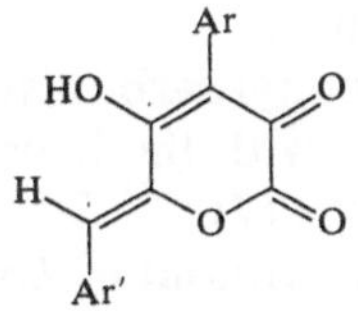

(7)

The pattern of hydroxylation in the aromatic rings of the grevillins
may be determined by ^{1}H-n.m.r. spectroscopy. Due to restricted rotation, the protons of ring A experience less deshielding by the pyrandione nucleus than the corresponding protons in ring C and consequently appear consistently at higher field (461). The benzylidene meth-
ine proton gives rise to a characteristic singlet near δ 6.9.

Table 1. *Occurrence of Grevillins*[a]

Grevillin-A (**1**)	*Suillus grevillei* [2×10^{-4}%] (*603*); TLC: *S. aeruginascens* var. *bresadolae* (= *S. bresadolae*) (*92*), *S. americanus, S. luteus, S. placidus* (*87*), *S. punctatipes* (*127*)
Grevillin-B (**2**)	*Suillus granulatus* [5×10^{-4}%] (*101*), *S. grevillei* [1.4×10^{-2}%] (*603*), *S. grevillei* var. *badius* (*204*), *S. luteus, S. placidus* (*101*), *S. tridentinus* (*97*); TLC: *Gastroboletus laricinus* (*87*), *Suillus aeruginascens, S. aeruginascens* var. *bresadolae* (*92*), *S. amabilis* (= *S. lakei*), *S. americanus, S. brevipes* (*127*), *S. collinitus* (*87*), *S. pictus* (*448*), *S. punctatipes* (*127*), *S. sibiricus* (*87*)
Grevillin-C (**3**)	*Suillus grevillei* [7×10^{-3}%] (*603*), *S. tridentinus* (*97*); TLC: *S. aeruginascens* (*127*), *S. aeruginascens* var. *bresadolae* (*92*), *S. amabilis, S. americanus* (*127*), *S. collinitus* (*87*), *S. granulatus* (*86*), *S. luteus,* (*87*), *S. pictus* (*448*), *S. placidus* (*87*), *S. punctatipes* (*127*), *S. sibiricus* (*87*)
Grevillin-D (**4**)	Widely distributed in *Suillus* Subsect. *Latiporini* and *Angustiporini* (*122, 127, 132*): e.g. *S. americanus* (*127, 622*), *S. granulatus* [5×10^{-3}%], *S. placidus* (*101*); TLC: *S. acidus, S. albidipes, S. brevipes* (*127*), *S. collinitus, S. flavidus* (*92*), *S. leptopus* (*86*), *S. pictus* (*448*), *S. pseudobrevipes, S. punctatipes, S. riparius* (*127*), *S. umbonatus* (*87*)
Anhydro-grevillin-D (**5**)	*Suillus americanus* (*622*), *S. granulatus* (*101*). Also detected in most fungi which produce (**4**), probably as an artefact formed during the isolation procedure

[a] Yields refer to percentage of fresh weight of fungus.

The ^{13}C-n.m.r. data for the grevillins-A to -D (*460, 461*) are in full accord with the 5-oxo structures (**1**) to (**4**). Thus, the signal from the ketonic carbonyl carbon at δ 176.7 appears as a doublet ($J = 2.7$ Hz) due to coupling with the benzylidene proton, thereby excluding 3-oxo tautomers of type (**7**).

All grevillins without *ortho*-hydroxy groups in ring A exhibit typical fragmentation patterns in the mass spectrum. Starting from the molecular ion, the consecutive loss of a carboxyl group and two molecules of carbon monoxide is observed and has been rationalised by the mechanism shown in Scheme 1 (*461, 463*).

Grevillin-D (**4**) shows an abundant $[M - H_2O]^+$ ion due to the formation of an ether bridge (*101*). The ease with which water is eliminated from the pigment (**4**) is also demonstrated by the formation of anhydrogrevillin-D (**5**) from (**4**) merely on standing in organic solvents in the presence of traces of acid (*101*). The structure (**5**) of anhydrogrevillin-D was established by analysis of the fully coupled ^{13}C-n.m.r. spectrum of the pigment (*461*). A dehydro derivative has also been obtained from grevillin-C (**3**) on standing. This was tentatively assigned the structure (**6**) which could arise by oxidation of ring

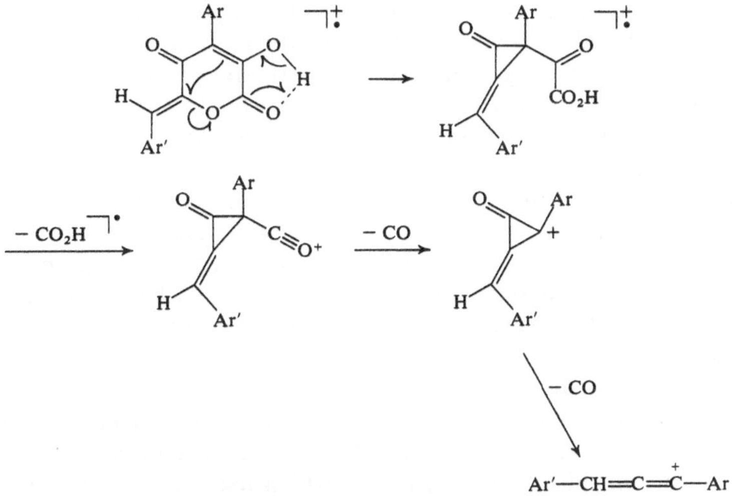

Scheme 1. Fragmentation of grevillins on electron impact

A in grevillin-C followed by conjugate addition of the hydroxy group to the resulting *ortho*-quinone (*599*).

Some properties of the grevillins are given in Table 2. It will be appreciated that with many colouring matters the colour of a pure pigment is not necessarily reflected in the colour of the substance either in solution or on a chromatography plate. Colours quoted in tables refer to the pure pigments unless stated otherwise.

Pigment-B$_1$ isolated from *Suillus grevillei* and from *S. grevillei* var. *badius* by EDWARDS and GILL (*204*) is identical with grevillin-B (**2**). These workers deduced the 4-(4-hydroxyphenyl)-6-(3,4-dihydroxyben-

Table 2. *Properties of Grevillins*

Pigment	Colour	TLC R$_f$[a]	Ultraviolet/Vis. [λ$_{max}$ (log ε), methanol]	Peracetyl Derivative M.p.	References
(**1**)	Orange	0.49	268 (4.27), 389 nm (4.23)	197–199°	(*461*)
(**2**)	Orange	0.42	280 (4.20), 396 nm (4.20)	184–186°	(*204, 461*)
(**3**)	Red-orange	0.23	287.3 (4.24), 405.6 nm (4.23)	175–178°	(*461, 603*)
(**4**)	Red-orange	0.20	287.6 (4.14), 408.9 nm (4.18)	–	(*461*)
(**5**)	Dark red	0.44	246.2 (4.13), 277.8 (3.92), 345 (sh., 3.67), 439.4 nm (4.35)	–	(*461*)

[a] Silica gel Merck 60 F$_{254}$ (benzene:ethyl formate:formic acid = 10:5:3).

zylidene)pyrandione structure from spectroscopic data and from the isolation of 4-hydroxy- and 3,4-dihydroxybenzoic acid after chromium trioxide oxidation of the tetra-acetate derivative of the pigment, but they invoked a tautomeric structure of type (7) to explain differences between the infrared spectrum of their pigment (v_{CO} 1736 cm^{-1}) and grevillin-B (2) (v_{CO} 1705–1715 cm^{-1}) (603). It is now clear that solvate formation is responsible for these differences since further experience with natural and synthetic grevillins has demonstrated significant variations in the positions of infrared bands in different preparations (460). This inconsistency makes comparison of grevillins based on infrared spectra, as in other cases of polyphenolic compounds, an unreliable method of identification.

The grevillins are easily converted into crystalline yellow peracetyl derivatives with acetic anhydride and a trace of concentrated sulphuric acid (Table 2) (204, 461, 603). On catalytic hydrogenation the grevillins yield colourless dihydro derivatives in which reduction of the exocyclic double bond has taken place (603). Conjugate addition of nucleophiles to the exocyclic enone system leading to colourless by-products has also been observed during acetylation of these pigments (101) and in their reaction with diazomethane in methanol (276, 461).

The structures of the grevillins (1)–(4) have been confirmed by a convergent synthesis (Scheme 2) in which the key step is the condensation of an aromatic aldehyde with a pyrandione of type (8) (461, 463). In this manner both the natural products and a variety of analogues have been prepared. By using an aromatic aldehyde incorporating a [^{13}CHO] label, grevillins specifically labelled at the benzylidene carbon may be conveniently obtained (see Section 2.1.2).

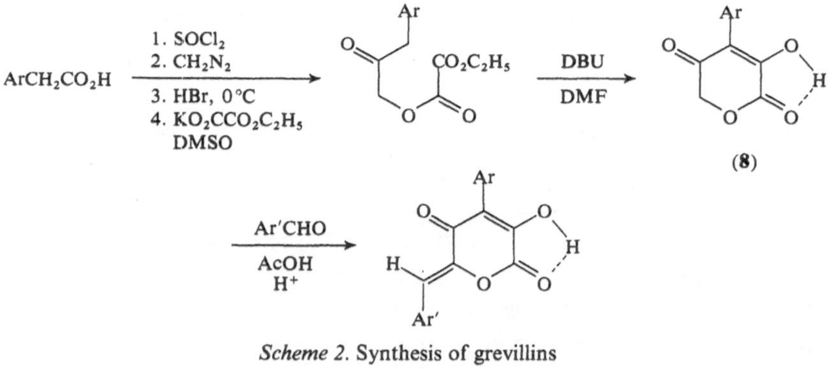

Scheme 2. Synthesis of grevillins

A possible biosynthesis of the grevillins from p-hydroxyphenylpyruvate (≡tyrosine) is depicted in Scheme 3. Preliminary feeding experiments have made it clear that DL-phenyl[3-^{13}C]alanine is not incorporated into the grevillins when given to fruit bodies of *Suillus grevillei* (*366*). In contrast, according to mass spectrometry, DL-[3'-^{13}C]tyrosine is efficiently incorporated into grevillin-A (12.8 atomic % excess) and grevillin-B (9.6%) by this mushroom. The conversion of ^{13}C-labelled grevillin-A (**1**) into grevillin-B (**2**) (29.4%) indicates that further hydroxylation of the aromatic ring in (**1**) occurs after formation of the pyrandione nucleus. DL-[3'-^{13}C]Tyrosine is also incorporated into grevillin-D (**4**) (2.5%) by fruit bodies of *S. luteus* (*366*). To explain the formation of the unusual 2,5-dihydroxylation pattern in (**4**) an oxidative rearrangement similar to the well known *p*-hydroxyphenylpyruvic acid → homogentisic acid rearrangement (*347*) is suggested.

The close connection between the grevillins and the terphenylquinone → pulvinic acid pathway (see Sections 2.1.2 and 2.1.3) is evident from the co-occurrence of pigments from both groups in sporocarps of some *Suillus* species. For example, *S. aeruginascens* contains the grevillins-B and -C, variegatorubin and thelephoric acid, *S. amabilis* (= *S. lakei*) contains the same two grevillins together with variegatic acid, while *S. grevillei* produces the grevillins-A, -B, and -C together with cyclovariegatin. BRESINSKY and coworkers (*86, 122*) reported the interesting finding that many *Suillus* species (e.g. *S. flavidus*, *S. grevillei*, *S. placidus* and *S. tridentinus*) produce xerocomic acid instead of the grevillins when grown in mycelial culture. Even *S. granulatus* and *S. luteus* which produce grevillin pigments in fresh mycelial culture change their metabolism in favour of xerocomic acid production on ageing. Obviously the mode of cyclisation of the intermediate (**9**) (Scheme 3, C-C versus C-O bond formation) ultimately determines which type of pigment is formed.

Closely related to grevillin-B (**2**) is pigment-A which was isolated from *S. grevillei* and *S. grevillei* var. *badius* after precipitation of the phenolics as lead salts and subsequent decomposition of the latter with methanolic hydrogen chloride (*204*). Pigment-A forms red needles [UV/ vis. (ethanol): λ_{max} (log ε) = 257 (4.08), 273 (infl., 4.03) and 444 nm (4.21)] which yield a cherry red colouration with concentrated sulphuric acid and a blue, changing to red-violet, colour when treated with dilute aqueous sodium hydroxide. From its spectroscopic data and the oxidation of its triacetyl derivative with chromium trioxide, which gave the acetates of 4-hydroxy- and 3,4-dihydroxybenzoic acid, structure (**10**) was derived for this pigment. The 3(2*H*)-furanone nucleus of the pigment could conceivably arise from the pivotal intermediate (**9**) of grevillin and terphenylquinone biosynthesis by dehydration involving a mode of C-O bond formation alternative to that which produces the grevillins. However, when the pigments from *S. grevillei* were isolated by

(9)

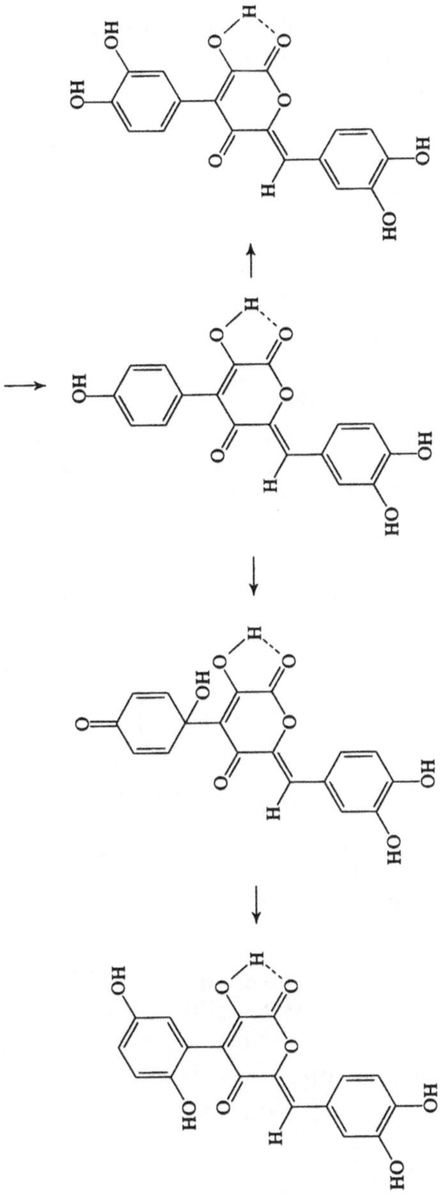

Grevillin-D

Grevillin-B

Grevillin-C

Scheme 3. Possible biosynthesis of grevillins

mild chromatographic methods only the grevillins (1)–(3) were obtained (603) and it is possible that the pigment (10) is an artefact formed from grevillin-B (2) during the lead acetate-methanolic hydrogen chloride treatment.

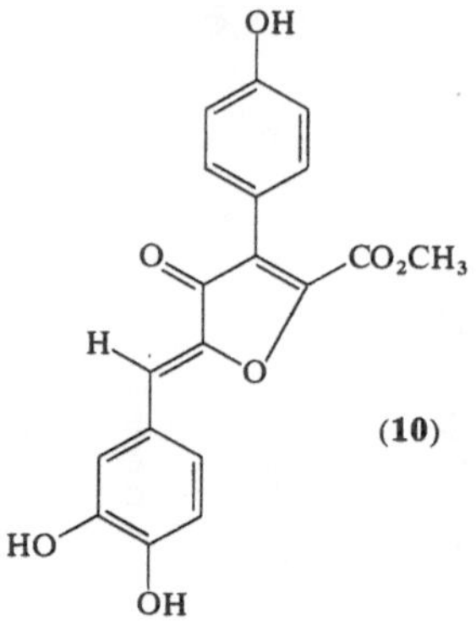

(10)

2.1.2. Terphenylquinones

A. Simple Terphenylquinones

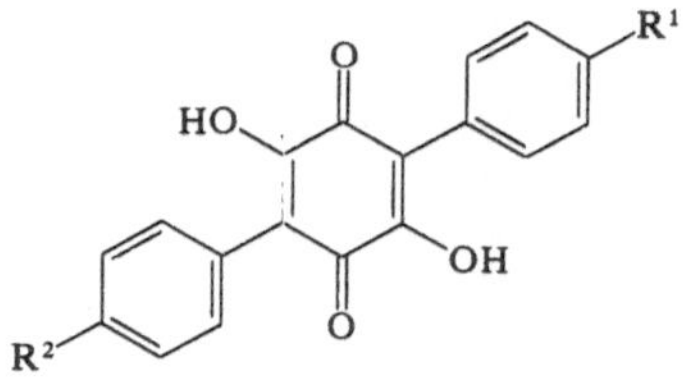

	R¹	R²
(11)	H	H
(12)	OH	H
(13)	OH	OH

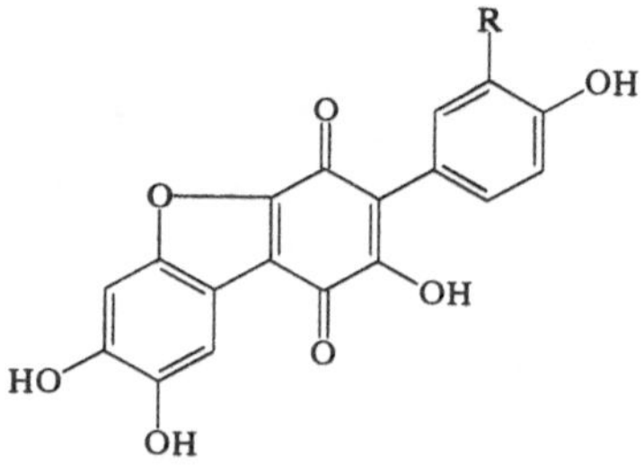

(14) R = H
(15) R = OH

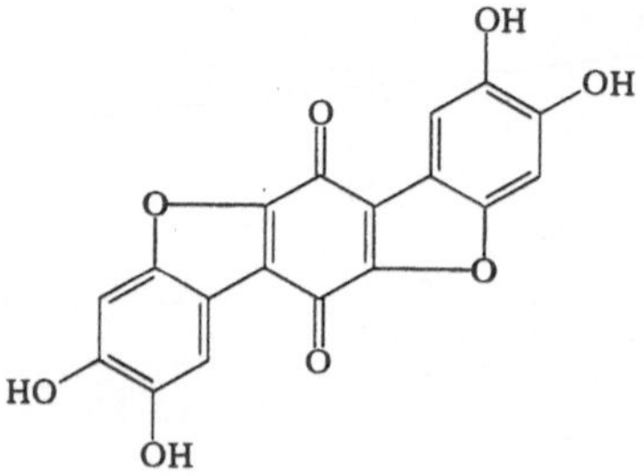

(16)

The isolation of polyporic acid (11) [STAHLSCHMIDT, 1877 (*591*, *592*)], atromentin (13) [THÖRNER, 1878 (*653*)] and thelephoric acid (16) [ZOPF, 1889 (*730*)] marked the start of the chemical investigation of fungal pigments. The structure elucidation and chemistry of these compounds has been amply covered in THOMSON's book (*655*) and does not need to be repeated here. In recent years several papers dealing with ultraviolet (*311*), infrared (*208*) and mass spectra (*304, 465*) and with the chromatography (*267*) and detection (*274*) of simple terphenylquinones have been published. Details of their known distribution in fungi are given in Table 3 and their physical properties are summarised in Table 4.

Table 3. *Occurrence of Simple Terphenylquinones*

Polyporic acid (11)	*Hapalopilus rutilans* (= *Polyporus nidulans* = *P. rutilans*) [43.5%] (*591*), (*63*), [18%] (*423*), [23%] (*248*), [cultures] (*495*), *Lopharia papyracea* [11.4%] (*395*), *Phanerochaete filamentosa* (= *Peniophora filamentosa*) [5%] (*501*)
Ascocorynin (12)	*Ascocoryne sarcoides* [c] (*534*)
Atromentin (13)	*Albatrellus cristatus* (*87*), *Hydnellum aurantiacum* (= *Hydnum aurantiacum*), *H. auratile*, *H. caeruleum* (*132*), *H. ferrugineum* (*321*), *H. peckii* (= *H. diabolus*) [1.43%] (*226*), (*409*), *H. suaveolens* (*132*), *Lampteromyces japonicus* [c] (*397*), *Leccinum eximium* (*127*), *Leucogyrophana olivascens* (*89*), *Omphalotus olearius* [c] (*397*), *O. subilludens* (= *Clitocybe subilludens*) [0.18%] (*638*), [c] (*637*), *Paxillus atrotomentosus* [1.4%] (*654*), [2%] (*273, 431*), [3.6%] (*80*), (*433, 653*), [c] (*397*), *P. involutus* [c] (*132*), *P. panuoides* [1.2%] (*273*), [c] (*397*), *Rhizopogon roseolus* [c] (*132*), *Suillus bovinus* [1.1 × 10⁻³%][a] (*70*), *Xerocomus subtomentosus* [c] (*122*)
Cycloleucomelone (14)	*Anthracophyllum archeri*, *A. discolor* [artefact?] (*385*), *Boletopsis leucomelaena* (= *Polyporus leucomelas*) ['leucomelone'] (*7, 386*), *Paxillus atrotomentosus* (*365*)
Cyclovariegatin (15)	*Suillus grevillei* [flesh], *S. grevillei* var. *badius* [cap skin?] (*204, 205*)
Thelephoric acid (16)	Characteristic pigment of the Thelephoraceae (*132*): *Bankera fuligineo-alba*, *B. violascens* (*132*), *Boletopsis leucomelaena* (*132, 386*), *Hydnellum aurantiacum* (*310*), *H. auratile* (*132*), *H. caeruleum* (*636*), *H. compactum* (*132*), *H. concrescens* (= *H. zonatum*) (*636*), *H. cruentum* (*132*), *H. ferrugineum* (= *Hydnum ferrugineum*) (*132, 727*), *H. mirabile* (*132*), *H. peckii*, *H. scrobiculatum* (*636*), *H. suaveolens* (= *Hydnum suaveolens*) [0.6%] (*313*), (*132, 636*), *H. velutinum*, *Lenzitopsis oxycedri* (*132*), *Phaeodon amarescens* (= *Hydnum amarescens*) (*556*), *Phellodon confluens* (*132*), *Ph. melaleucus* (= *Hydnum graveolens*) (*132, 555, 556*), *Ph. niger* (= *Calodon niger* = *Hydnum nigrum*) [1.7%][a] (*556*), (*132*), *Ph. tomentosus* (= *Calodon cyathiformis* = *Hydnum cyathiforme*) (*132, 556*), *Polyozellus multiplex* (= *Cantharellus multiplex*) (*555*), *Pseudotomentella*

Table 3 (*continued*)

---

mucidula (= *Tomentella mucidula*) (*132*), *Sarcodon aspratus* (= *Hydnum aspratum*) (*555*), *S. imbricatus* (= *Hydnum imbricatum*) (*466, 555, 556*), *S. scabrosus* (= *Hydnum scabrosum*) (*555*), *Thelephora anthocephala* (*132*), *Th. caryophyllea*, *Th. mollissima* (= *Th. intybacea*) (*730*), *Th. palmata* [0.5%] (*730*), [0.9%] (*51*), (*430*), *Th. terrestris* (*730*), *Tomentellina fibrosa* (= *Kneiffiella bombycina* = *Caldesiella ferruginosa*) (*132*); Corticiaceae s. lat.: *Punctularia strigoso-zonata* (= *Phlebia strigoso-zonata*) (*556*); Polyporaceae s. lat.: *Trametes multicolor* (= *Coriolus zonatus*) (*92*), *T. versicolor* (= *Polyporus versicolor* = *Polystictus versicolor*) (*555, 580*); Boletales: *Lampteromyces japonicus* [c] (*397*), *Omphalotus olearius* [c] (*128*), *O. illudens* (= *Clitocybe illudens*) (*127*), *Omphalotus subilludens* [c] (*637*), *Paxillus atrotomentosus* [7×10^{-4}%] [a] (*273*, see however *92*), *Rhizopogon colossus*, *Rh. hawkeri*, *Rh. parksii* [trace], *Rh. subareolatus* [trace] (*127*), *Suillus aeruginascens* (*87*), *S. grevillei* var. *badius* [cap skin; probably an artefact formed from cyclovariegatin (**15**)] (*204, 205*), *S. tridentinus* (*87*)

---

[a] Refers to isolated yield from fresh fungus. Other yields refer to dried material.

Many terphenylquinones have been synthesised by arylation of 2,5-dichlorobenzoquinone (**17**) with *N*-nitrosoacetanilides (*8, 9, 10, 80, 149, 150*) or diazotised aromatic amines (*9, 80, 148, 149, 150, 205, 206*) followed by alkaline hydrolysis. The procedure is illustrated in Scheme 4 for the preparation of leucomelone (**19**) (*205*).

This approach in general gives only moderate overall yields, especially in the case of unsymmetrically substituted terphenylquinones. It is interesting to note, however, that the use of the methoxymethyl ether protecting group allowed conversion of the dichloroquinone (**18**) into the corresponding dihydroxy compound with alkali prior to deprotection of the phenolic hydroxy groups, which was subsequently achieved under mild acidic conditions in high yield [see also (*206*)]. This expedient avoids the formation of cycloleucomelone (**14**) and decomposition products which were observed in earlier attempts to synthesise the quinone (**19**) (*9, 80*).

Recently a new general synthesis of terphenylquinones has been developed which makes use of the methoxide catalysed rearrangement of grevillins (Scheme 5) (*462*). Because the starting materials (**20**) are easily prepared from pyrandiones of type (**8**) (Scheme 2) this method offers an efficient and versatile route to unsymmetrically substituted and specifically carbon-labelled terphenylquinones (*366*). Besides polyporic acid (**11**), ascocorynin (**12**), and leucomelone (**19**) a number of analogues containing heterocyclic or styryl residues have been prepared (*462*). The enolate (**21**) involved in this transformation is similar to

Table 4. *Properties of Terphenylquinones*

Pigment	Colour	TLC R_f	Ultraviolet/Vis. [λ_{max} (log ε)]	Peracetate M.p.	Leucoperacetate M.p.	References
(11)	Bronze leaflets	0.15[a] 0.75[b]	256 (4.63), 262 (4.63), 330 (sh., 4.06), 465 nm (2.60) [ethanol]; 261 (4.45), 332 nm (3.86) [dioxan]	209°	267–268°	(311, 395, 423, 579, 655)
(12)	Green needles	0.42–0.56[c]	262 (4.59), 353 (3.71), 470 nm (2.67) [dioxan]	–	198–200°	(462, 534)
(13)	Bronze leaflets	0.21[a] 0.67[b]	268 (4.55), 385 nm (3.66) [dioxan]	242–245°/ 290–294°	235–236°	(307, 311, 431)
(14)	Brown leaflets	0.53[d]	258 (4.27), 293 (4.14), 420 nm (3.60) [methanol]	226–227°	203–204°	(7, 9, 385, 386)
(15)	Dark red needles	0.38[e]	259 (4.39), 301 (4.44), 340 (infl., 4.01), 445 nm (3.84) [ethanol]	Oil	Oil	(204, 205)
(16)	Dark violet prisms	0.63[a] 0.10[b]	217 (4.33), 264 (4.27), 305 (4.30), 390 (sh., 3.48), 483 nm (3.86) [ethanol]	330–335° (dec.)	380–390° (dec.)	(310, 313, 655)

[a] Silica gel G (methyl ethyl ketone:water:formic acid = 250:25:1) (272).
[b] Silica gel G (methanol:acetic acid = 3:1) (636).
[c] Silica gel G (benzene:ethyl formate:formic acid = 10:5:3) (462).
[d] Silica gel 60 F_{254} (toluene:ethyl formate:formic acid = 10:5:3) (385).
[e] Silica gel Merck PF$_{254+366}$ (benzene:ethyl formate:formic acid = 50:49:1) (204).

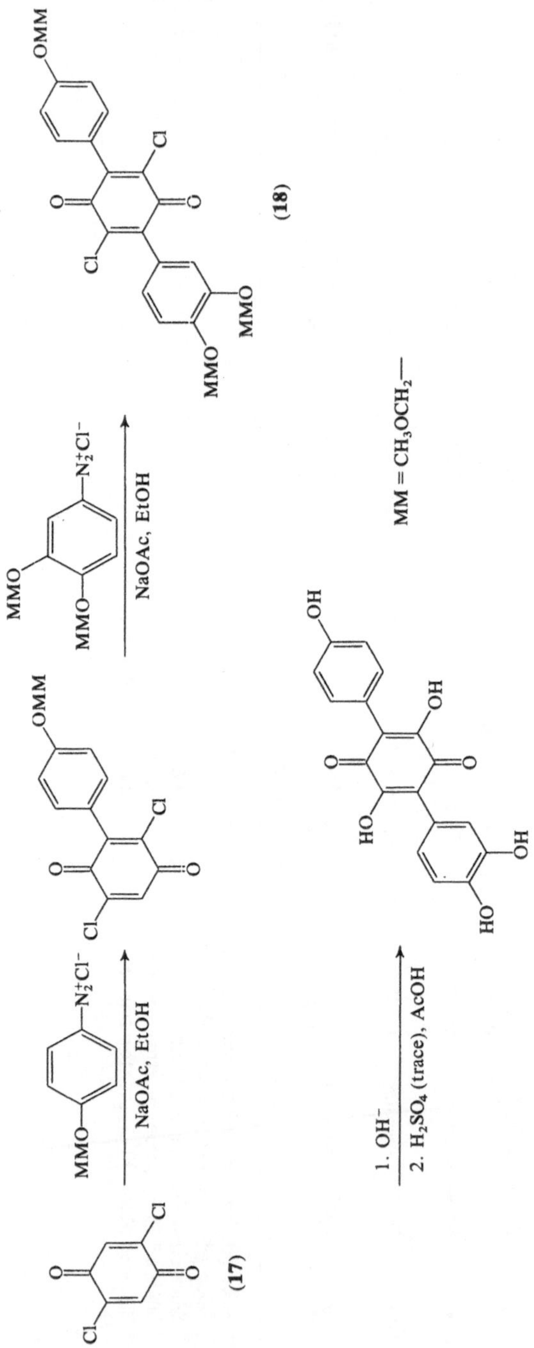

Scheme 4. Synthesis of leucomelone

MM = CH₃OCH₂—

$MM = CH_3OCH_2-$

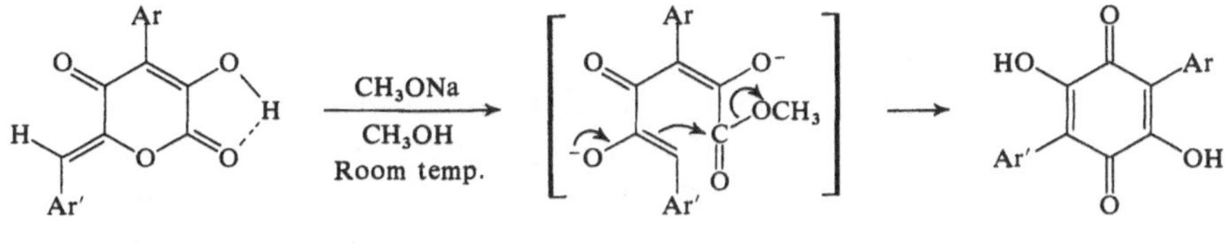

Scheme 5. Rearrangement of grevillins to terphenylquinones

the putative intermediate (9) in the biosynthesis of grevillins and ter-phenylquinones (Scheme 3).

The initial suggestion that the naturally occurring terphenylqui-nones (11), (13), and (16) are assembled biosynthetically by condensa-tion between two molecules of an unbranched phenylpropanoid precur-sor was due to READ and VINING (540) who subsequently obtained the first experimental proof of this pathway (541). These authors ob-served the efficient incorporation of L-phenyl[1-^{14}C]alanine, DL-phe-nyl[2-^{14}C]lactic acid and DL-[1'-^{14}C]-m-tyrosine into the unique ter-phenylquinone pigment volucrisporin (22) (185, 186) when the acids were fed to cultures of Volucrispora aurantiaca. Later it was shown that DL-phenyl[3-^{13}C]alanine serves as a precursor of the terphenyl ring system of phlebiarubrone (23) (113), the carbon atom of the methy-lene bridge coming from formate or from methionine (17). Recently the incorporation of L-phenyl[U-^{14}C]alanine into polyporic acid (11) by young fruit bodies of Hapalopilus rutilans was demonstrated [incor-poration rate: 17.3%] (366).

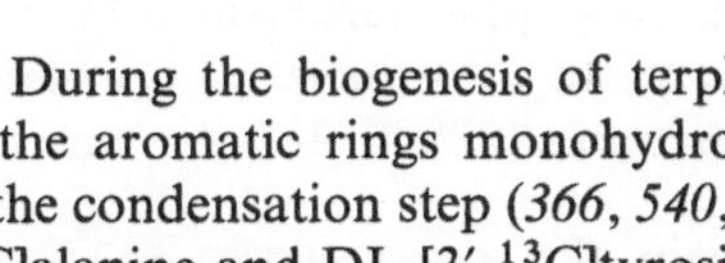

(22) 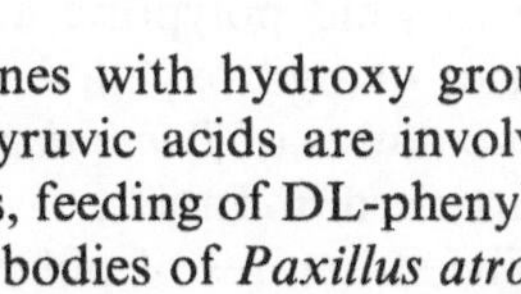   (23)

During the biogenesis of terphenylquinones with hydroxy groups in the aromatic rings monohydroxyphenylpyruvic acids are involved in the condensation step (366, 540, 541). Thus, feeding of DL-phenyl[3-^{13}C]alanine and DL-[3'-^{13}C]tyrosine to fruit bodies of Paxillus atroto-mentosus resulted only in incorporation of the latter amino acid into atromentin (13) (366). In accord with the proposed biosynthesis via p-hydroxyphenylpyruvic acid (540) all of the label was subsequently found at C-3 and C-6 in the quinone (13) by ^{13}C-n.m.r. spectroscopy (1.7 atomic % excess).

Atromentin (13) is a key intermediate for further conversions, e.g. to more highly hydroxylated terphenylquinones and to pulvinic acids (see Section 2.1.3). Whereas polyporic acid (11) and ascocorynin (12) have not been observed as co-metabolites of more highly hydroxylated terphenylquinones, atromentin or its derivatives often co-occur with cycloleucomelone (14) (e.g. in *Paxillus* and *Anthracophyllum*) and with thelephoric acid (16) (e.g. in *Hydnellum*). In cultures of *Omphalotus subilludens* (= *Clitocybe subilludens*) the relative proportion of atromentin to thelephoric acid decreases with ageing, which provides a strong indication that thelephoric acid is derived from atromentin (*637*).

The discovery that polyporic acid (11) exhibits antileukemic activity (*146*) induced the synthesis of numerous analogues (*148, 149, 150*). The pigment also causes contraction of isolated rabbit ileum suspended in oxygenated Tyrode solution (*395*). Similarly, atromentin (13) shows significant smooth muscle stimulant activity (*638*) and proves to be an effective anticoagulant (*226, 409*). Polyporic acid lacks this latter property completely (*409*).

The simple terphenylquinones occur in fungi often in the form of derivatives of the quinones themselves or of the corresponding hydroquinones. These derivatives will now be discussed according to the 'parent' terphenylquinone system.

B. Polyporic Acid and Derivatives

Terphenylquinones with unsubstituted phenyl rings have a limited distribution in fungi. Polyporic acid (11) occurs in a few wood inhabiting Aphyllophorales where it is produced in surprisingly high concentrations (Table 3). Phlebiarubrone (23) has been isolated from cultures of *Punctularia strigoso-zonata* (= *Phlebia strigoso-zonata*) (*478, 480*). This red *ortho*-quinone formed a leuco-diacetate and with *o*-phenylene-diamine afforded a quinoxaline derivative. The protons of the methylenedioxy group give rise to a singlet at δ 6.26 in the ^{1}H-n.m.r. spectrum of phlebiarubrone and the acetal function was hydrolysed with aqueous base to yield polyporic acid (11). Recently, phlebiarubrone (23) and a series of hydroxylated derivatives, (24) to (26), have been isolated from cultures of *Punctularia atropurpurascens* (*22*); their properties are summarised in Table 5.

The structures of phlebiarubrone (23) and its dihydroxy derivative (25) were confirmed by synthesis involving methylenation of the potassium salts of polyporic acid (11) and atromentin (13), respectively, with methylene sulphate in the presence of sodium bicarbonate (*22,*

	R¹	R²	R³
(24)	OH	H	H
(25)	OH	H	OH
(26)	OH	OH	OH

Table 5. *Properties of the Phlebiarubrone Group*

Pigment	Colour	TLC R_f [a]	Ultraviolet/Vis. [λ_{max} (log ε)]	References
Phlebiarubrone (23)	Red needles	0.51	268 (4.48), 332 (3.64), 465 nm (3.54) [ethanol]	(478)
4'-Hydroxy-phlebiarubrone (24)	Violet crystals	0.37	273 (4.22), 352 (2.87), 500 nm (3.13) [methanol]	(22)
4',4''-Dihydroxy-phlebiarubrone (25)	Dark violet crystals	0.26	276 (4.33), 382 (3.19), 530 nm (3.38) [methanol]	(22)
3',4',4''-Trihydroxy-phlebiarubrone (26)	Black-violet crystals	0.16	277 (4.51), 543 nm (3.58) [methanol]	(22)

[a] Silica gel Merck 60 F_{254} (benzene:ethyl formate:formic acid = 10:5:3).

316). Phlebiarubrone and polyporic acid have also been synthesised beginning with 1,2-dimethoxy-4,5-(methylenedioxy)benzene (178). The co-occurrence of pigments (23) to (26) is of biogenetic interest in that it suggests a pattern of consecutive hydroxylations beginning with phlebiarubrone which has no parallel with polyporic acid itself.

From mycelial mats of *Phlebiopsis gigantea* (= *Peniophora gigantea*) the colourless 2, 3, 5-trimethoxy-*p*-terphenyl (27) has been isolated (134) which appears to be biogenetically related to polyporic acid.

C. Ascocorynin

Ascocorynin (12) and 4'-hydroxyphlebiarubrone (24) represent the only known monohydroxylated terphenylquinones found to date in nature.

D. Atromentin and Derivatives

	R¹	R²
(28)	H	H
(29)	(epoxy-hexenoyl)	H
(30)	(epoxy-hexenoyl)	(epoxy-hexenoyl)

Since Thörner's statement (653) that atromentin (13) occurs in the fruit bodies of *Paxillus atrotomentosus* principally in the form of the corresponding hydroquinone, the presence of leuco-atromentin (28) in the colourless flesh of this fungus has been widely assumed. However, a recent reinvestigation of the colourless constituents of *P. atrotomentosus* after extraction of freeze-dried fruit bodies with acetone has revealed that they consist mainly of leucomentin-3 (29), accompanied by some leucomentin-4 (30) (365). The leucomentins are tri- and tetraesterified derivatives of leuco-atromentin (28) with (2Z,4S,5S)-4,5-epoxy-2-hexenoic acid. The positions of the acyl residues in leucomentin-3 (29) followed from the ¹H-n.m.r. spectrum of the triacetyl derivative in which two acetyl signals appear coincident at δ 2.35. This is compatible only with acetoxylation at the 4'- and 4''-positions of the peripheral phenyl rings. The third acetyl group (δ 1.78) must therefore be located, like the epoxyhexenoyl residues, in the central ring. The (4S,5S)-absolute configuration at the oxirane ring stereocentres in leucomentin-3 (29) followed from its conversion into O-acetyl-(+)-osmunda lactone (31) (Scheme 6) and tetra-O-acetylatromentin on acid catalysed acetylation with acetic anhydride. The formation of (31) has been explained by the mechanism shown in Scheme 6 in which cleavage of the epoxide ring in leucomentin-3 is effected intramolecularly with concomitant inversion of the configuration at C-5 in the side chain.

An alternative mechanism for degradation of leucomentin-3 to the acetate (31) of (+)-osmunda lactone would involve initial acid catalysed addition of acetic acid to the epoxide followed by intramolecular displacement of the hydroquinone moiety and would lead, significantly,

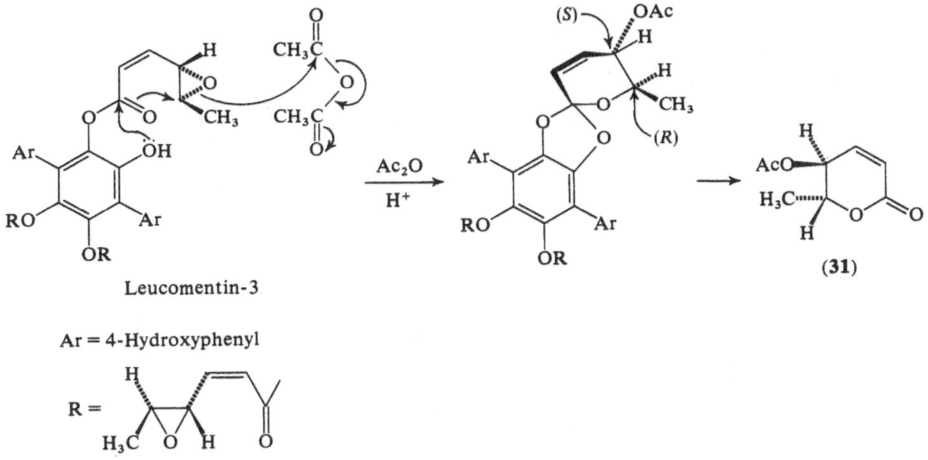

Scheme 6. Degradation of leucomentin-3

from (4 R, 5 R)-leucomentin-3 (the enantiomer of the natural product) to compound (31) again possessing the (4 S, 5 R)-stereochemistry. The two possible mechanisms have been clearly differentiated by a labelling experiment in which acetylation was brought about using [^{18}O$_2$]acetic anhydride. In this case O-acetyl-(+)-osmunda lactone (31) was obtained bearing only a single ^{18}O-label in the 4-acetoxy substituent, a result compatible only with the type of mechanism depicted in Scheme 6.

The leucomentins (29) and (30) are easily hydrolysed to atromentin (13) under conditions which parallel the isolation procedure and may therefore be considered as the genuine immediate precursors of the terphenylquinone (13) in *Paxillus atrotomentosus*. The leucomentins may also be regarded as the likely precursors of a number of orange and violet pigments which have been isolated in small amounts from fruit bodies of *Paxillus atrotomentosus* and *P. panuoides* (273, 365). The structures (32)–(39) of these compounds, called flavomentins and spiromentins, reveal the close chemical relationships which exist between individual members of the group.

The structures of the flavomentins and spiromentins were determined using mainly ^{1}H-n.m.r. spectroscopy (365). Interestingly, in the ^{1}H-n.m.r. spectra of the monoacylated terphenylquinones (33), (34), and (35) (among others) the protons of the two aryl residues appear to be equivalent due to a rapid acyl shift between the oxygen atoms in the quinonoid ring (364).

The distribution and some properties of the flavomentins and spiromentins are summarised in Table 6 (365).

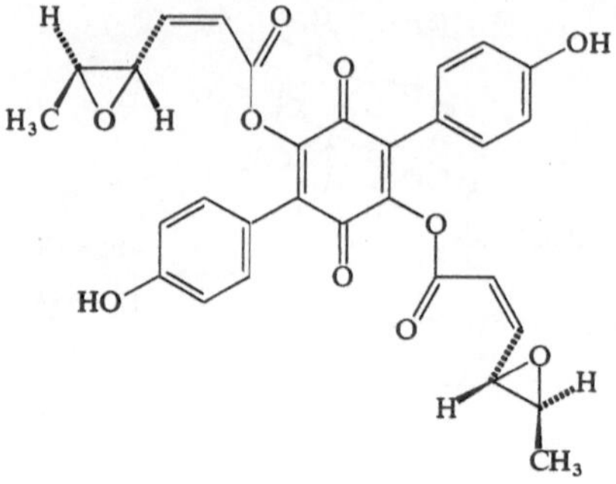

M. GILL and W. STEGLICH:

(32)

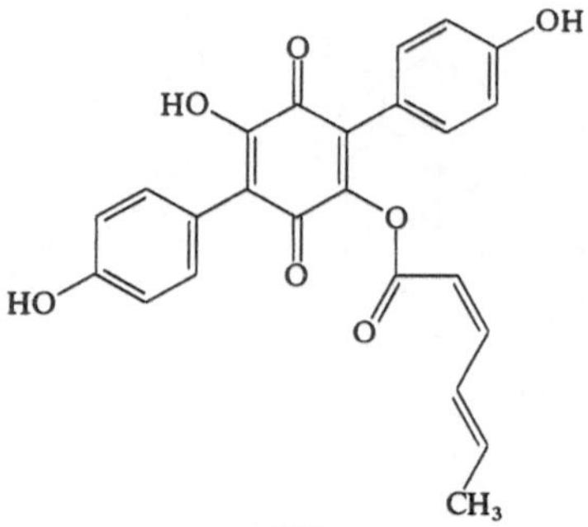

(33)

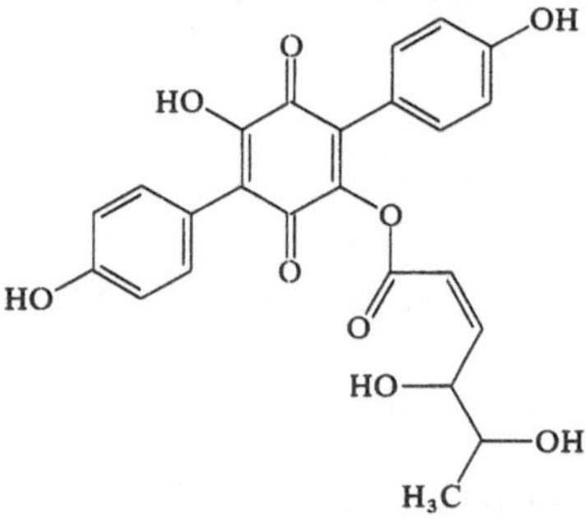

(34)

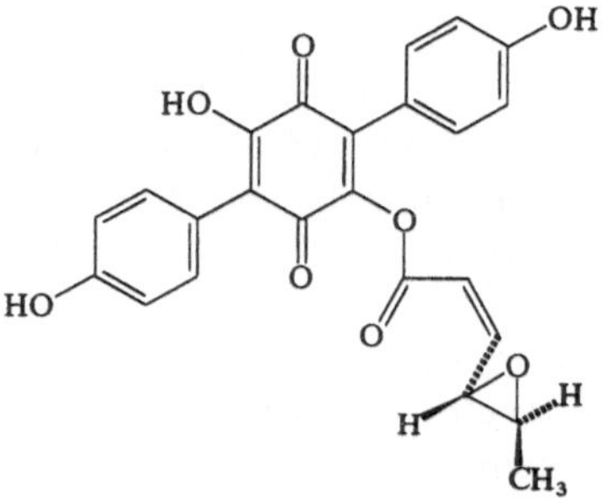

(35)

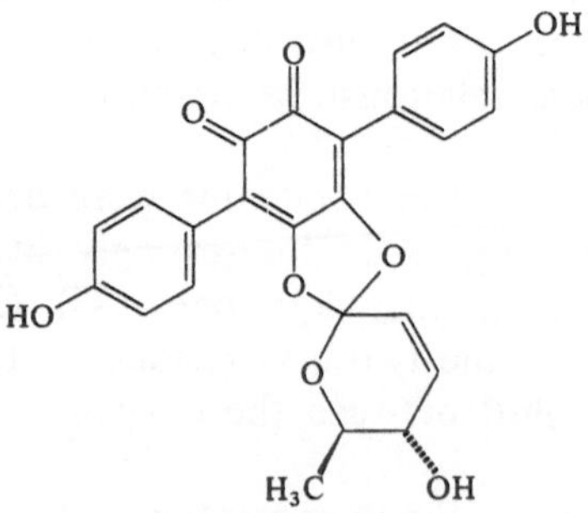

(36)

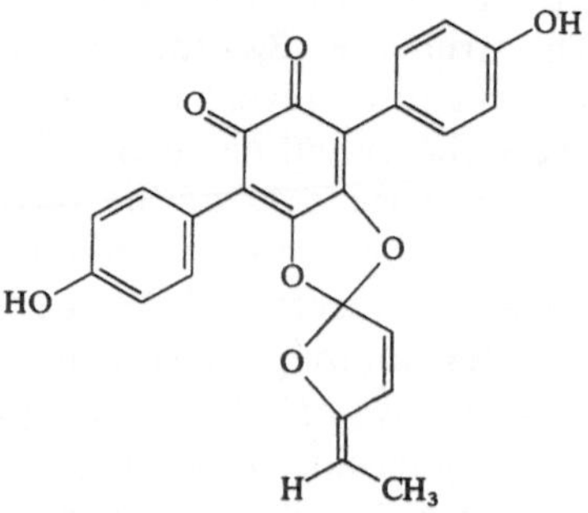

(37)

(38)   (39)

Table 6. *Flavomentins and Spiromentins*

Pigment	Occurrence[a]	Colour on TLC	TLC R_f[b]	Ultraviolet/Vis. [λ_{max}, qualitative, ethanol]
Flavomentin-A (32)	*Paxillus atrotomentosus*	Orange-yellow	0.48	265, 348 nm
Flavomentin-B (35)	*Paxillus atrotomentosus, P. panuoides* var. *ionipus*	Orange[c]	0.44	266, 371 nm
Flavomentin-C (33)	*Paxillus atrotomentosus, P. panuoides*	Orange[c]	0.48	–
Flavomentin-D (34)[d]	*Paxillus panuoides*	Orange[c]	0.28	–
Spiromentin-A (37)	*Paxillus atrotomentosus, P. panuoides* [cultures]	Violet	0.38	268, 382, 550 nm
Spiromentin-B (38)	*Paxillus atrotomentosus*	Violet	0.25	270, 382, 552 nm
Spiromentin-C (36)	*Paxillus atrotomentosus*	Violet	0.27	272, 380, 555 nm
Spiromentin-D (39)	*Paxillus atrotomentosus*	Violet	0.32	270, 385, 560 nm

[a] All occurrences are cited in ref. *365*. Flavo- and spiromentins have also been detected chromatographically in cultures of *Leucogyrophana olivascens* (*89*).
[b] Silica gel Merck 60 F_{254} (benzene:ethyl formate:formic acid = 10:5:3).
[c] Violet on exposure to ammonia vapour.
[d] Probably an artefact.

The structures and inter-relationships of the various flavomentins and spiromentins have been supported by chemical studies (Scheme 7) (*365*). The synthesis of 4′,4″-di-*O*-methylflavomentin-C (**41**) was achieved by reaction of the cyclic carbonate (**40**) with (2*Z*, 4*E*)-sorbic acid. Epoxidation of the ester (**41**) with *m*-chloroperbenzoic acid then led to the racemic flavomentin-B derivative (42), which on dissolving in trifluoroacetic acid cyclised to the spiromentin-B and spiromentin-C derivatives (43) and (44), respectively. This reaction sequence appears to mimic the likely biosynthesis of these compounds, although the possibility that they are (at least partially) artefacts formed from the leucomentins during ageing of the fruit bodies, or during isolation and work up, cannot be completely ruled out.

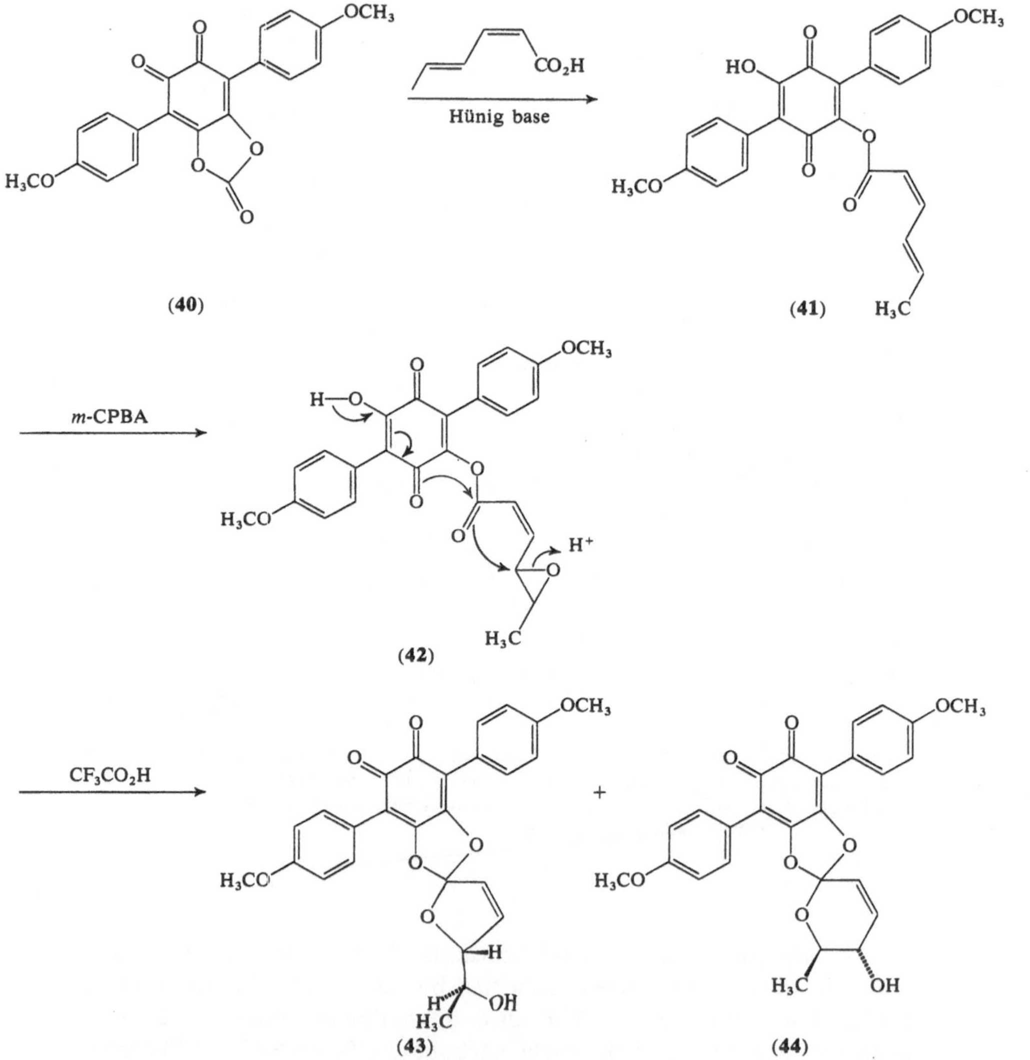

Scheme 7. Synthesis of flavomentin and spiromentin derivatives

Table 7. *Further Mono- and Diacyl Derivatives of Atromentin*

Pigment	Occurrence	Colour	TLC R_f	Ultraviolet/Vis. [λ_{max} (log ε)]
2-*O*-Acetyl-atromentin (**45**)	*Albatrellus cristatus* (*90*), *Omphalotus olearius* [cultures], *Paxillus atrotomentosus* (*397*), *P. panuoides* (*365, 397*), [c] (*397*)	Orange	0.39[a]	–
2,5-Di-*O*-acetyl-atromentin (**46**)	*Anthracophyllum archeri* (*385*)	Orange	0.61[a]	252 (4.42), 397 nm (3.69) [methanol]
Aurantiacin (**47**)	*Hydnellum aurantiacum* (= *Hydnum aurantiacum*) [1–2%] (*307*), *H. auratile* (*132*), *H. caeruleum* (*492*), *H. ferrugineum* (*132*), *H. scrobiculatum* (*636*), *H. velutinum* (*132*)	Dark red	0.85[b]	240 (4.63), 405 nm (3.82) [dioxan]
Dihydro-aurantiacin-1,4-dibenzoate (**48**)	*Hydnellum aurantiacum* (*310*), *H. caeruleum* (*492, 636*)	Colourless		234 (4.84), 270 nm (infl., 4.43) [dioxan]

[a] Silica gel Merck 60 F_{254} (benzene:ethyl formate:formic acid = 10:5:3) (*365*).
[b] Silica gel G (methyl ethyl ketone:water:formic acid = 250:25:1) (*272*).

Besides the flavomentins several other mono- and diesters of atromentin (**13**) have been isolated from fungi (Table 7). The structure of 2-*O*-acetylatromentin (**45**) was confirmed by synthesis of the 4′, 4″-di-*O*-methyl ether derivative using the cyclic carbonate (**40**) and acetic acid (*365*).

Aurantiacin (**47**) is accompanied in *Hydnellum* species by its leuco-dibenzoate (**48**) (*310*). The structure of aurantiacin (**47**) followed from its almost instantaneous hydrolysis to atromentin and benzoic acid when treated with aqueous alkali, its insolubility in sodium carbonate

	R^1	R^2
(**45**)	Ac	H
(**46**)	Ac	Ac
(**47**)	PhCO	PhCO

(**48**)

solution, and by the formation from (47) of a dimethyl ether identical with synthetic 4′,4″-dimethoxypolyporic acid dibenzoate (307). Aurantiacin itself has been synthesised unambiguously by arylation of 2,5-dichlorobenzoquinone with diazotised p-(methoxymethoxy)aniline followed, sequentially, by hydrolysis with dilute alkali, benzoylation of the newly introduced hydroxy groups, and detachment of the protecting acetal groups with dilute acid (206).

From sporophores of *Hydnellum ferrugineum* and *H. concrescens* (= *H. zonatum*) Gripenberg (321) has isolated an interesting yellow quinonoid pigment, hydnuferruginin, which is present to the extent of 1.2% of the dry weight in both species. The pigment shows absorption maxima at 288 (log ε = 4.21) and 372 nm (2.72) in ethanol, which is in good agreement with the values reported for 2,5-dimethoxy-1,4-benzoquinone. In the ¹H-n.m.r. spectrum aromatic proton signals are absent, and from the presence of characteristic signals in the aliphatic and olefinic regions the highly symmetrical structure (50) was deduced. The structure of hydnuferruginin was confirmed by a single crystal X-ray analysis (321) which also defined the stereochemistry as that shown.

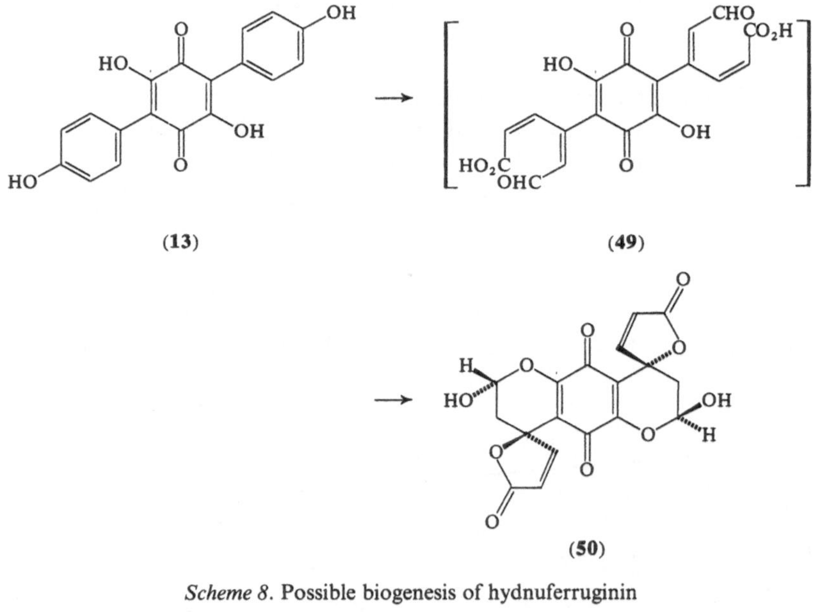

(13)   (49)

(50)

Scheme 8. Possible biogenesis of hydnuferruginin

Hydnuferruginin could be biogenetically derived from atromentin (13) by assuming a dioxygenase cleavage of both aromatic rings at the 3,4-position. The resulting dialdehydo-diacid **(49)** (Scheme 8) is then at the oxidation level of the pigment and may undergo direct cyclisation to give hydnuferruginin **(50)**. Alternatively, the quinone **(50)** may arise *via* intra-diol cleavage of the 3,4-dihydroxyphenyl rings of variegatin **(51)** *(321)*.

(51)

E. Cycloleucomelone and Derivatives

In 1942 AKAGI isolated a terphenylquinone, 'leucomelone', from fruit bodies of the edible black mushroom *Boletopsis leucomelaena* (= *Polyporus leucomelas*) for which he proposed the formula **(19)** *(7)*. The quinone formed a peracetyl derivative, m.p. 226–227° C *(9)*, and was accompanied in the fungus by its leuco-peracetate, 'protoleucomelone' **(52)**, m.p. 204–205° C. A synthesis of 'leucomelone' by sequential arylation of 2,5-dichloro-1,4-benzoquinone with diazotised 4-methoxyaniline and 3,4-dimethoxyaniline followed by demethylation and replacement of the chlorine substituents by hydroxyl with aqueous alkali apparently confirmed the structure *(9)*. Later syntheses, however, gave the quinone **(19)**, its peracetate and leuco-peracetate, all of which differed in their melting points from the values reported for the natural product and its derivatives by AKAGI *(205, 462)*.

A reinvestigation of the constituents of *B. leucomelaena* obtained by extraction of the fungus with acetone has afforded principally the leuco-peracetates **(53)** and **(54)** of cycloleucomelone and thelephoric acid, respectively *(386)*. The leuco-peracetate **(53)** has the same melting point as AKAGI's 'protoleucomelone' and we consider these compounds to be one and the same.

Minor colourless constituents accompanying the leuco-peracetate **(53)** in the acetone extracts of *B. leucomelaena* were the series of acetates **(55)** to **(58)** which have lost one, two, and three acetyl groups, respectively, from the peripheral hydroxy substituents. The position of the

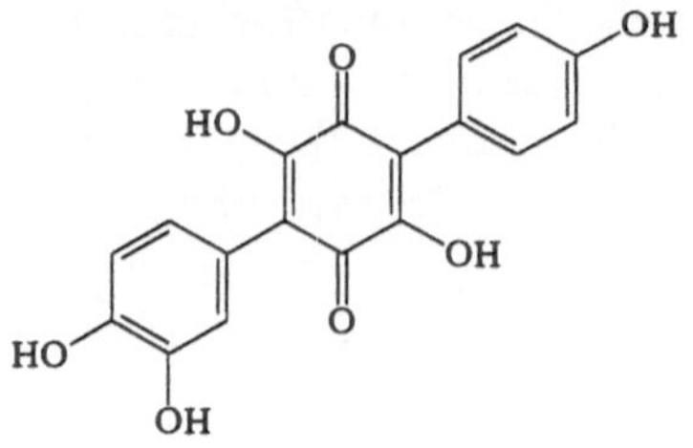

(19)

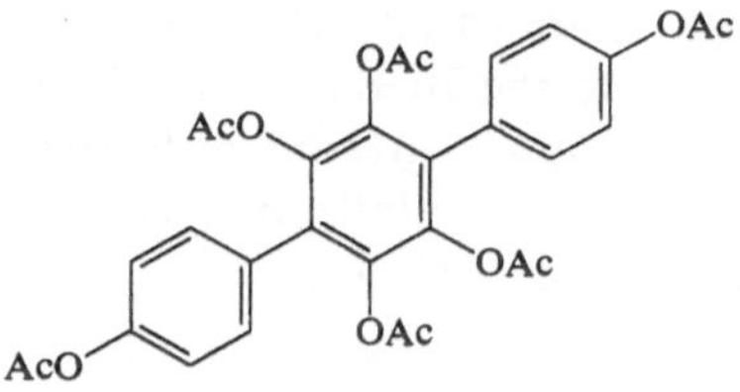

(52)

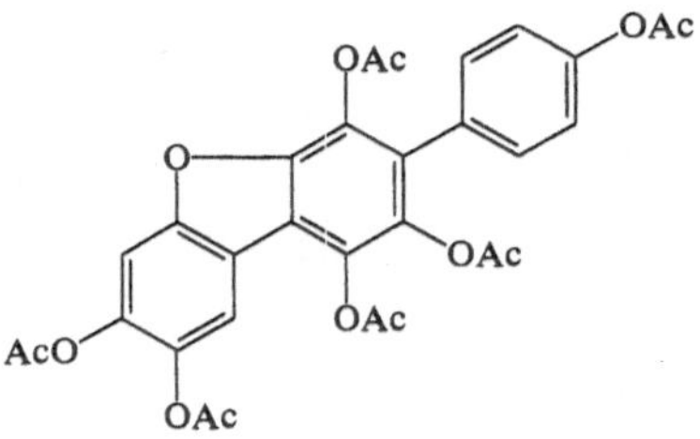

(53)

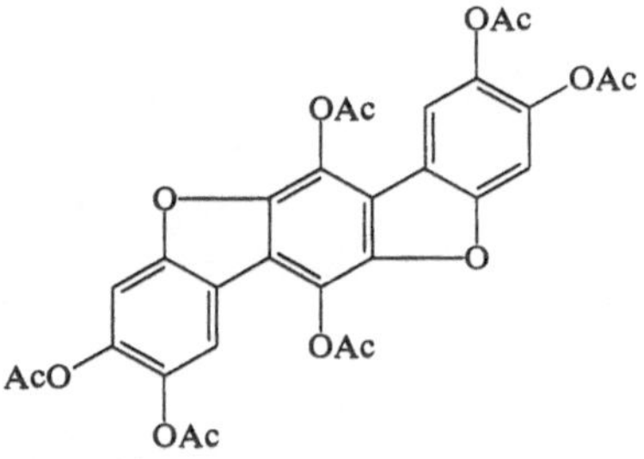

(54)

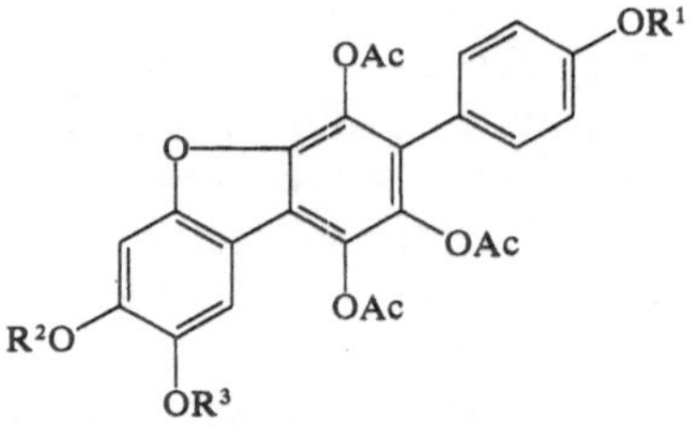

	R¹	R²	R³
(55)	H	Ac	Ac
(56)	Ac	H	H
(57)	H	H	Ac
(58)	H	H	H

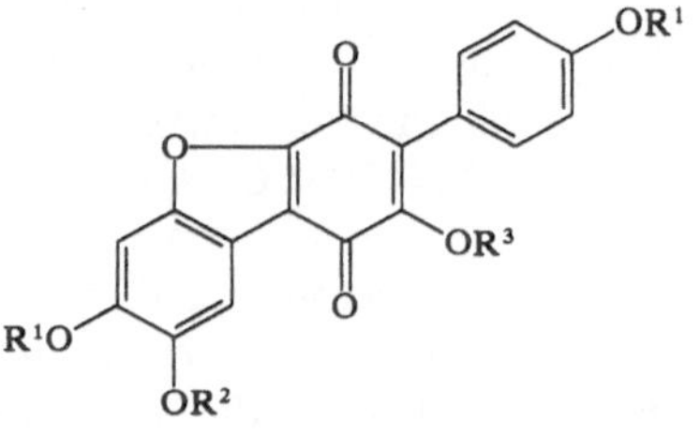

	R¹	R²	R³
(59)	Ac	Ac	Ac
(60)	H	Ac	Ac
(61)	H	H	Ac
(62)	H	Ac	H

acetate groups which remain could be determined by their influence on the chemical shifts of the aromatic protons in the ^{1}H-n.m.r. spectrum. It cannot be excluded that these minor constituents are artefacts formed during the extraction and separation procedure.

On extraction of *B. leucomelaena* with methanol a blue-green solution resulted from which cycloleucomelone (14), the leuco-triacetate (58) and a blue pigment have been obtained (386). Because cycloleucomelone (14) and its tetra-acetyl derivative (59) have the same melting points as AKAGI's 'leucomelone' and 'leucomelone peracetate', respectively, and since the identity of 'protoleucomelone' with the leucoperacetate (53) of cycloleucomelone is established, we have no doubt that the Japanese author was also dealing with the cyclised compounds. Synthetic leucomelone (19) is easily oxidised to cycloleucomelone (14), e.g. with Fetizon's reagent, and it can be confidently assumed that in the final step of AKAGI's 'leucomelone' synthesis (9) the ether bridge was formed on exposure of leucomelone (19) to alkali [see also (205)]. The only authentic natural terphenylquinone known with the hydroxylation pattern of leucomelone is 3',4',4''-trihydroxyphlebiarubrone (26) (22) in which case the cyclisation process is prohibited by methylenylation of the hydroxy groups in the central ring.

The typical green colour reaction shown by fruit bodies of *Anthracophyllum* species when touched with aqueous alkali is caused by the presence of cycloleucomelone derivatives (385). From *Anthracophyllum discolor* and *A. archeri* a yellow pigment, anthracophyllin [UV/vis. (methanol): $\lambda_{max} = 254$ and 410 nm], has been isolated which constitutes up to 7% of the dry weight of these fungi. By analysis of spectral data, especially the chemical shifts of the aromatic protons in the ^{1}H-n.m.r. spectrum, the formula (60) was deduced. Anthracophyllin (60) is accompanied by small amounts of a mixture of the monoacetates (61) and (62) together with free cycloleucomelone (14).

Sporophores of *Hydnellum ferrugineum* and *H. concrescens* contain, besides hydnuferruginin (50) (321), a second ring-opened pigment, hydnuferrugin, which was isolated in yields of 0.8 and 1.8%, respectively, of the weight of the fungi (319, 321). Hydnuferrugin forms dark violet crystals [UV/vis. (dioxan): λ_{max} (log ϵ) = 255 (4.35), 293 (4.06), 322 (3.71), 410 (sh., 3.37) and 485 nm (3.47)] and yielded a yellow triacetate and a colourless leuco-penta-acetate. On treatment of its methanolic solution with concentrated hydrochloric acid a monomethyl derivative was formed which suggested the presence in hydnuferrugin of a hemiacetal moiety. From close inspection of the infrared and ^{1}H-n.m.r. spectra the structure (64) was proposed (319). The stereochemistry of hydnuferrugin (64) is suggested from the X-ray structure analysis of its co-metabolite, hydnuferruginin (50) (321). No optical activity has been detected for hydnuferrugin but this may be due to the low solubility of this quinone.

Biosynthetically, hydnuferrugin (64) may arise *via* dioxygenase fission of the 4-hydroxyphenyl ring in cycloleucomelone (14), followed by cyclisation of the resulting aldehydo-acid (63) (Scheme 9). Alternatively, hydnuferrugin may be derived *via* intradiol cleavage of cyclovariegatin (15) (319, 321).

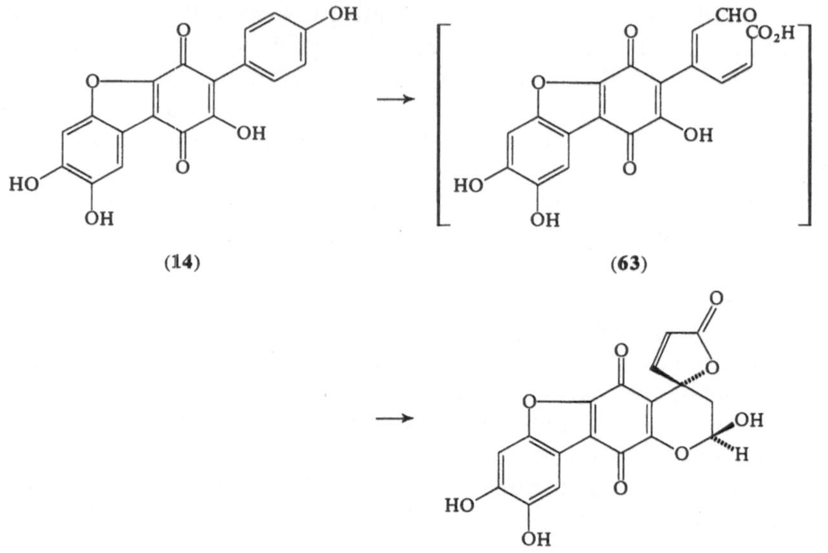

(14)   (63)

(64)

Scheme 9. Possible biogenesis of hydnuferrugin

F. Cyclovariegatin and Derivatives

EDWARDS and GILL (*204, 205*) isolated from the flesh of *Suillus grevillei* and *S. grevillei* var. *badius* pigment-C_2, a precursor of thelephoric acid. Even on standing in acetone solution this dark red quinone is slowly converted into thelephoric acid (**16**). This behaviour is readily accounted for by the formula (**15**), which followed from spectroscopic data, mass spectral comparison with variegatin (**51**) and leucomelone (**19**), and conversion of the pigment into a penta-acetate and a leuco-hepta-acetate. Since the probable biogenetic precursor (**51**) of the quinone (**15**) has been named variegatin (*205*) we propose the name cyclovariegatin for (**15**) by analogy with cycloleucomelone.

It has been suggested that cyclovariegatin (**15**) is responsible for pigmentation of the dark brown cap skins of *S. grevillei* var. *badius* from which thelephoric acid was earlier isolated (*204*). The quinone (**15**) may in fact enjoy a wide distribution as an unstable precursor of thelephoric acid in other fungi which are reported to contain the latter pigment.

During an investigation of a species of *Anthracophyllum* collected in New Zealand, besides thelephoric acid and anthracophyllin (**60**),

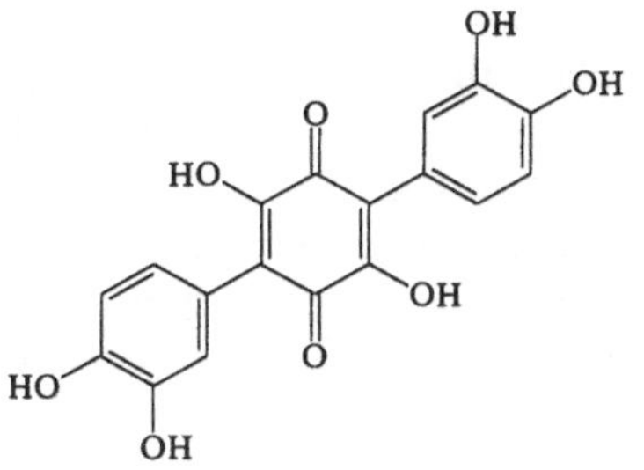

(51)

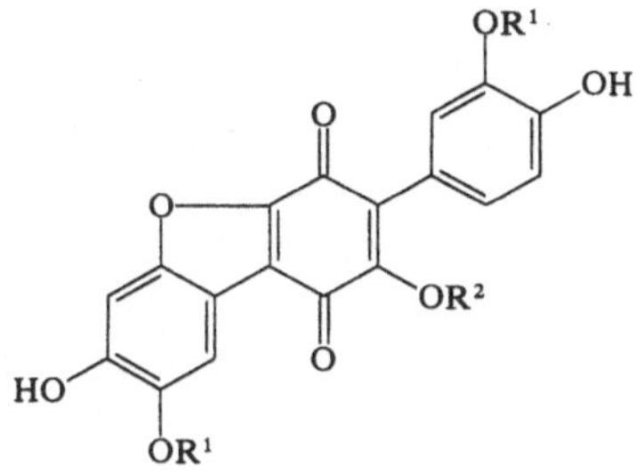

R¹ R²
(15) H   H
(65) H   Ac
(66) Ac Ac

two new pigments were isolated which were shown to be 2-*O*-acetylcy-clovariegatin (65) and 2,3′,8-tri-*O*-acetylcyclovariegatin (66) (*385*). On dissolving in methanol the quinone (65) is rapidly converted into thelephoric acid.

G. Thelephoric Acid and Derivatives

The terphenylquinone structure (16) for thelephoric acid emerged from the meticulous work of GRIPENBERG (*313*) following suspicions by others (*6, 539*) that KÖGL's original report (*430*) was incorrect. Thelephoric acid is widely distributed in Thelephoraceae and may be considered as an important taxonomic character (*132*). It can be synthesised by methods developed by GRIPENBERG (*313*) and others (*464, 700*) which are discussed in THOMSON's book on naturally occurring quinones (*655*).

Recently, the leuco-peracetate (54) of thelephoric acid has been isolated from sporophores of *Boletopsis leucomelaena* (*386*). The leuco-permethyl ether (67) of thelephoric acid occurs in minute amounts in the indigo blue coloured mycelial mats of *Pulcherricium caeruleum* (= *Corticium caeruleum*) and its structure has been established by an X-ray crystal analysis (*581, 711*). The chemical nature of the blue pigment from this organism is a matter of dispute. BRIGGS *et al.* (*135*) obtained blue solutions ($\lambda_{max} = 565$ nm) on extraction of the mycelium with chloroform. These solutions turned green and then brown on exposure to alkali, a phenomenon which was ascribed to the presence of quinhydrones. Either by direct chromatographic separation of the crude pigments or by preliminary conversion into the acetates or leucoacetates followed by thin layer chromatography, these authors were able to purify three compounds which they named the corticins-A, -B, and -C. Mainly based on ¹H-n.m.r. evidence, formulas (68), (69), and (**70**) were suggested for these compounds and structure (**69**) was confirmed by synthesis. Of these, formula (**70**) for corticin-C appears most improbable on biogenetic grounds and needs confirmation by synthesis. Independent work on *P. caeruleum* by a French group (*511, 512*) had earlier detected the presence of leuco-thelephoric acid derivatives and further concluded on the basis of mass spectral analysis that the blue pigments were polymeric in nature. Clearly, the chemistry of these interesting pigments needs still further clarification.

(**67**)   (**68**)

(**69**)   (**70**)

2.1.3. Pulvinic Acid Derivatives

Hydroxylated pulvinic acids are responsible for the yellow and red colours of most boletes. They also cause the characteristic blueing of the flesh which occurs when the fruit bodies of many species are cut or bruised. Because the distribution of the hydroxylated pulvinic acids such as variegatic acid (**73**) and xerocomic acid (**72**) is restricted to the order Boletales, their isolation from *Gomphidius* (*608*), *Omphalotus* (*128, 397, 584*), *Hygrophoropsis* (*91, 126, 127, 198*), *Coniophora* (*86, 89, 122, 134*), *Leucogyrophana* (*89*), *Rhizopogon* (*618*) and *Serpula* (*89, 90, 121*) provides strong evidence for the inclusion of these genera in Boletales.

A. Common Pulvinic Acid Derivatives

The blue stain produced in the fruit bodies of many boletes when the flesh is damaged has long attracted the attention of chemists. In 1885, BOEHM (108) isolated a red crystalline compound 'Luridussäure' from *Boletus luridus*. The pigment formed yellow solutions in water which on addition of sodium carbonate changed to an emerald green and finally an indigo blue colour. Later, BERTRAND (84, 85) named the blueing principle 'boletol' and showed that the colour change in the fungus was brought about by the action of oxygen in the presence of oxidase enzymes. BERTRAND noted that the flesh of boletes which lack these enzymes, e.g. *Xerocomus subtomentosus*, remained yellow after bruising despite the fact that they contained 'boletol'. BERTRAND's suggestion that 'boletol' may be a quinone seemed to find support in the work of KÖGL and DEIJS (427, 428) who proposed an hydroxylated anthraquinone carboxylic acid structure for the compound. Unfortunately, the experiments on which this conclusion was based could not be repeated by later workers. Thus, all attempts to detect anthraquinones in boletes have failed (73, 607) and, furthermore, synthetic materials corresponding to KÖGL's formulation did not develop blue colours with oxidases or with chemical oxidising agents (117, 256, 607).

The problem was finally solved by EDWARDS and co-workers (73, 200) who showed that the blueing of *Suillus variegatus*, *S. bovinus*, *Boletus erythropus*, and *B. appendiculatus* is caused by a tetrahydroxypulvinic acid (73), which they named variegatic acid. Shortly afterwards STEGLICH, FURTNER, and PROX (607) characterised a second blueing principle, xerocomic acid (72), from fruit bodies of *Xerocomus chrysenteron* and several other boletes. Variegatic acid (73) and xerocomic acid (72) are identical with 'boletol' and 'isoboletol' which had been extracted from various boletes and separated chromatographically by GABRIEL (265). Subsequent chromatographic work by BRESINSKY's group (92, 127, 131, 132) and by SCHMITT (562) has demonstrated the wide distribution of these and related pigments in Boletales (Table 8).

	R¹	R²
(71)	H	H
(72)	OH	H
(73)	OH	OH

(74)

Table 8. *Occurrence of Common Pulvinic Acid Derivatives*[a]

Atromentic Acid (71)	*Leccinum aurantiacum* [cultures] *(130)*, *Omphalotus illudens* (= *Clitocybe illudens*) [c] *(584)*, *Paxillus atrotomentosus* [c] *(272)*, *P. panuoides* [c] *(273)*, *Xerocomus chrysenteron (607)*; TLC: *Boletellus emodensis (281)*, *B. mirabilis (127)*, *Boletinus cavipes (131)*, *Boletus edulis*, *B. erythropus (257)*, *Chalciporus piperatus (131, 609)*, [c] *(87)*, *Coniophora puteana* (= *C. cerebella*) [c] *(86)*, *Gomphidius glutinosus (131, 257)*, *G. maculatus*, *G. roseus (92)*, *Hygrophoropsis aurantiaca* [c] *(126)*, *Lampteromyces japonicus* [c] *(397)*, *Leucogyrophana mollusca* [c], *L. olivascens* [c], *L. pinastri* [c], *L. romellii* [c] *(89)*, *Omphalotus olearius* [c] *(128)*, *Paxillus filamentosus* [c] *(397)*, *Rhizopogon roseolus* [c] *(122)*, *Serpula lacrimans (89)*, [c] *(86, 121)*, *Suillus flavidus* [c], *S. granulatus* [c], *S. luteus* [c] *(122)*, *S. placidus (87)*, [c] *(122)*, *S. plorans (92)*, [c] *(122)*, *Xerocomus badius (257)*, *X. parasiticus (131)*, *X. subtomentosus* [c] *(122)*, present also in traces in most fungi which produce (72) or (73).
Xerocomic Acid (72)	*Chroogomphus helveticus (606)*, *Ch. rutilus* (= *Gomphidius rutilus*) $[2 \times 10^{-3}\%]$ *(72)*, *Coniophora arida* var. *suffocata* (= *C. suffocata*) [c], *C. tomentella* [c] *(134)*, *Gomphidius glutinosus* [stipe base, $4 \times 10^{-2}\%$] *(608)*, *G. maculatus (689)*, *Paxillus atrotomentosus* [c] *(272)*, *P. panuoides* [c] *(273)*, *Phaeogyroporus portentosus (281)*, *Suillus luteus (276)*, *Xerocomus chrysenteron* $[9 \times 10^{-4}\%]$ *(607)*; TLC: *Boletellus mirabilis (127)*, *B. obscurecoccineus (281)*, *B. russellii (127)*, *Boletinus cavipes (131)*, [c] *(122)*, *B. paluster*, *Boletus auripes (127)*, *B. calopus (607)*, *B. edulis (257)*, *B. erythropus (562, 607)*, *B. fechtneri (92)*, *B. fibrillosus*, *B. frostii* ssp. *floridanus*, *B. gertrudiae (127)*, *B. impolitus*, *B. junquilleus*, *B. luridus (92)*, *B. miniatoolivaceus*, *B. pallidus*, *B. rubellus* ssp. *dumetorum*, *B. rubricitrinus (127)*, *B. satanas (92)*, *Chalciporus piperatus (131, 265, 609)*, [c] *(122)*, *Ch. pseudorubinus* (= *Ch. pseudorubinellus*) *(127)*, *Ch. rubinellus* [c] *(87)*, *Chroogomphus tomentosus (127)*, *Coniophora arida* var. *arida* [c], *C. marmorata* [c], *C. olivacea* [c] *(89)*, *C. puteana* [c] *(86, 122)*, *Gastroboletus turbinatus (127)*, *G. xerocomoides (87)*, *Gomphidius subroseus (127)*, *Gyrodon lividus* [c] *(92)*, *Hygrophoropsis aurantiaca* [c] *(126)*, *Leccinum rubropunctum*, *L. subglabripes* ssp. *corrugis (127)*, *Leucogyrophana arizonica* [c, trace], *L. mollusca* [c], *L. olivascens* [c], *L. pinastri* [c], *L. romellii* [c] *(89)*, *Omphalotus olearius* [c] *(128)*, *Pulveroboletus auriporus*, *P. hemichrysus (127)*, *P. lignicola* [c], *Rhizopogon luteolus (86)*, *S. himantioides* [c], *S. incrassata* [c] *(89)*, *Serpula lacrimans (89)*, [c] *(86, 121)*, *Suillus aeruginascens (87)*, [c] *(122)*, *S. amabilis* (= *S. lakei* = *Boletinus lakei*) [c], *S. bovinus* [c] *(122)*, *S. collinitus* [c] *(86)*, *S. flavidus* [c], *S. granulatus* [c] *(122)*, *S. grevillei (87)*, [c] *(122)*, *S. grisellus*, *S. hirtellus* ssp. *thermophilus (127)*, *S. luteus (87)*, [c] *(122)*, *S. placidus (87)*, [c] *(122)*, *S. plorans (92)*, [c] *(122)*, *S. serotinus (127)*, *S. sibiricus* [c] *(86)*, *S. spectabilis*, *S. tomentosus (127)*, *S. tridentinus* [c], *S. variegatus* [c] *(122)*, *Truncocolumella rubra (127)*, *Xerocomus badius (131, 257)*, *X. illudens* var. *xanthomycelinus (127)*, *X. parasiticus (131, 265)*
Variegatic Acid (73)	*Boletus appendiculatus (73)*, *B. calopus* $[5 \times 10^{-3}\%]$ *(607)*, *B. edulis (257)*, *B. erythropus* $[4 \times 10^{-2}\%]$ *(607)*, *(73)*, *Chalciporus piperatus (609)*, *Phaeogyroporus portentosus*, *Phylloporus hyperion*, *P. rhodoxanthus (281)*, *Rhizopogon roseolus (618)*, *Suillus bovinus* $[3 \times 10^{-3}\%]$ *(70, 73)*, *S. luteus (276)*, *S. variegatus (73, 200)*, *Xerocomus badius (257)*, *X. chrysenteron* $[4 \times 10^{-3}\%]$ *(607)*; TLC: *Boletellus ananas (127)*, *B. emodensis (281)*,

Table 8 *(continued)*

	B. mirabilis (127), *B. obscurecoccineus (281)*, *B. russellii*, *B. zelleri (127)*, *Boletinus cavipes (131, 265)*, *B. cavipes* f. *aureus (131)*, *B. paluster (127)*, *Boletus aestivalis* (= *B. reticulatus*) *(265)*, *B. auripes (127)*, *B. bicolor (127)*, *B. coniferarum (87)*, *B. edulis (127, 131, 265)*, *B. fechtneri (92)*, *B. fibrillosus*, *B. fraternus*, *B. frostii* ssp. *floridanus*, *B. gertrudiae (127)*, *B. impolitus (92)*, *B. junquilleus (131)*, *B. luridus (127, 265, 562)*, *B. miniatoolivaceus (127)*, *B. morrisii (364)*, *B. pallidus (127)*, *B. pinicola (132, 265)*, *B. pulverulentus (127, 131, 265, 562)*, *B. queletii (86)*, *B. radicans (131, 562)*, *B. regius (132)*, *B. rhodoxanthus (131, 265)*, *B. rubellus* ssp. *dumetorum*, *B. rubricitrinus (127)*, *B. rubripes (87)*, *B. satanas (92)*, *B. speciosus (87)*, *B. splendidus* ssp. *splendidus* (= *S. satanoides*) *(131)*, *B. subglobosus (281)*, *Chalciporus piperatus* [c] *(87)*, *Ch. pseudorubinus (127)*, *Coniophora puteana (89)*, [c] *(86)*, *Gastroboletus subalpinus (87)*, *G. turbinatus (127)*, *G. xerocomoides (87)*, *Gomphidius subroseus (127)*, *Gyrodon lividus (131)*, *G. merulioides*, *Hygrophoropsis aurantiaca (127)*, [c] *(126)*, *Leccinum rubropunctum*, *L. subglabripes* ssp. *corrugis (127)*, *Leucogyrophana mollusca* [c], *L. olivascens* [c], *L. pinastri* [c] *(89)*, *Omphalotus olearius* [c] *(128)*, *Paxillus panuoides* [c] *(397)*, *Pulveroboletus auriporus* (= *Boletus auriporus*) *(87, 127)*, *P. cramesinus (474)*, *P. hemichrysus (127)*, *P. lignicola (131)*, *Rhizopogon luteolus (86)*, *S. himantioides* [c] *(89)*, *Serpula lacrimans (89)*, [c] *(86, 121)*, *Suillus aeruginascens* [young fruit bodies] *(132)*, *S. amabilis (127)*, *S. bovinus* [c], *S. collinitus* [c] *(122)*, *S. grevillei (87)*, *S. grisellus (127)*, *S. hirtellinus (87)*, *S. hirtellus* ssp. *thermophilus (127)*, *S. luteus (87)*, *S. pictus (127)*, *S. placidus (87)*, *S. plorans (131)*, *S. punctipes (87)*, *S. serotinus*, *S. spectabilis*, *S. tomentosus (127)*, *S. tridentinus* [c] *(122)*, *Truncocolumella citrina*, *T. rubra (127)*, *Xerocomus badius (131, 562)*, *X. illudens* var. *xanthomycelinus (127)*, *X. moravicus*, *X. parasiticus (131)*, *X. rubellus (87)*, *X. subtomentosus (131, 562)*
Variegatorubin (74)	Present in almost all species which contain variegatic acid (73), in many cases detected as an artefact formed from (73) during work up. The few species quoted contain higher concentrations of (74): *Boletus erythropus (595)*, *Chalciporus piperatus (609)*, *Rhizopogon roseolus (618)*; TLC: *Boletellus emodensis*, *B. obscurecoccineus (281)*, *Boletus frostii* ssp. *floridanus (127, 595)*, *B. luridus (92, 609)*, *B. rhodoxanthus (131, 609)*, *B. rubellus* ssp. *dumetorum (127)*, *B. satanas (92)*, *B. splendidus* ssp. *splendidus (131)*, *B. subglobosus (281)*, *Suillus aeruginascens (92)*, *Xerocomus chrysenteron (607)*, *X. rubellus* (= *Boletus rubellus*) *(282)*

[a] Yields refer to percentage of fresh weight of fungus.

The hydroxylated pulvinic acids such as variegatic acid (73) may be characterised in the form of their dilactone peracetates, e.g. (75) (Scheme 10), which exhibit sharp melting points, in contrast to the parent pigments, and give rise to diagnostic infrared absorption above 1800 cm^{-1} (73). Dilactone formation also occurs on electron impact and the mass spectral fragmentation of pulvinic dilactones has proved valuable in structural elucidation (452, 453, 607, 608). On treatment with diazomethane the hydroxypulvinic acids are converted into the

Table 9. *Properties of Common Pulvinic Acid Derivatives*

Pig-ment	Colour on TLC	TLC R_f[a]	Colour on TLC +K$_3$FeCN$_6$: NaHCO$_3$	Ultraviolet/Vis. [λ_{max} (log ε), ethanol]	Permethyl Derivative M.p.	Per-acetate M.p.	References
(71)	Yellow	0.36	No change	258 (4.12), 387 nm (3.81)	170–171°	271°	(425, 584)
(72)	Yellow	0.23	Blue	261 (4.03), 411 nm (3.83)	148–149°	221–223°	(202, 607)
(73)	Yellow	0.18	Blue	275 (4.02), 415 nm (3.88)	145°	282° (dec.)	(73, 607)
(74)	Red	0.41	Grey-blue	510 nm (4.34)	–	260–261°	(609)

[a] Silica gel Merck Kieselgel HF$_{254}$ (benzene:ethyl formate:formic acid = 13:5:4) (131).

corresponding permethyl derivatives (73) (Table 9). Methyl variegatate (76) was obtained on treatment of 3,3′,4,4′-tetra-acetoxypulvinic lactone (75) with methanolic potassium hydroxide (73).

The blue colour formed on oxidation of variegatic acid (73) by oxidases or by alkaline potassium ferricyanide is due to the formation of delocalised hydroxyquinone methide anions of type (77) (Scheme 10) (607). The electronic spectrum of anions such as (77) is in close accord with that of the blue anion (78) generated by oxidation of catechol in the presence of Meldrum's acid (700). Due to its nature as a vinylogous carboxylic acid a substantial proportion of the hydroxyquinone methide (77) is present in the form of its anion even at neutral pH.

The common red pigment of boletes, variegatorubin (74), is formed by oxidation of variegatic acid (73). This facile reaction takes place even when variegatic acid is allowed to stand in solution, especially if traces of acid are present (609), and has been carried out on a preparative scale with hydrogen peroxide as oxidant in the presence of either a copper catalyst (609) or ammonium tungstate (202). Under these conditions a 1,2-benzoquinone moiety is generated in ring C followed by intramolecular addition of the carboxyl group. The formation of the second lactone ring in variegatorubin (74) brings the extended chromophore into coplanarity thus explaining the dramatic bathochromic shift observed when one compares the electronic spectra of the pigments (73) and (74). The out-of-plane twisting of the aryl ring situated near the carboxyl group in the pulvinic acids themselves has been proved by an X-ray structural analysis of the lichen metabolite vulpinic acid (357). Some of the chemistry of variegatic acid is summarised in Scheme 10.

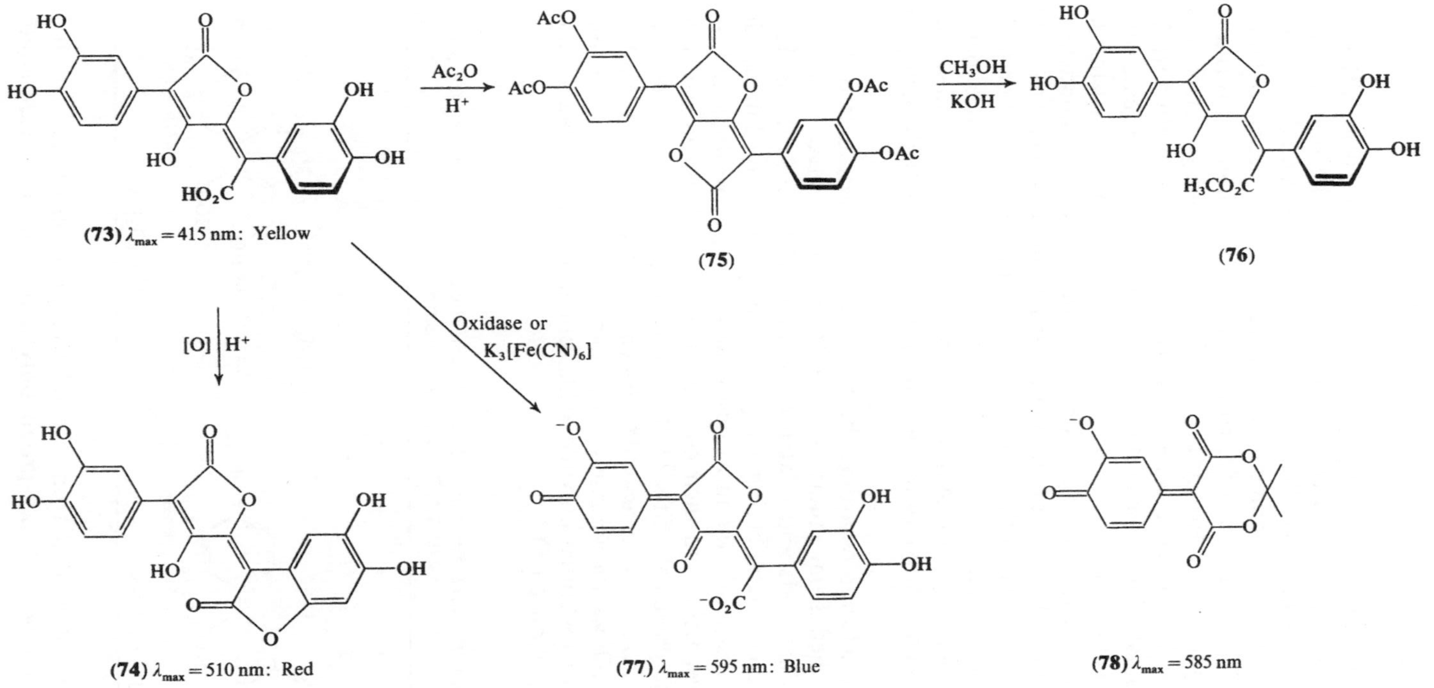

Scheme 10. Some chemistry of variegatic acid

When both aryl rings of the pulvinic acid nucleus bear different substituents structure elucidation is complicated by the potential for positional isomerism. Complications of this kind presented immense problems during the assignment of structures to the lichen pulvinic acids [cf. (4, 5)]. In the case of xerocomic acid the validity of formula (72) was established by EDWARDS and GILL by chemical correlation of the pigment from *Chroogomphus rutilus* with synthetic methyl tri-O-methylxerocomate (79) (Scheme 11). The structure of the synthetic substance (79) was proved by its degradation to methyl 4-methoxybenzoyl-formate (81) on ozonolysis (202). A superior degradative method which involves alkaline cleavage of permethylated pulvinic acids was later developed by the same authors. In the case of methyl tetra-O-methyl-xerocomate (80), obtained from the natural pigment with diazometh-ane, the products were 2-(3,4-dimethoxyphenyl)-3-methoxymaleic an-hydride (82) and methyl (4-methoxyphenyl)acetate (83) (Scheme 11) (203).

The same conclusion regarding the substitution pattern in xerocomic acid was reached from studies of the ¹H-n.m.r. spectra of several isomeric pulvinic acids and related compounds (276, 604). Thus, H-2 and H-6 in the aromatic ring (ring A) which abuts the butenolide nucleus are consistently the most deshielded aromatic protons and re-sonate at lower field than their counterparts in ring C of the isomeric structure. The chemical shifts of the aromatic protons of xerocomic acid (72) and isoxerocomic acid (84) shown in Table 10 are illustrative. This method allows the unambiguous assignment of structure to un-symmetrically substituted pulvinic acids, pulvinones and related com-pounds (see Section 2.1.5).

Table 10. ¹H-n.m.r. Data for Xerocomic Acid (72) and Isoxerocomic Acid (84) (δ values, with TMS as internal standard)

(72) [D₆] acetone   (84) [D₆] acetone

As can be anticipated from the relative position of the catechol ring and the enolic hydroxy group only xerocomic acid (72) could

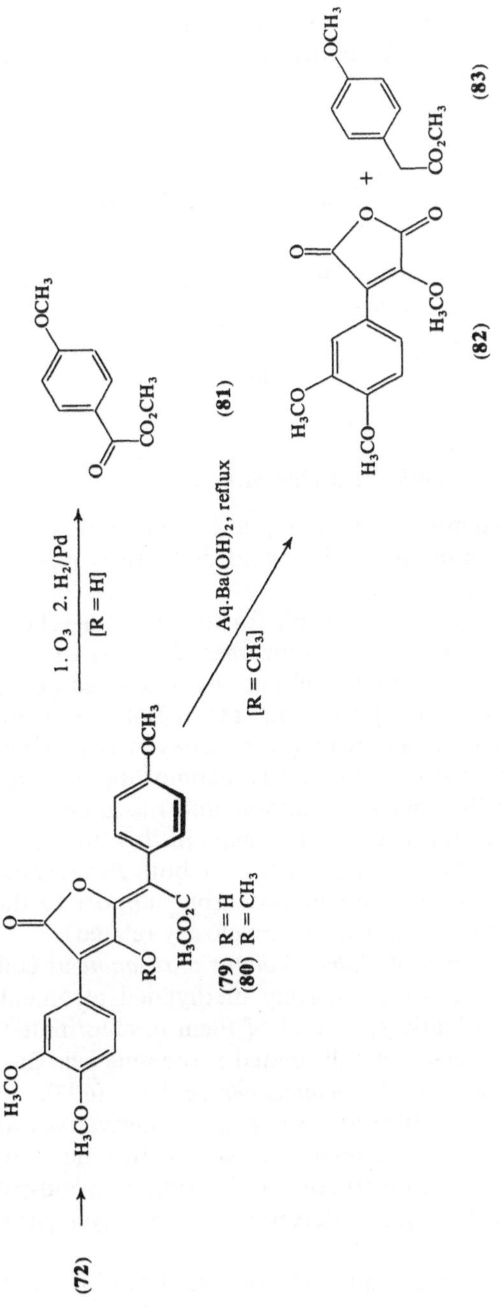

Scheme 11. Degradation of xerocomic acid derivatives

be oxidised to a blue hydroxyquinone methide anion analogous to (77). On the other hand, only isoxerocomic acid (84) was transformed into xerocomorubin (85) by hydrogen peroxide in the presence of cupric or tungstate ion (202, 604).

(85)

B. Less Common Pulvinic Acid Derivatives

Besides atromentic acid, xerocomic acid, and variegatic acid a number of less common pulvinic acid derivatives have been isolated from Basidiomycetes (Table 11).

Vulpinic acid (86) is responsible for the greenish-yellow appearance of the fruit bodies of *Pulveroboletus ravenelii* (474) and of *Boletus sub-globosus* (281). The occurrence of a nonhydroxylated pulvinic acid bio-genetically related to polyporic acid (11) in Boletales is most unusual. The co-occurrence of (86) with the hydroxylated pigments variegatic acid (73), variegatorubin (74), and xerocomorubin (85) in fruit bodies of *B. subglobosus* is indeed remarkable and raises questions about the biogenetic relationship between pigments in this fungus. The isolation of vulpinic acid as the principal pigment in both *Pulveroboletus ravenelii* and *Boletus subglobosus* adds considerable support to the suggestion by Singer (582) that these species are closely related.

The fruiting bodies of *Pulveroboletus auriflammeus* collected in Japan contain a series of extensively methylated atromentic acid and xerocomic acid derivatives, several of them in chlorinated form (474). The presence of traces of a chlorinated xerocomic acid had been noted before in sporophores of *Xerocomus chrysenteron* (607).

Isoxerocomic acid (84) and some of its derivatives are produced by cultures of *Serpula lacrimans* (89, 90). Variegatic acid (73) is also produced by this most destructive and insidious wood-rotting fungus and may be used for on-site detection by thin layer chromatography (125).

A series of variegatic acid derivatives (93) to (95) specifically methylated at the 3-hydroxy group is formed in cultures of *Hygrophoropsis*

Table 11. *Structures and Distribution of Less Common Pulvinic Acid Derivatives*

Pigment	R^1	R^2	Name	Occurrence
	—	—	(86) Vulpinic acid	*Boletus subglobosus* (281), *Pulveroboletus ravenelii* [*ca.* 8%, dry weight]; TLC: *P. frians* (474)
	H	H	(87) Methyl 4,4'-di-*O*-methyl-atromentate	*Pulveroboletus auriflammeus* (474)
	H	Cl	(88) Methyl 3'-chloro-4,4'-di-*O*-methylatromentate	*P. auriflammeus* (474)
	Cl	Cl	(89) Methyl 3,3'-dichloro-4,4'-di-*O*-methylatromentate	*P. auriflammeus* (474)
	H	—	(79) Methyl 3,4,4'-tri-*O*-methyl-xerocomate	*P. auriflammeus* (474)
	Cl	—	(90) Methyl 3'-chloro-3,4,4'-tri-*O*-methylxerocomate	*P. auriflammeus* (474)

Table 11 (continued)

Pigment	R¹	R²	Name	Occurrence
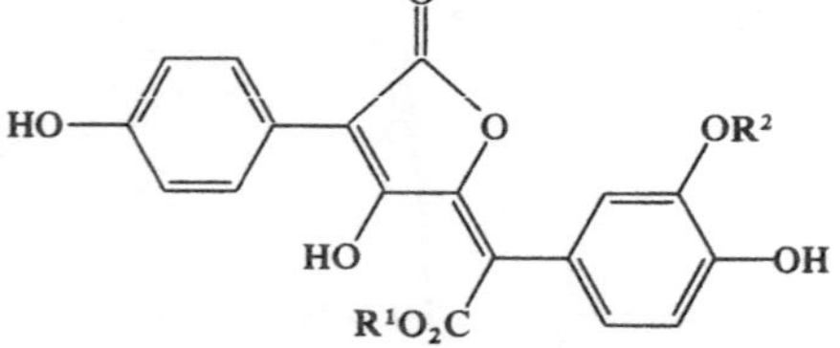	H	H	(84) Isoxerocomic acid	Serpula lacrimans [cultures] (89, 90)
	CH₃	H	(91) Methyl isoxerocomate	Rhizopogon roseolus (90)
	H	CH₃	(92) 3'-O-Methylisoxerocomic acid	Serpula lacrimans [c] (90)
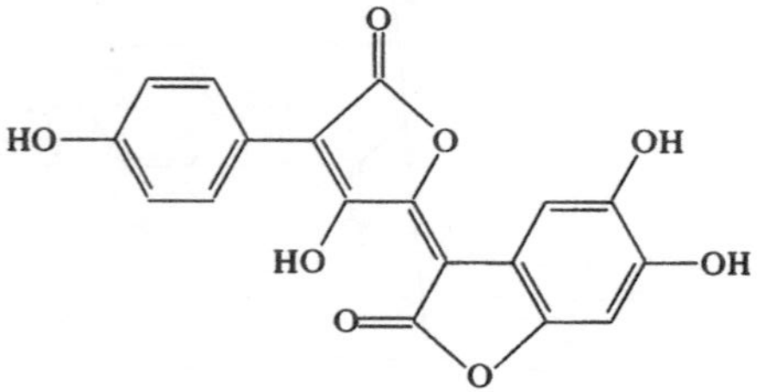	—	—	(85) Xerocomorubin	Serpula lacrimans [c] (89, 90); TLC: Boletus subglobosus (281)
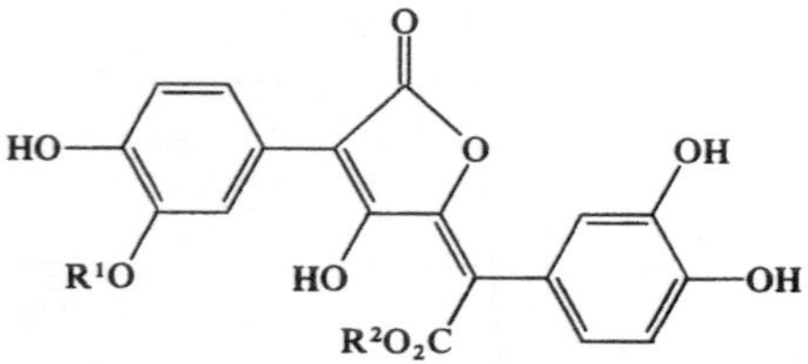	H	CH₃	(76) Methyl variegatate	Hygrophoropsis aurantiaca [1.7 × 10⁻³%, fresh fungus] (198), [c] (91); TLC: Leucogyrophana mollusca [c] (89); (76) has been detected by TLC as an artefact formed from (73) during extraction with acidified methanol (127)
	CH₃	H	(93) 3-O-Methylvariegatic acid	Hygrophoropsis aurantiaca [c] (91)
	CH₃	CH₃	(94) Methyl 3-O-methylvariegatate	Hygrophoropsis aurantiaca [c] (91)

References, pp. 253–286

(95) 3-*O*-Methylvariegatorubin   *Hygrophoropsis aurantiaca* [c] *(91)*

(96) Methyl per-*O*-methyl-variegatate   *Hygrophoropsis aurantiaca* [trace] *(198)*

(97) Gomphidic acid   *Gomphidius glutinosus (608)*

(98) Methyl bovinate   *Suillus bovinus* [c] *(99)*

aurantiaca (91). The methyl ester and permethylated derivatives (76) and (96), respectively, of variegatic acid are present in the orange fruit bodies of this same fungus (*198*).

Gomphidius glutinosus contains xerocomic acid and the novel pulvinic acid derivative gomphidic acid (97) in its yellow stalk base (*608*). The hydroxylation pattern in gomphidic acid was assigned from the [1]H-n.m.r. spectrum of the pigment which exhibits chemical shifts for the protons of the AA'BB'-system of ring C identical with those observed in the spectrum of xerocomic acid (Table 10) (*604*).

A unique pulvinic acid derivative, methyl bovinate (98), which contains an extra carbon atom in the form of the carbonyl group of an additional δ-lactone ring has been isolated by BESL *et al.* from cultures of *Suillus bovinus (99)*. Its structure followed from the spectroscopic data, particularly the proton coupled ^{13}C-n.m.r. spectrum.

During preliminary studies on the North American bolete *Boletus morrisii* a polar pigment has been isolated from dried fungal material for which a glycosidically bound variegatic acid structure is probable (*364*).

C. Synthesis and Biosynthesis of Pulvinic Acid Derivatives

Most of the hydroxylated pulvinic acids from Basidiomycetes were first synthesised (*73, 200, 607, 608, 609*) using the classical methods of VOLHARD (*687*) and ASANO (*52*). For unsymmetrically substituted pulvinic acids such as xerocomic acid (72) the procedure is shown in Scheme 12 (*202*). The use of sodium hydride as base in the second condensation step is recommended (*11*) in order to reduce contamination of the dinitrile (99) with symmetrical by-products which arise *via* retro-aldol processes when sodium ethoxide is used. This route has also been followed for the synthesis of ^{14}C-labelled compounds (*514*).

In the cases of xerocomic acid and gomphidic acid the classical route afforded mixtures of the natural pigments and their regioisomers. EDWARDS and GILL (*202*) were able to separate the mixture of (72) and (84) obtained from the dilactone (100) by fractional crystallisation. Interestingly, when the mixture of (72) and (84) was heated with hydroiodic acid in acetic acid for longer than one hour exclusive formation of xerocomic acid (72) was observed.

Symmetrically substituted pulvinic dilactones are conveniently obtained from terphenylquinones by oxidative ring cleavage. This may be brought about with hydrogen peroxide (*425*), with lead tetra-acetate (*248*) or best with dimethyl sulphoxide and acetic anhydride (*493, 715*). Because of the ease with which terphenylquinones themselves may be derived by methoxide catalysed rearrangement of grevillins (*462*)

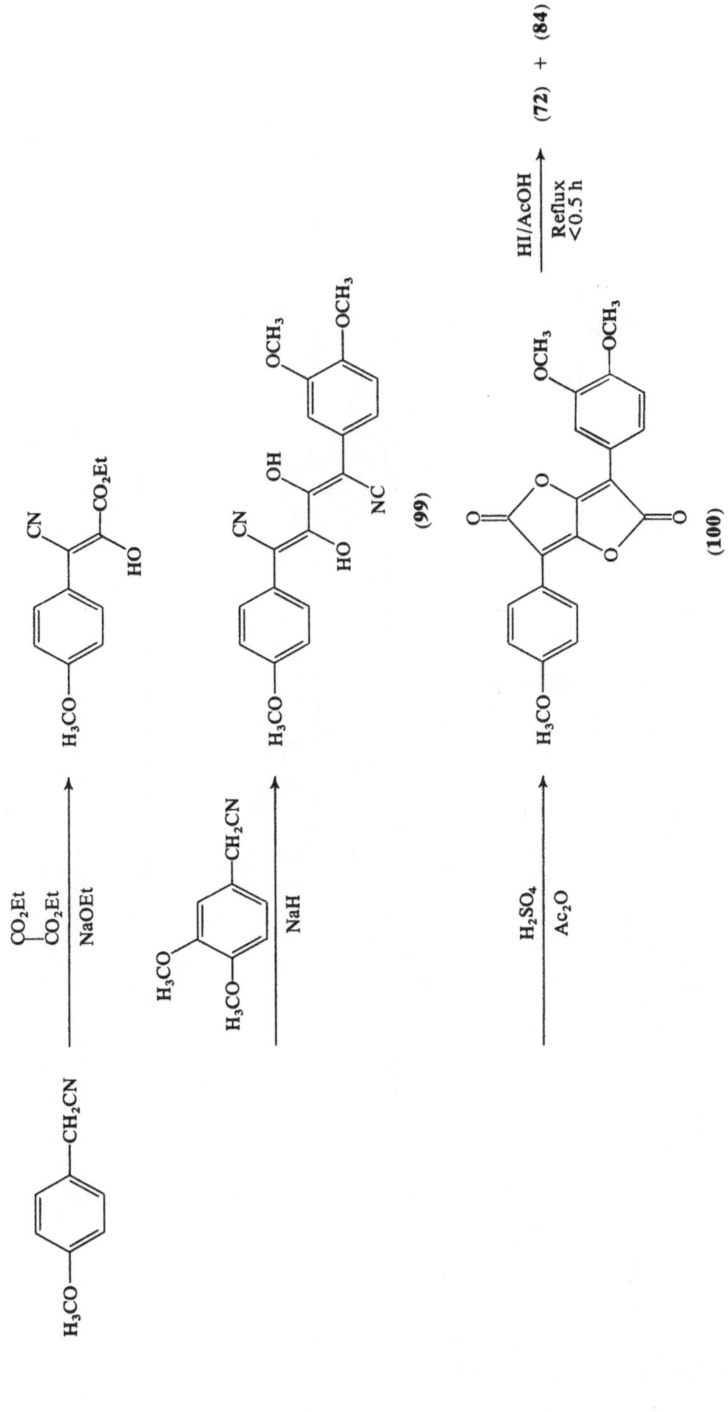

Scheme 12. Synthesis of xerocomic acid

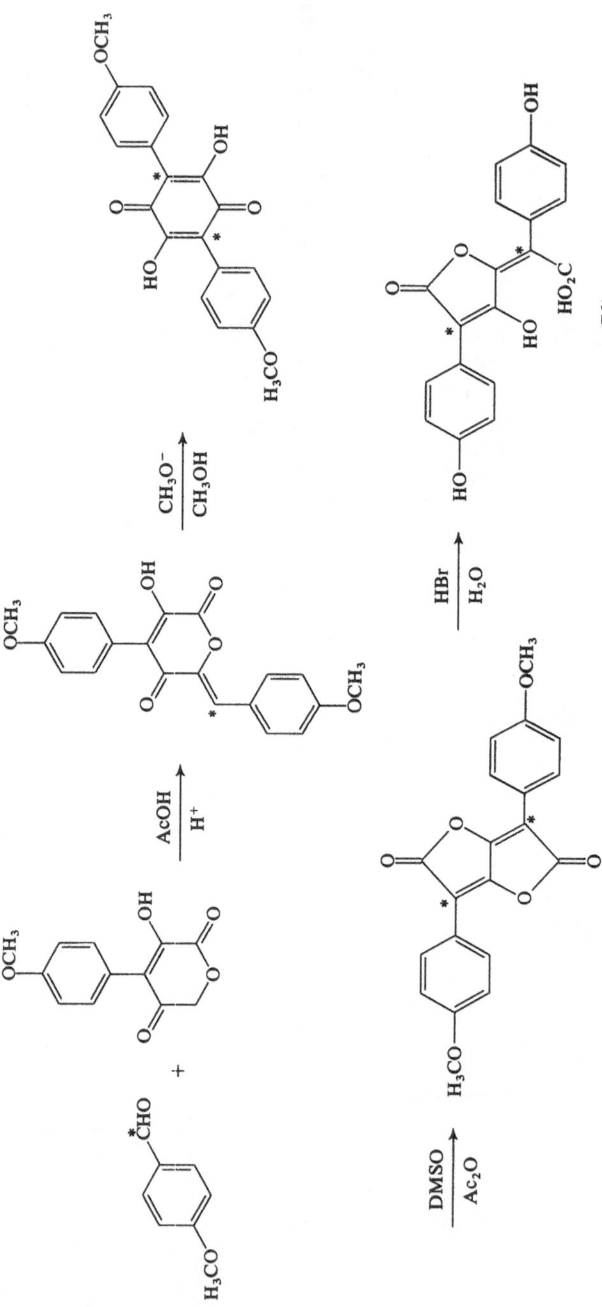

Scheme 13. Synthesis of ^{13}C-labelled atromentic acid

(Scheme 5) this procedure too constitutes an attractive alternative to the classical route to pulvinic acids, e.g. for the synthesis of specifically ^{13}C-labelled atromentic acid (Scheme 13) (366).

Recently, several regiospecific syntheses of unsymmetrically substituted pulvinic acid derivatives have been developed. KNIGHT and PATTENDEN (420, 422) have obtained 3-methoxybutenolides of type (102) from the readily available maleic anhydrides (101) (202). Metallation of the butenolides at $-78°$ C followed by reaction with an aroylformate ester then gave a tertiary alcohol of type (103) which was dehydrated with a suspension of phosphorous pentoxide in dry benzene (Scheme 14, route A). Whereas dehydration of the carbinol (103; $Ar^1 =$ 4-methoxyphenyl, $Ar^2 =$ phenyl) gave O-methylpinastric acid (104) possessing the natural (E,E)-geometry, the carbinol (103; $Ar^1 = 3,4,5$-trimethoxyphenyl, $Ar^2 = 4$-methoxyphenyl), in contrast, afforded only permethylgomphidic acid with the unnatural (E,Z)-geometry shown in structure (107) (422, 521). Nevertheless, the ^{1}H-n.m.r. spectrum of the permethyl derivative (107) was considered to offer strong support for the structure (97) for natural gomphidic acid. PATTENDEN and coworkers (521) have synthesised gomphidic acid (97) with the correct geometry commencing again from the maleic anhydride (101; $Ar^1 = $ 3,4,5-trimethoxyphenyl). In this second approach (Scheme 14, route B) the anhydride was reacted with the zinc enolate derived from methyl (4-methoxyphenyl)acetate which afforded the intermediate carbinol (106) (40%) as a single diastereoisomer. The assignment of the geometry shown assumes a chair-like conformation of the zinc chelate intermediate (105) in which the most bulky groups occupy equatorial orientations. Elimination of the elements of water from (106) was brought about via the mesylate derivative and produced, in a 5:1 ratio, the photolabile (E,Z)-isomer (107) and the (E,E)-derivative (108) of natural gomphidic acid. Permethylgomphidic acid (108) when exposed to iodotrimethylsilane in chloroform yielded the free pigment (97).

RAMAGE and coworkers (537) have developed a biomimetic synthesis of pulvinic acids which relies for its success on the facility with which dioxolanones of type (110) undergo nucleophilic attack at the lactone carbonyl group with subsequent extrusion of cyclohexanone. In the synthesis of xerocomic acid (Scheme 15) the dioxolanone (110), obtained as the predominant isomer from reaction between the phosphorane (109) and methyl (3,4-dibenzyloxyphenyl)glyoxalate, was cleaved with the lithium enolate of t-butyl (4-benzyloxyphenyl)acetate. The intermediate dianion (111) probably exists at first as the chelate (112) which is then broken down on aqueous work up and subsequently cyclised specifically at the less hindered carbonyl group to produce the ester (113). The dianion (111) is analogous to the hypothetical

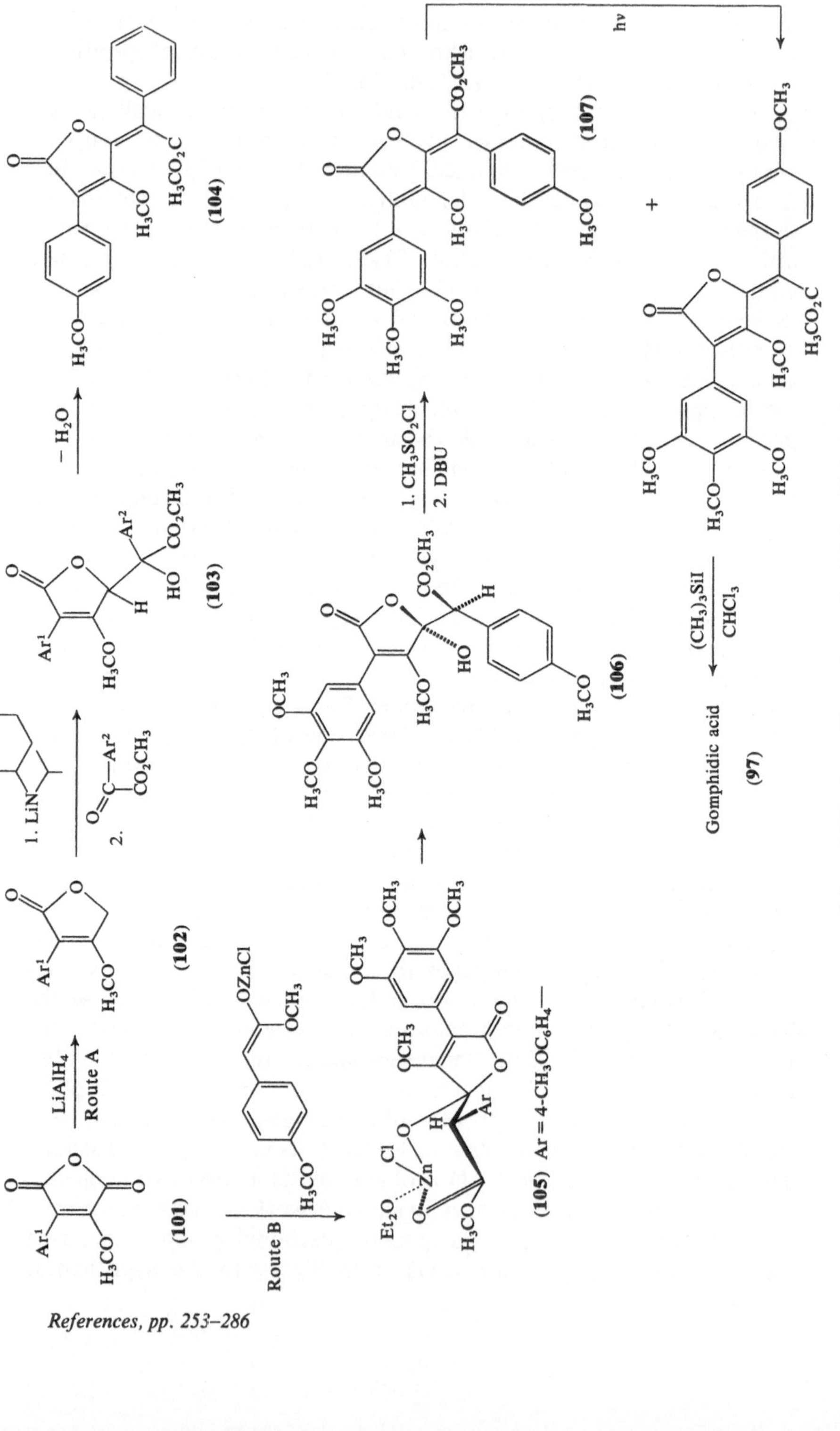

Scheme 14. Regiospecific syntheses of pulvinic acid derivatives

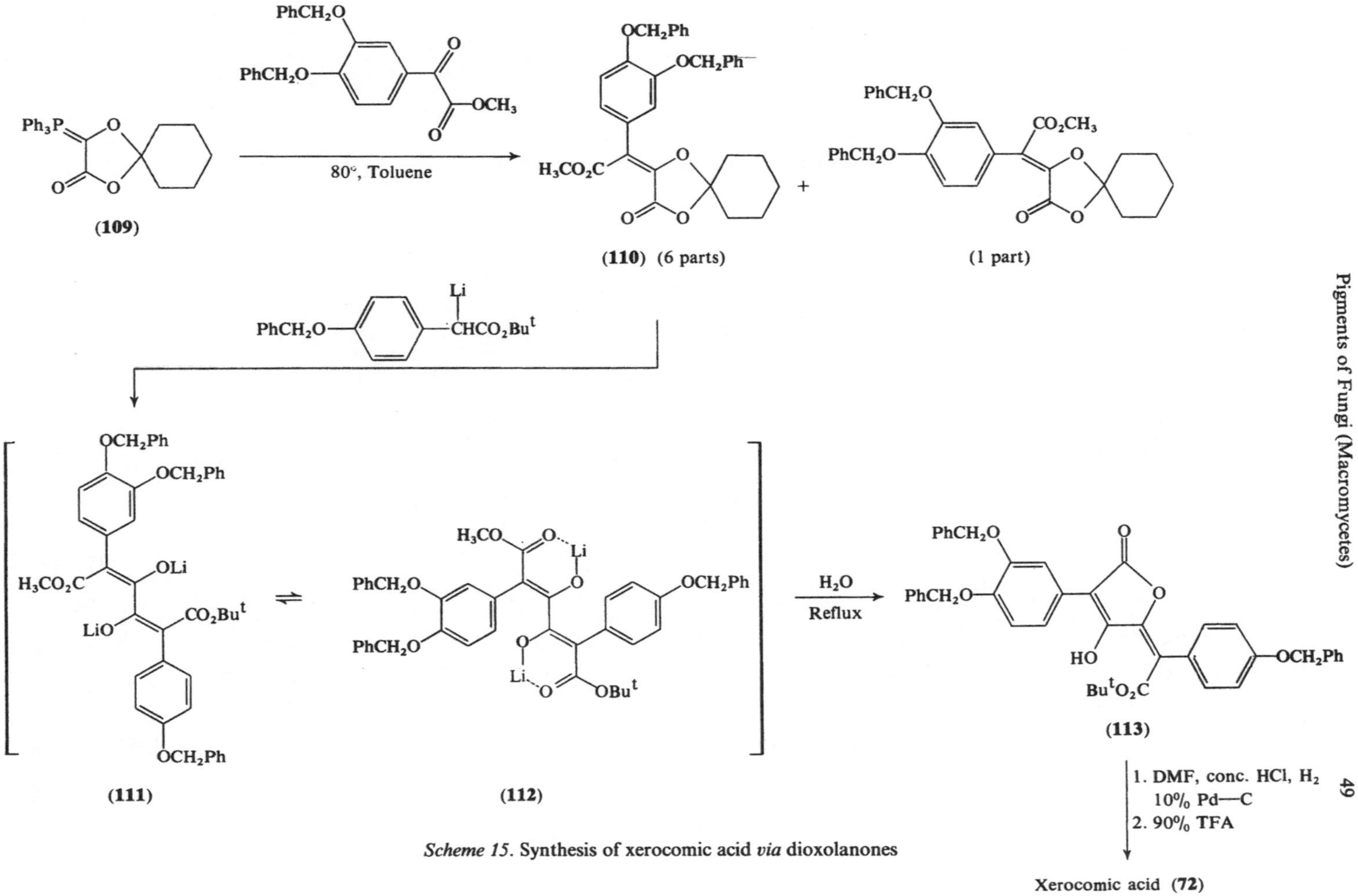

Scheme 15. Synthesis of xerocomic acid *via* dioxolanones

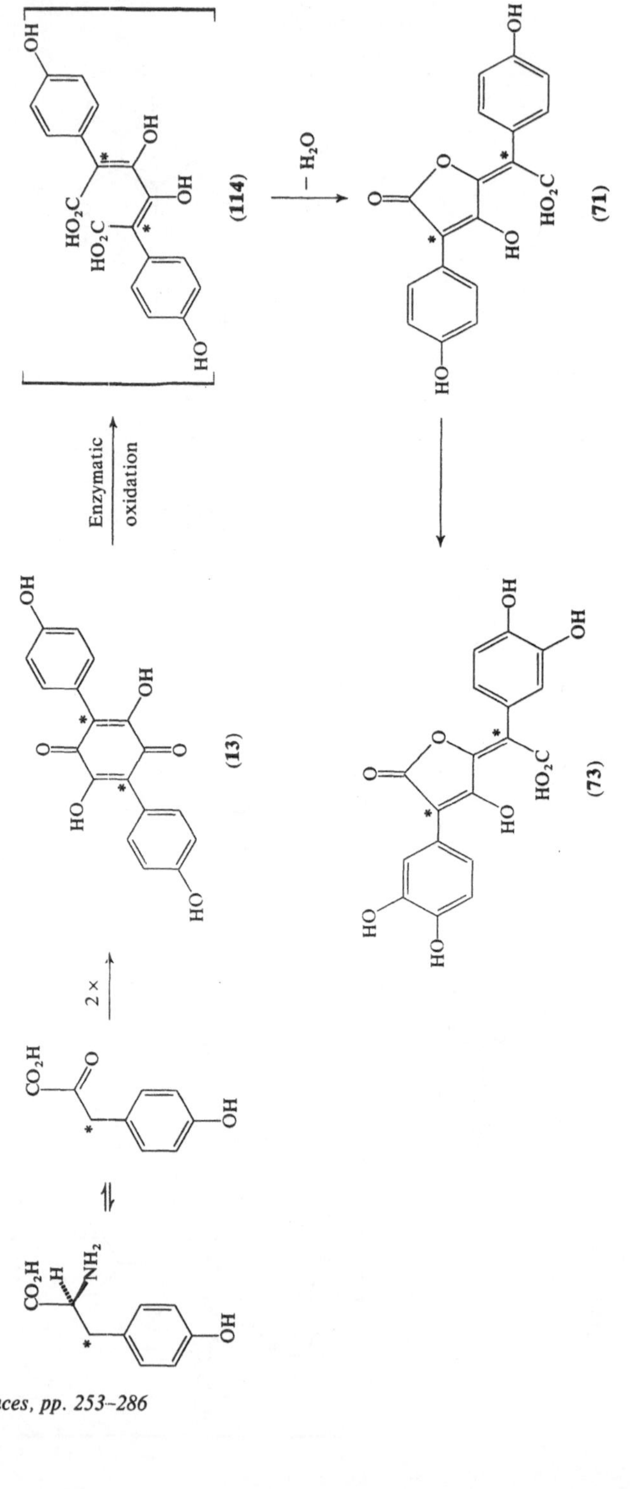

M. GILL and W. STEGLICH:

Scheme 16. Biogenesis of fungal pulvinic acids

intermediate (114) involved in the biosynthesis of pulvinic acids from terphenylquinones (see Scheme 16). After unmasking the phenolic hydroxy groups by hydrogenolysis of the benzyl ethers in (113) and cleavage of the t-butyl ester with trifluoroacetic acid, xerocomic acid (72) identical with the natural product was obtained.

The biosynthesis of fungal pulvinic acid derivatives follows the same general course as has been established for lichens (467, 468, 469, 494). Thus, DL-[3'-^{13}C]tyrosine was efficiently incorporated into variegatic acid (73) by Boletus erythropus and the role of atromentin (13) as an intermediate was proved by feeding it in ^{13}C-labelled form to fruit bodies of the same fungus. The high incorporation of both ^{13}C-labelled tyrosine and ^{13}C-labelled atromentic acid (71) into variegatic acid (73) is in accord with the biogenesis shown in Scheme 16 (366).

Xerocomic acid exhibits moderate antibiotic activity against several bacteria (533, but compare 76). In recent years numerous pulvinic acid derivatives have been synthesised, and many have been shown to possess useful anti-inflammatory properties (710).

D. Pulvinones

Closely related to the pulvinic acids is pigment-B$_3$ which was isolated from Suillus grevillei and S. grevillei var. badius by EDWARDS and GILL (204). The pale yellow pigment [UV/vis. (ethanol): λ_{max} (log ε) = 243 (4.12), 308 (infl., 3.96), 343 (4.10) and 378 nm (4.14)] was identified as a hydroxylated derivative of pulvinone from its infrared and ^{1}H-n.m.r. spectra, and from the formation of tetramethyl and tetra-acetyl derivatives. Oxidation of the latter compound with chromium trioxide gave the acetates of 4-hydroxy and 3,4-dihydroxybenzoic acid which thus established the hydroxylation pattern in the aromatic rings. A strong inference that the pigment possessed the 3',4',4-trihydroxypulvinone structure (115) was drawn from the ^{1}H-n.m.r. spectrum in which the H-2 and H-6 protons appear most deshielded (δ 7.94) indicating their close proximity to the butenolide ring (276, 604). The substitution pattern in the pulvinone (115) was confirmed by correlation of the permethyl derivative of the pigment with the methoxycyclopentenedione (116) which had been derived unambiguously from isoxerocomic acid (84) (202, 204) (see Scheme 25).

The unambiguous synthesis of 3',4',4-trihydroxypulvinone (115) has more recently been reported by RAMAGE and coworkers (Scheme 17) (536). By cleavage of the dioxolanone (117) with the lithium enolate of methyl (4-benzyloxyphenyl)acetate at −78° C the bright yellow carboxylic acid (118) was obtained in hydrated form after work up. Attempts to purify (118) brought about efficient lactone formation. Final-

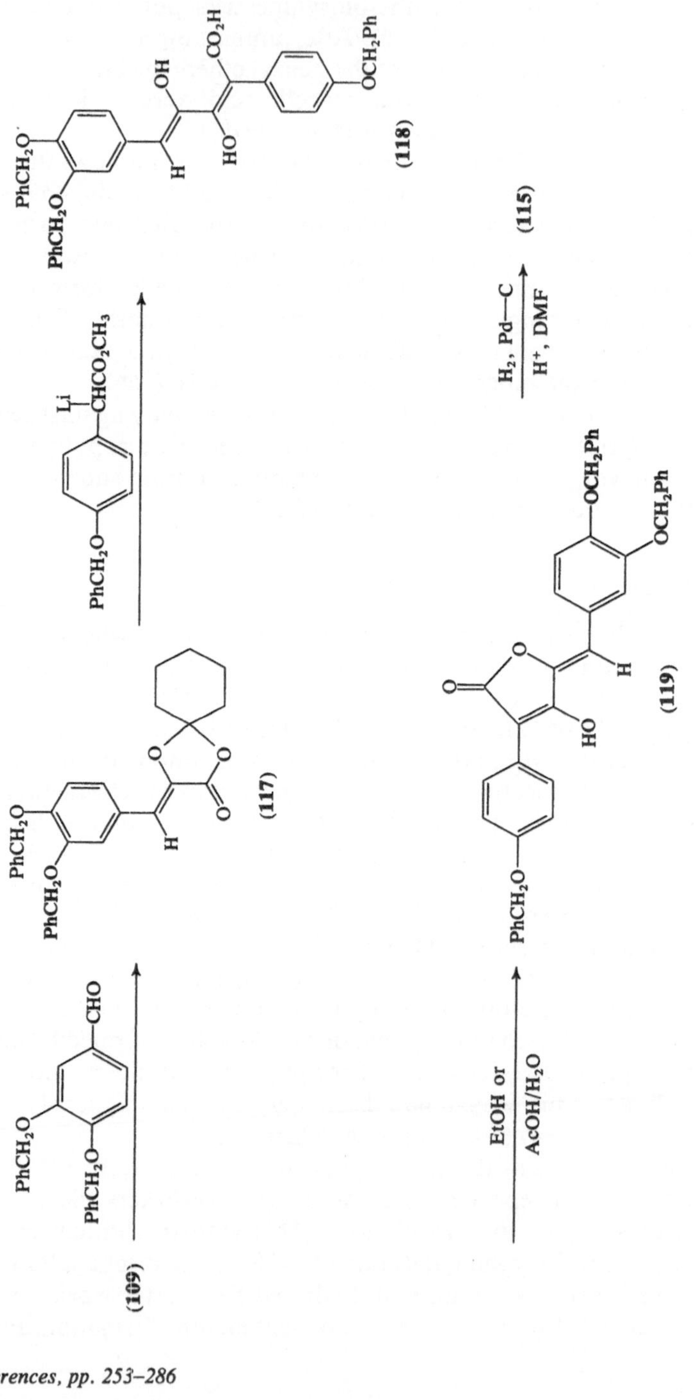

Scheme 17. Synthesis of 3′,4′,4-trihydroxypulvinone

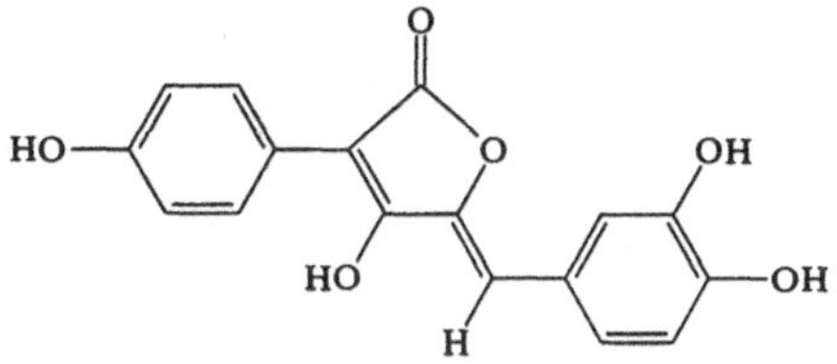

(115)

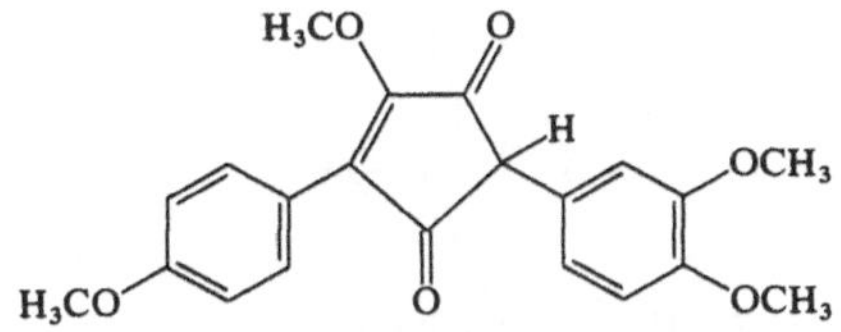

(116)

ly, the benzyl ether groups in the pulvinone **(119)** were removed by catalytic hydrogenolysis in acidic dimethylformamide to afford pulvinone **(115)** in 96% yield.

Earlier, KNIGHT and PATTENDEN (*419, 421*) had synthesised the permethyl derivative **(121)** of the pigment **(115)** from the metallated tetronic acid **(120)** and 3,4-dimethoxybenzaldehyde as shown in Scheme 18. Subsequently, the anti-inflammatory properties of several pulvinone derivatives have been recognised and a further synthesis of (*E*)- and (*Z*)-pulvinones has been developed (*153*).

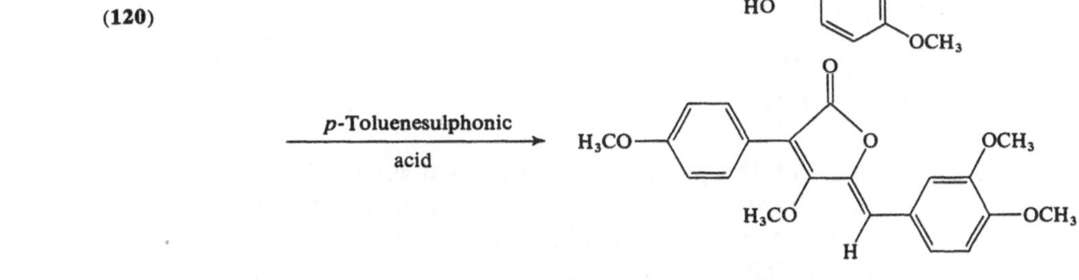

Scheme 18. Synthesis of *O*-methyl-3′,4′,4-trimethoxypulvinone

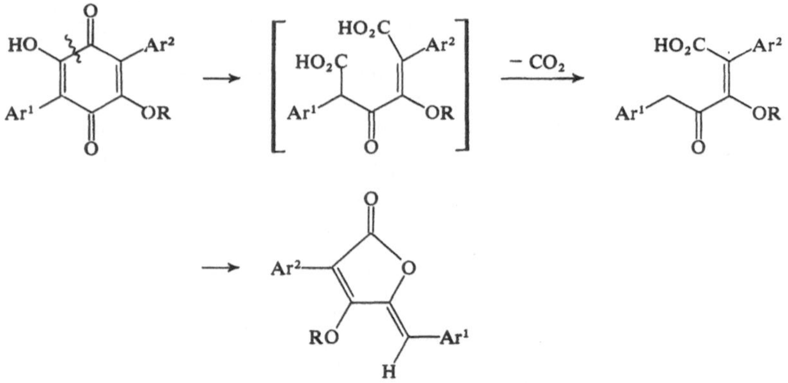

R = H, or protecting group

Scheme 19. Hypothetical biosynthetic pathway to the pulvinone nucleus

Studies of the early stages of aspulvinone biosynthesis have shown that labelled tyrosine is efficiently incorporated into aspulvinone-E (4,4′-dihydroxypulvinone) (*574*). It seems highly probable, therefore, that the later stages follow closely those described above for the fungal pulvinic acids. Thus, oxidative cleavage of an appropriate phenylpropanoid derived terphenylquinone followed by decarboxylation could furnish the pulvinone skeleton (Scheme 19) (*520*). In this regard it is interesting to note that when 3-hydroxy-6-methoxy-2,5-diphenyl-1,4-benzoquinone (**122**) is exposed to dimethyl sulphoxide and acetic anhydride (see Scheme 13) the product in high yield is *O*-methylpulvinone (**123**) (*493, 715*).

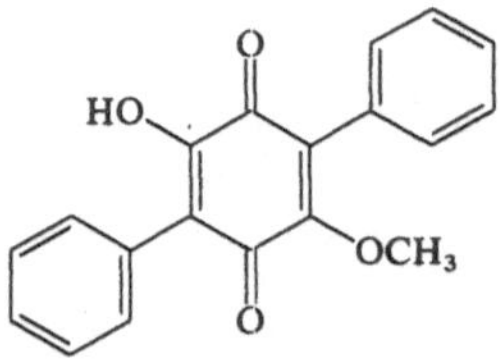

(122) 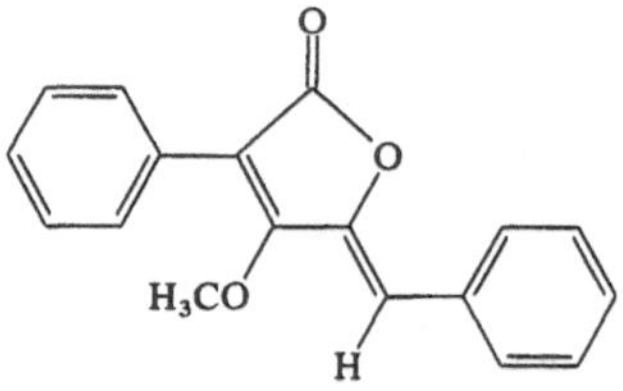   (123)

Alternatively, the pulvinone nucleus of pigment (**115**) could derive *via* oxidative decarboxylation of the hypothetical intermediate (**9**) already implicated in the biosynthesis of grevillins, terphenylquinones (and hence pulvinic acids), and the pigment (**10**).

In conclusion, it must be noted that since 3′,4′,4-trihydroxypulvinone (**115**) was isolated from *Suillus grevillei* after precipitation of the

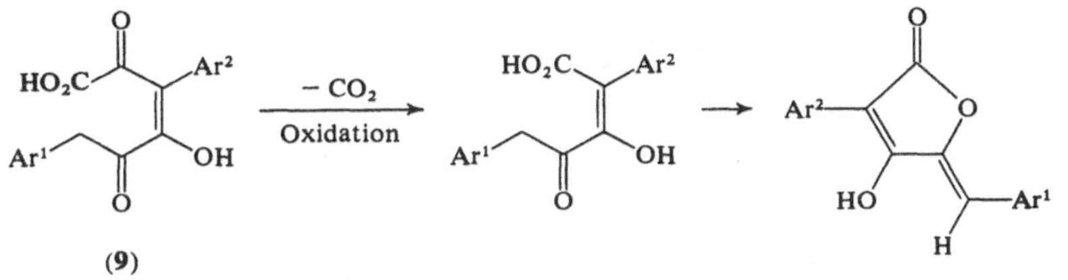

Scheme 20. Alternative biosynthesis of pulvinones

phenolic constituents in the form of their lead salts (204) the possibility that this compound is an artefact formed by oxidative degradation of grevillin-B (2) cannot be excluded.

2.1.4. Badione Group

The pigments responsible for the chocolate brown colour of the cap of the 'bay bolete' (*Xerocomus badius*) and other boletes have been investigated (594). Extraction of the cap skins with a mixture of acidified methanol and acetone yielded a reddish-brown solution from which the pigments could be separated and purified by chromatography on Sephadex LH-20.

The main pigment, badione-A (124), is easily recognised by the presence in the ^{1}H-n.m.r. spectrum ([D$_6$]DMSO) of two doublets ($J = 1.5$ Hz) occurring at unusually low field (δ 9.06 and 9.26). From the n.m.r. data the presence of two benzylidenetetronic acid moieties could

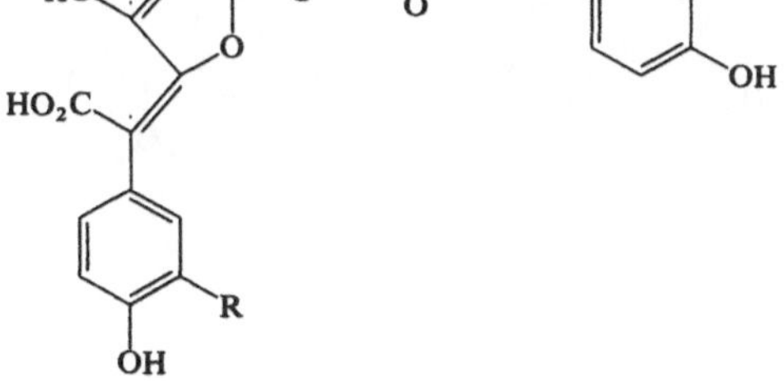

(124) R = H
(125) R = OH

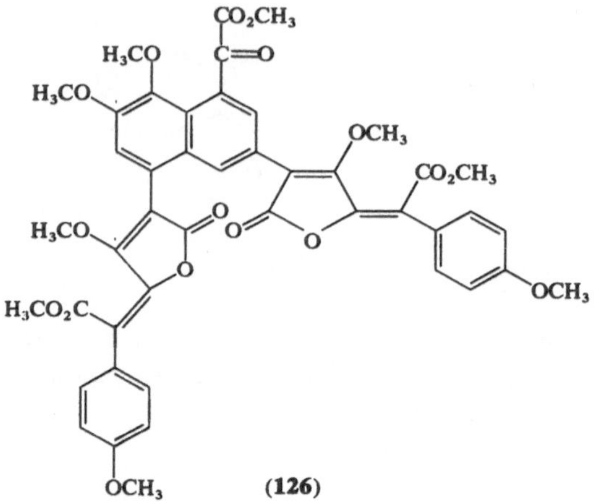

(126)

be deduced and, following systematic decouplings in the proton coupled ^{13}C-n.m.r. spectrum, structure (124) was established. On methylation with dimethyl sulphate and potassium carbonate in acetone, badione-A formed the nonamethyl derivative (126). Obviously, the presence of methanol during the methylation process has led to opening of the lactone ring. Badione-A (124) occurs mainly as a very stable potassium salt from which the metal can be removed only with difficulty. In the cap skins of *Boletus erythropus* badione-A is replaced by the more highly hydroxylated analogue badione-B (125) (*594*).

Badione-A is accompanied in the cap skins of *X. badius* by the closely related pigments norbadione-A (127), bisnorbadioquinone-A (128) and pulviquinone-A (129). The structures of these naphthalenoid pulvinic acids followed from detailed analysis of the ^{1}H- and ^{13}C-n.m.r. spectra and from their close chemical and spectroscopic relationship

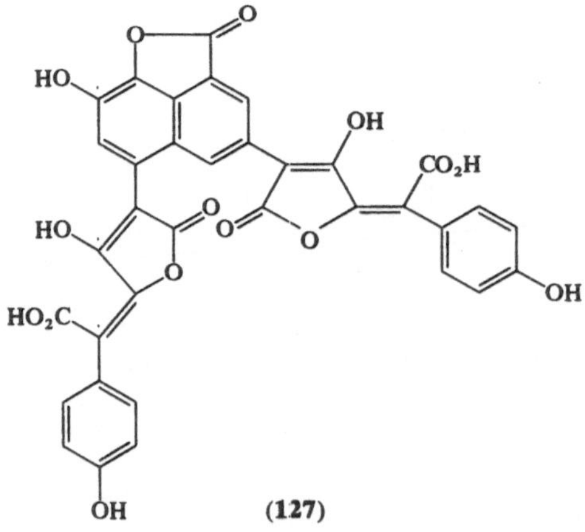

(127)

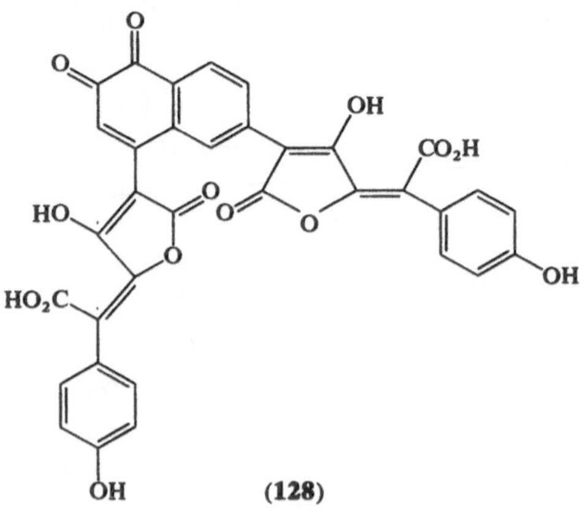

(128)

(129) R = H
(130) R = CH₃

with badione-A **(124)**. Pulviquinone-A was isolated in the form of its *O*-methyl derivative **(130)** which is probably an artefact formed during its isolation with acidified methanol.

Norbadione-A **(127)** occurs in large quantities in the important ectomycorrhizal fungus *Pisolithus arhizus* (= *P. tinctorius*) (*282*). Extraction of the fresh fruit bodies of this gasteromycete with acetone furnished directly norbadione-A as its crystalline potassium salt. The ¹H-n.m.r. spectrum of the crude acetone extracts of *P. arhizus* revealed essentially pure norbadione-A as its potassium salt, which constitutes greater than 25% of the dry weight of the fungus! The stability of this salt may be explained by effective chelation from the two neighbouring pulvinic acid chains, but its precise role in the fungus is not yet known. The free pigment **(127)** may be liberated from its salt with dilute hydrochloric acid and was identified by direct spectroscopic and chromatographic comparison with norbadione-A from *Xerocomus badius* (*282, 594*). With acetic anhydride and a trace of acid, norbadione-A **(127)** formed the crystalline lactone triacetate **(131)** and with dimethyl sulphate and potassium carbonate in acetone yielded the hepta- and nonamethyl derivatives **(132)** and **(133)**, respectively, depending on the reaction conditions (*282, 594*). Small quantities of badione-A **(124)**

(131)

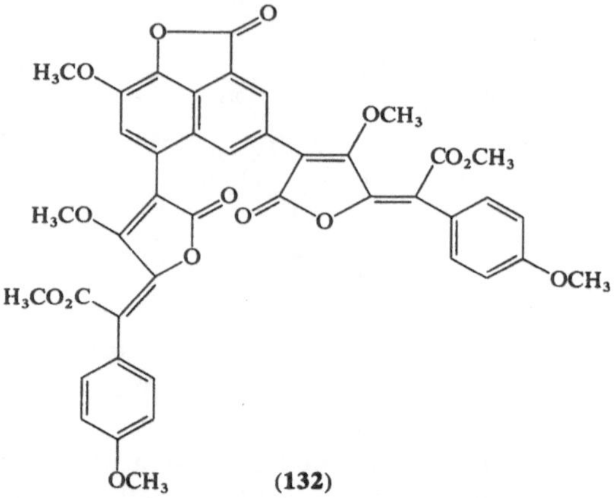

(132)

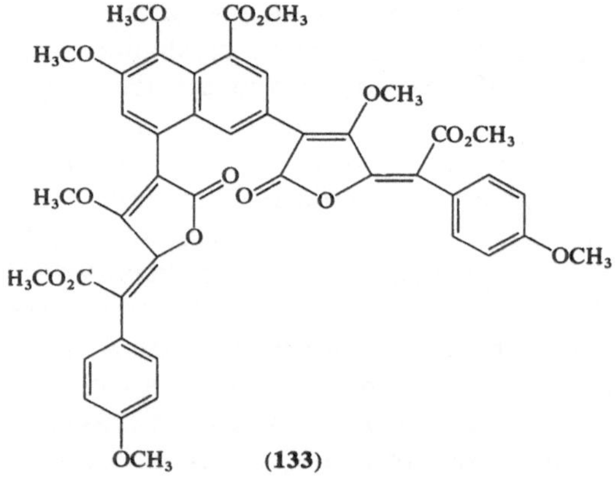

(133)

and bisnorbadioquinone-A (128) are also present in *Pisolithus arhizus* (283).

The isolation of pulvinic acid derivatives from a member of the Sclerodermataceae points to a close taxonomic relationship between these gastroid fungi and the boletes (286).

Some properties and the distribution of pigments of the badione group are summarised in Table 12.

The synthesis of bis-*O*-methylpulviquinone-A lactone (135) has been achieved by using successive grevillin-terphenylquinone-pulvinic lactone rearrangements starting from the naphthaldehyde (134) (Scheme 21) (561).

The biosynthesis of badione-A seems to occur *via* oxidative dimerisation of xerocomic acid (72). The feasibility of this hypothesis has been demonstrated *in vivo* by applying an aqueous solution of xerocomic acid to the cap of *Xerocomus badius* from which part of the

Table 12. *Pigments of the Badione Group*

Pigment	Occurrence[a]	R_f	TLC		Ultraviolet/Vis. [λ_{max} (methanol)]
			Colour	Colour + NH_3	
(124)	*Austroboletus gracilis, Boletus pinicola (594), Phylloporus rhodoxanthus (595), Pisolithus arhizus (= P. tinctorius) (283), Xerocomus badius (594), X. castanellus (595)*	0.83[b]	Brown	Green	260, 368, 482 nm (sh.)
(125)	*Boletus erythropus (594)*	0.76[b]	Yellow-brown	Green	–
(127)	*Pisolithus arhizus (282), Xerocomus badius (594)*	0.89[b]	Yellow-brown	Salmon	260, 367, 406 nm (sh.)
(128)	*Pisolithus arhizus (283), Xerocomus badius (594)*	0.83[b]	Brown	Violet	250 (sh.), 307, 356, 415 nm (sh.)
(129)[c]	*Xerocomus badius (594)*	0.24[d]	Yellow-brown	Red-violet	265 (sh.), 278, 312, 354, 422 nm

[a] Pigments of this group are concentrated in the cap skin of the various boletes cited.
[b] Aluminium oxide Merck 150 F_{254} impregnated with saturated ethanolic ascorbic acid solution (butan-2-one:formic acid:water = 15:3:2) (594).
[c] Isolated as the methyl derivative (130).
[d] Silica gel Merck 60 F_{254} (benzene:ethyl formate:formic acid = 10:5:3) (594).

brown skin had been removed. After several hours the yellow zone to which xerocomic acid had been applied had become brown, and the presence of badione-A (124) was established by chromatographic and n.m.r. comparison with authentic material (594). Badione-B (125) presumably arises by dimerisation of variegatic acid (73).

For the formation of the naphtho[1,8-*bc*]pyrandione nucleus in these pigments a mechanism involving Diels-Alder dimerisation of a 1,2-benzoquinone precursor can be envisioned (Scheme 22). The dimer (136) may be further oxidised to the intermediate (137) which could then undergo fragmentation to give (138) as indicated by the arrows. Tautomerisation and lactone ring formation then leads to the badiones-A or -B.

This hypothesis has been tested *in vitro* (Scheme 23) by treating the 'dimer' (139) of 4-methyl-1,2-benzoquinone with aqueous sodium hydroxide. Rapid acidification of the resulting solution gave the red naphtho[1,8-*bc*]pyrandione derivative (140) in high yield (610). The pyrandione (140) forms a blue anion and exhibits a ketone carbonyl

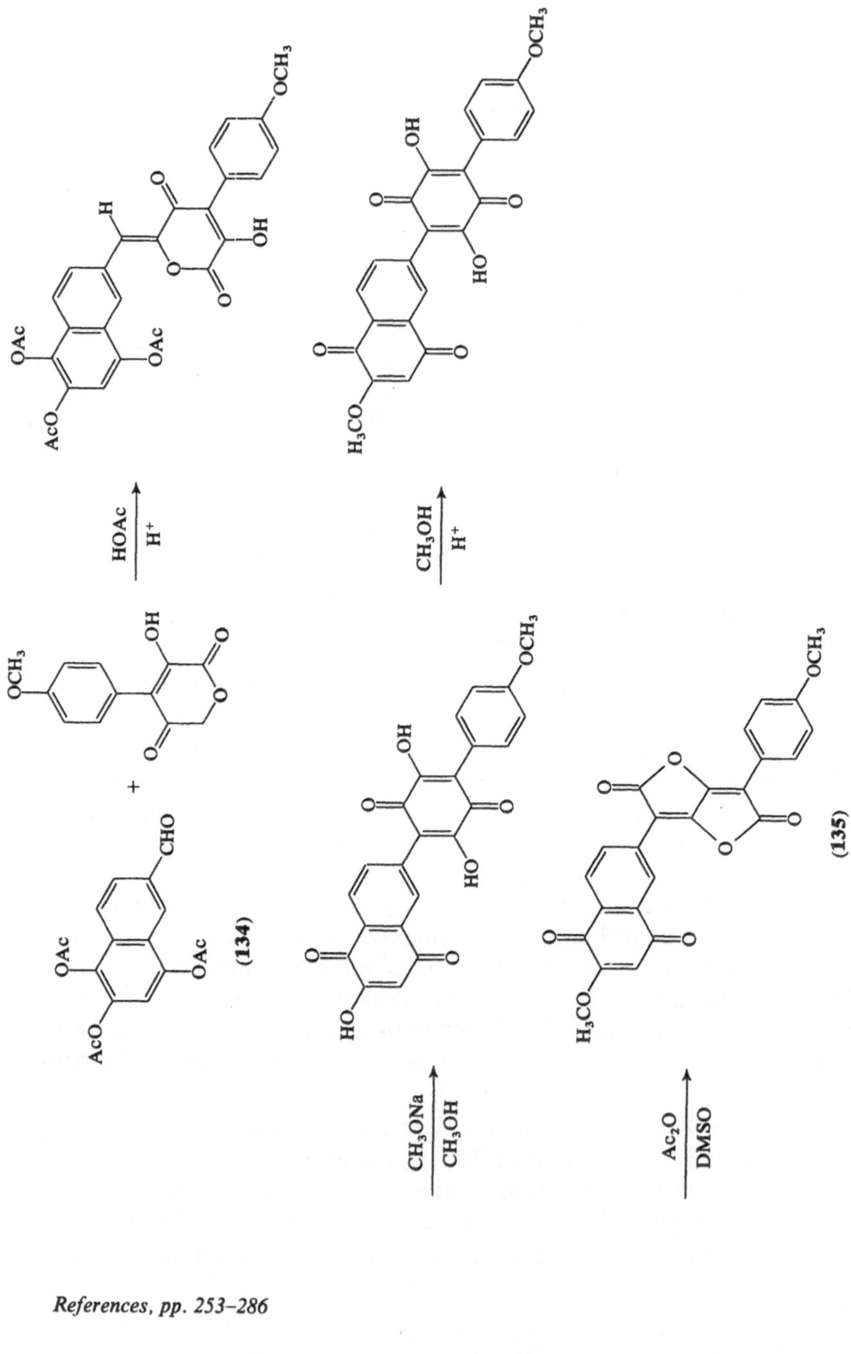

Scheme 21. Synthesis of a pulviquinone-A derivative

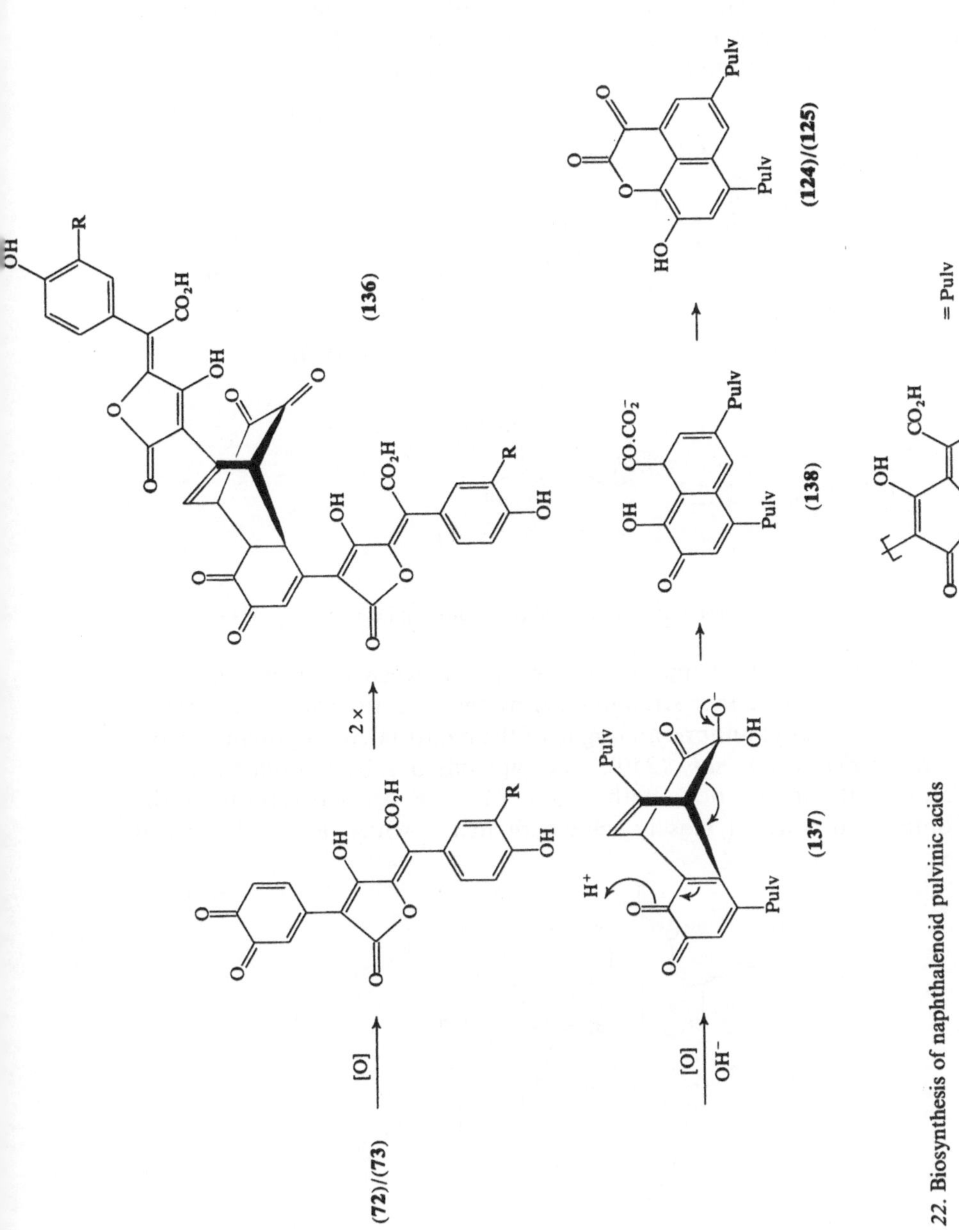

Scheme 22. Biosynthesis of naphthalenoid pulvinic acids

signal in the ^{13}C-n.m.r. spectrum at unusually high field, δ 172.8, in good agreement with badione-A (δ 173.0). This is taken to indicate a significant contribution to the ground state of the molecule from the alternative mesomeric form shown in Scheme 23.

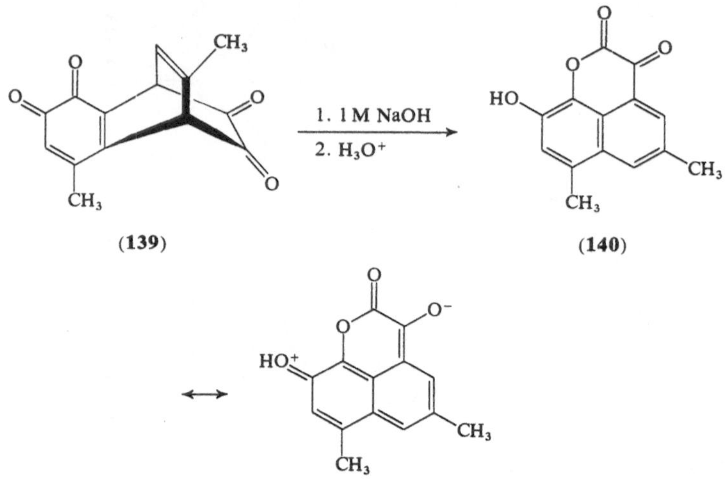

Scheme 23. Synthesis of the badione nucleus

The postulated chemotaxonomic link between Sclerodermataceae and Boletales is further strengthened by the detection of sclerocitrin-D (141), a terphenylquinone analogue of the naphthalenoid pulvinic acids, in *Scleroderma citrinum* (250). The pigment has been isolated in the form of its yellow nonamethyl derivative by permethylation of the complex mixture of closely related pigments which are to be found

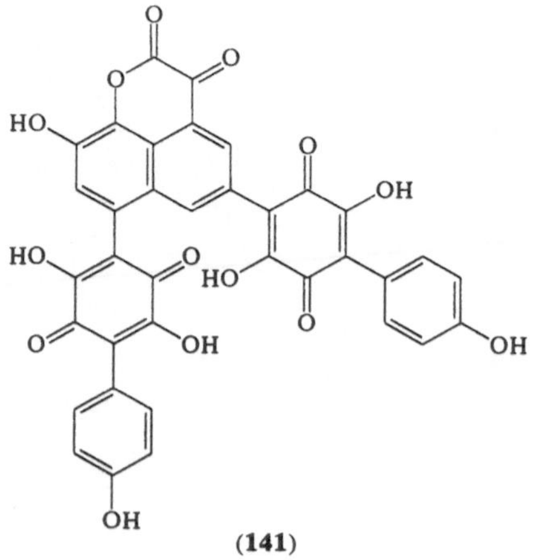

(141)

in this fungus. Extensive purification involving silica gel, Sephadex LH-20, and high performance liquid chromatography was necessary in order to obtain the substance in pure form. By these means, 7 mg of the nonamethyl derivative was obtained from 1.9 kg of fresh peridia. By assuming methanolysis of the ketolactone ring in sclerocitrin-D during methylation structure (141) has tentatively been proposed. It is supported by the similarity of the electronic and ^{1}H-n.m.r. spectra of the nonamethyl compound to those of permethylated badione-A and from its mass spectral fragmentations.

2.1.5. Cyclopentanoids

Several hydroxylated diarylcyclopentenones closely related to the pulvinic acids and pulvinones are found in some species of Boletales (Table 13). Their presence indicates the close relationship of *Gyrodon* to *Gyroporus* and the Paxillaceae and is strong evidence for inclusion of the gastroid *Chamonixia* in Boletales.

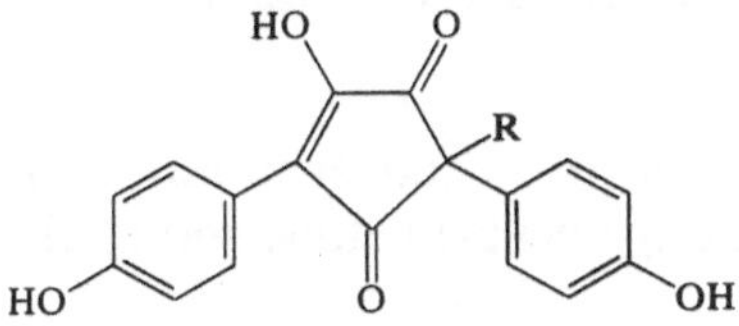

(142) R = H
(143) R = OH

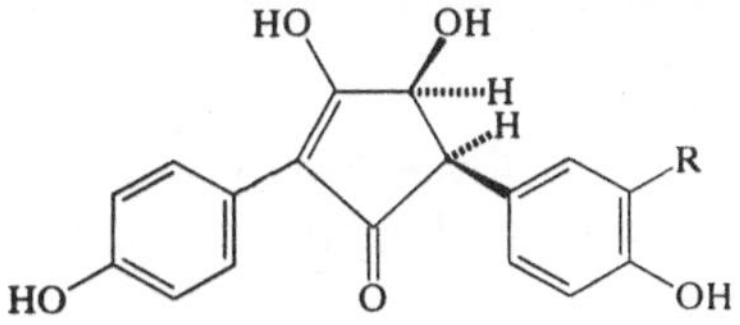

(144)* R = H
(145)* R = OH
* Relative configuration only

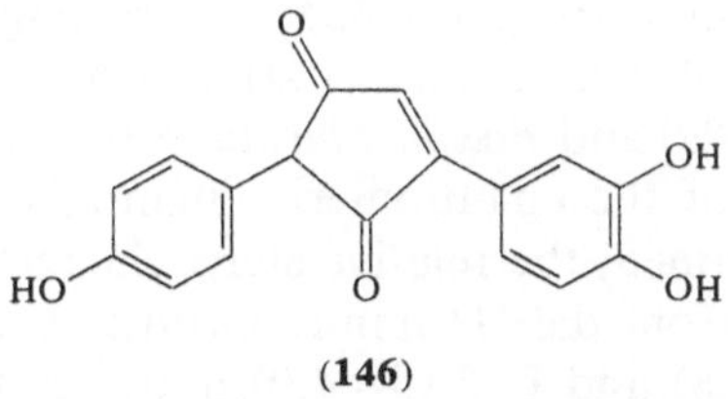

(146)

The blueing of injured sporophores of *Gyroporus cyanescens* is due to oxidation of gyrocyanin (142) to the blue anion (147) (Scheme 24) (95). The anion (147) is soluble in ethyl acetate and with sodium bicarbonate afforded the green dianion (148). Reduction of the blue anion (147) with ascorbic acid regenerated gyrocyanin (142). When the anion (147) was neutralised with dilute mineral acid a second cyclopentenone, gyroporin (143), was produced which is also present in fruit bodies of several fungi (Table 13). The cyclopentanoid structure (142) for gyro-

Table 13. *Occurrence of Cyclopentenones*

Gyrocyanin (**142**)	*Chamonixia caespitosa* [5×10^{-2}%] (*625*), *Gyroporus cyanescens* [0.1%][a] (*95*), *Lampteromyces japonicus* [cultures] (*397*), *Leccinum aurantiacum* (*625*, *692*)
Gyroporin (**143**)	*Chamonixia caespitosa* (*625*), *Gyroporus cyanescens* (*95*), *Lampteromyces japonicus* [c] (*397*), *Leccinum aurantiacum* [c] (*138*); TLC: *Albatrellus cristatus*, *Leccinum aurantiacum* (*87*), *L. eximium*, *L. oxydabile* (*127*), *L. scabrum* (*86*, *127*), *L. testaceoscabrum* (*87*), *Omphalotus illudens* (*127*), [c] (*128*), *O. olearius* [c] (*128*), *Suillus serotinus* (*127*)
Chamonixin (**144**)	(+)-Enantiomer: *Chamonixia caespitosa* [0.5%] (*625*); (−)-Enantiomer: *Gyrodon lividus* (*88*), *Paxillus involutus* (*625*); Undetermined, TLC: *Leucogyrophana pinastri* [c] (*89*), *Suillus aeruginascens*, *S. serotinus* (*87*)
Involutin (**145**)	(−)-Enantiomer: *Gyrodon lividus* (*88*), *Paxillus involutus* [ca. 3×10^{-2}%][a] (*201*), (*202*, *625*); Undetermined, TLC: *Gyrodon lividus* [c] (*87*), *Leucogyrophana pinastri* [c] (*89*), *Paxillus filamentosus* (*397*), [c] (*397*), *Suillus serotinus* (*87*)
Anhydroinvolutin (**146**)	*Paxillus involutus* (*600*)

[a] Refers to isolated yield from fresh fungus. Other yields refer to dried material.

cyanin followed analysis of spectroscopic data and synthesis of its 4,4′-di-*O*-methyl ether from 4,4′-dimethoxydibenzyl ketone and diethyl oxalate.

The blueing of fruit bodies of the gasteromycete *Chamonixia caespitosa* is also due to production of the anion (**147**). In this case the anion arises by oxidation of both gyrocyanin (**142**) and its dihydro derivative, (+)-chamonixin (**144**) (*625*). Interestingly, it is the laevorotatory enantiomer of chamonixin (**144**) which occurs in fruit bodies of *Gyrodon lividus* (*88*) and *Paxillus involutus* (*625*). Although the absolute configuration of the enantiomeric chamonixins and of involutin (**145**) remains undefined, the relative stereochemistry in the five membered ring is clear from the ^{1}H-n.m.r. spectra in which 7 Hz coupling between H-4 ($\delta \sim 4.8$) and H-5 ($\delta \sim 4.0$) indicates the *cis* arrangement between these protons (*95*, *202*, *625*).

Involutin (**145**) is responsible for the intense brown stain produced when the fruit body of *Paxillus involutus* is bruised. It formed a pentaacetate which, on oxidation with chromium trioxide or potassium permanganate solution, gave a mixture of 3,4-diacetoxybenzoic acid and 4-acetoxybenzoic acid. With dimethyl sulphate and sodium hydroxide the trimethyl ether (**151**) resulted (*201*, *202*). These observations, coupled with spectroscopic comparison with model compounds, established the nature of the cyclopentanoid nucleus and the level of hydrox-

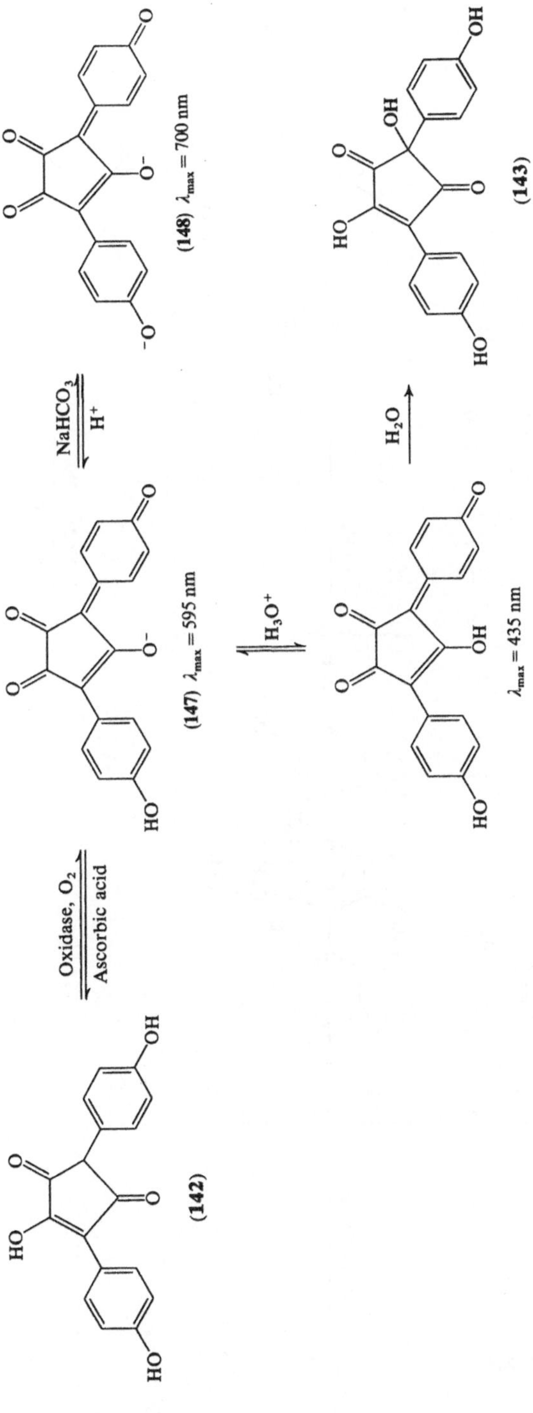

Scheme 24. Chromogenic cyclopentanoids

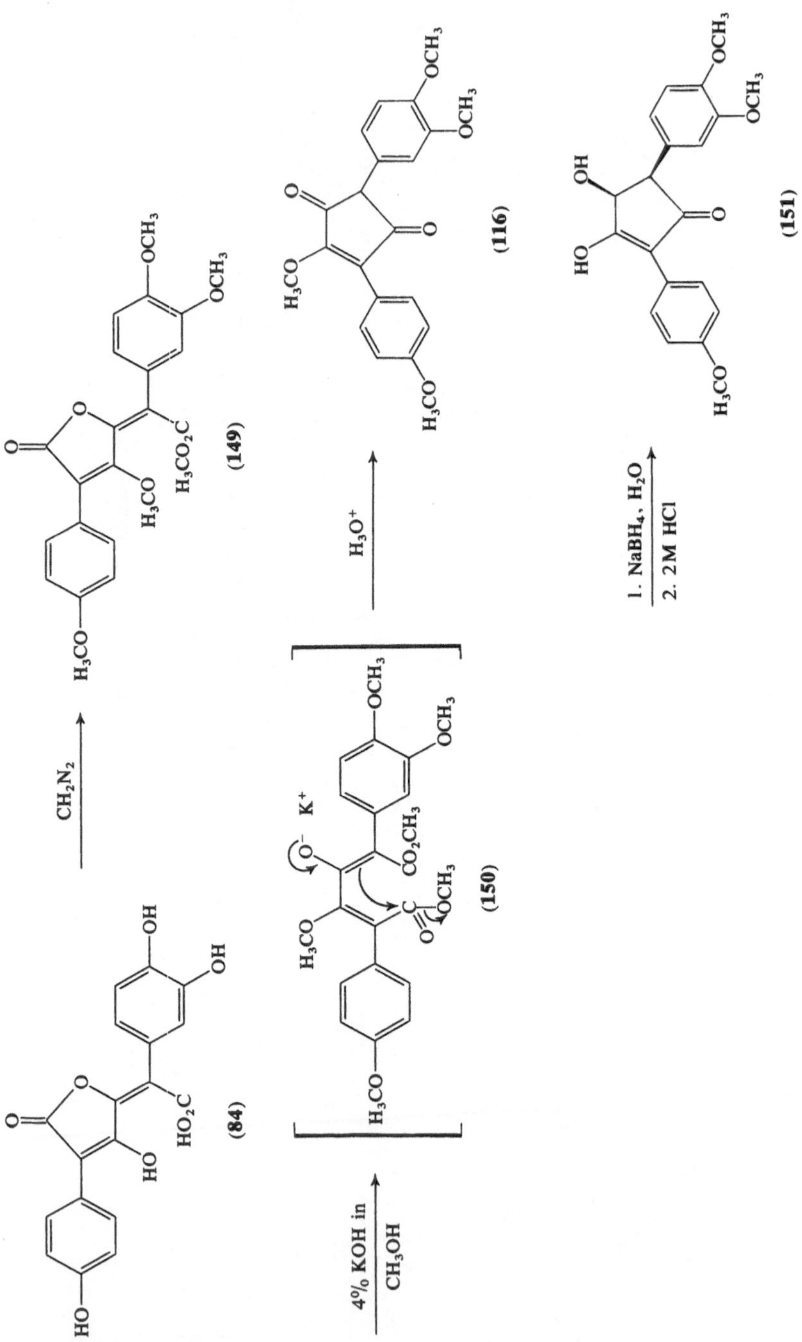

Scheme 25. Synthesis of involutin trimethyl ether

ylation in the aromatic rings but could not distinguish the structure (145) from the alternative in which the phenyl rings were interchanged. However, inspection of the ^{1}H-n.m.r. spectra of involutin and several of its derivatives (95, 202, 276) strongly suggested that the *para*-substituted phenyl ring in these compounds is subject to the deshielding effect of an adjacent hydroxyenone moiety. This conclusion was verified by synthesis of racemic involutin trimethyl ether (151) from isoxero-comic acid (84) (Scheme 25) (202). Thus, exposure of the permethyl derivative (149) of isoxerocomic acid to dilute methanolic potash produced a purple solution from which the cyclopentendione (116) could be obtained upon acidification. This rearrangement, first described by KöGL (426) for the transformation of methyl vulpinate into 4-methoxy-2,5-diphenylcyclopent-4-ene-1,3-dione, proceeds *via* Dieckmann cyclisation of the potassium enolate (150) followed by hydrolysis and decarboxylation. In the present case this led to the cyclopentendione (116) and thence to involutin trimethyl ether (151) in which the substitution pattern is unambiguously defined. The alternative trimethyl ether isomeric with (151) was produced in a similar way from xerocomic acid (72) and showed the expected trend in the ^{1}H-n.m.r. spectrum (202, 276).

Recently a yellow pigment present in *Paxillus involutus* has been isolated (600). It may be easily recognised by its characteristic colour reactions, e.g. with aniline (orange), ferric chloride (blue) and with ammonia (red-violet), and has been identified as anhydroinvolutin (146). The structure, derived from the spectroscopic data, was proved by the formation of this pigment from involutin (145) on treatment with pyridinium *p*-toluenesulphonate in benzene under reflux.

Some properties of the fungal cyclopentenones are given in Table 14.

Table 14. *Properties of Fungal Cyclopentenones*

Compound	TLC R_f [a]	Colour on TLC	Ultraviolet/Vis. [λ_{max} (log ε), methanol]	References
(142)	0.47	Lemon yellow	225 (4.18), 264 (4.28), 366 nm (4.08)	(95)
(143)	0.37	Blue-green [c]	226 (4.13), 273 (4.22), 371 nm (4.04)	(95)
(144)	0.25	Dark blue [b]	225 (4.27), 252 (4.29), 276 nm (sh., 4.14)	(88)
(145)	0.17	Red-brown [b]	200 (4.73), 254 (4.18), 279 nm (4.10)	(88, 201)
(146)	0.23	Orange-brown [b]	229, 271 (4.24), 397 nm (3.87)	(600)

[a] Silica gel Merck Kieselgel 60 (benzene:ethyl formate:formic acid = 10:5:3).
[b] After spraying with potassium ferricyanide: sodium bicarbonate.
[c] After spraying with concentrated sulphuric acid.

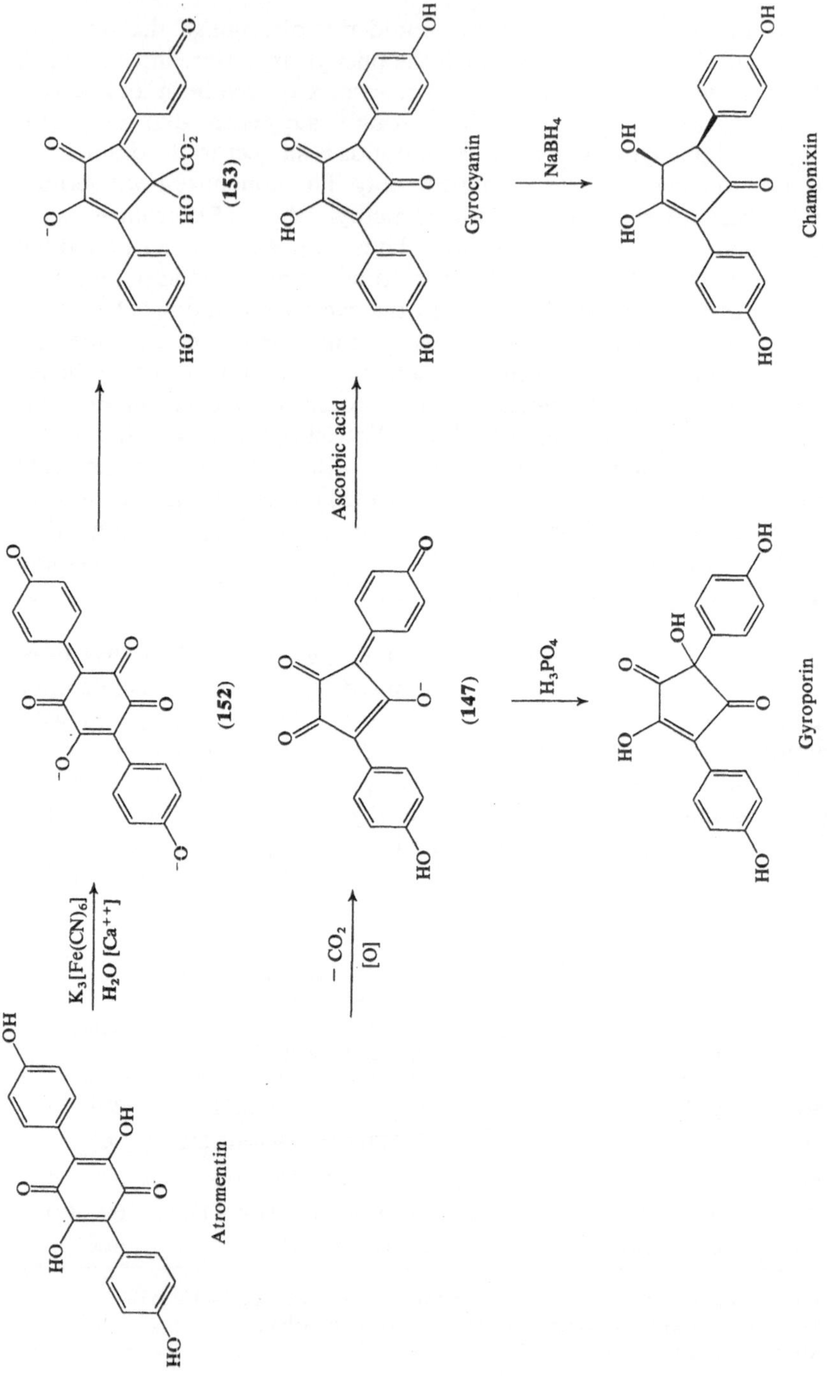

M. GILL and W. STEGLICH:

Scheme 26. Biomimetic synthesis of cyclopentenones

Several pathways for the biosynthesis of the diarylcyclopentenones may be envisioned. The possibility that the cyclopentenones arise *via* oxidative ring contraction from atromentin (13) is strongly supported by generation of the blue anion (147) when atromentin is oxidised by potassium ferricyanide in the presence of calcium ions (*95*). This ring contraction which presumably involves a benzylic acid rearrangement of the quinone methide (152) yielded the anion (147) after decarboxylation and oxidation of the carboxylate (153) (Scheme 26). Protonation of the anion (147) led to gyroporin in 19% yield from atromentin. Alternatively, reduction of the blue anion (147) by addition of ascorbic acid gave gyrocyanin which with sodium borohydride afforded chamonixin (*95, 625*).

A biogenesis of cyclopentenones from pulvinic acid precursors along the lines of Scheme 25 must also be considered (*202*). In this regard the rearrangement of *O*-methylpulvinone to 4-methoxy-2,5-diphenylcyclopent-4-ene-1,3-dione mediated by 4% methanolic potassium hydroxide (*426*) is noteworthy. Decarboxylation of an intermediate muconic acid derivative has also been invoked (*481*).

As a final alternative, the biosynthesis of fungal cyclopentenones may by-pass the terphenylquinone and pulvinic acid pathway altogether. Condensation of two 4-(hydroxyphenyl)pyruvate moieties, either directly or *via* the now familiar intermediate (9), could lead to the α-

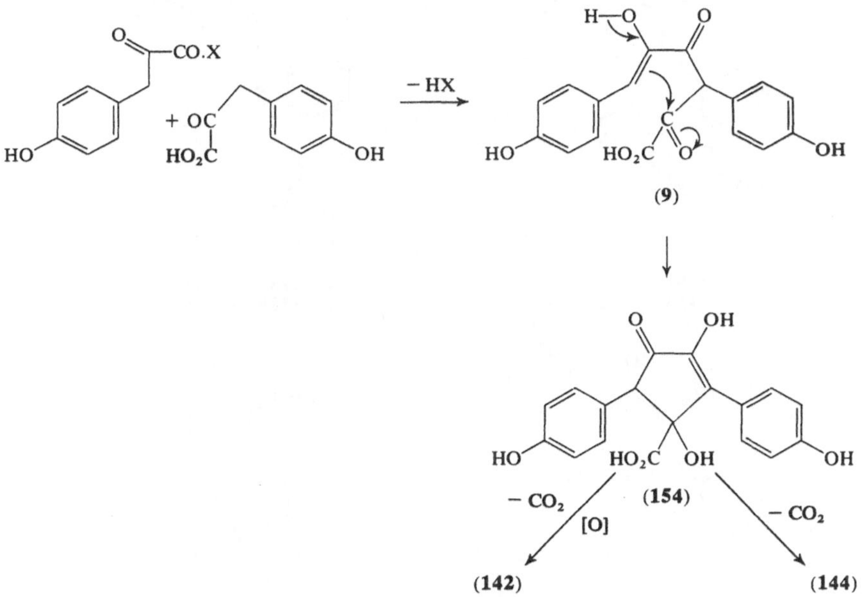

Scheme 27. Possible biogenesis of cyclopentenones

hydroxycarboxylate (154) and thence by decarboxylation to chamonixin (144) or by oxidative decarboxylation to gyrocyanin (142)
(Scheme 27).

2.1.6. Xylerythrin Group

A series of red quinone methide pigments closely related to the
terphenylquinones and the pulvinic acids has been isolated by GRIPEN-
BERG from wood infected by the fungus *Phanerochaete sanguinea* (=
Peniophora sanguinea). The main component, xylerythrin (155), exhibited characteristic lactone and chelated quinone carbonyl absorption
in the infrared spectrum and formed diacetyl and dimethyl ether derivatives (*315, 329*). Further reaction of the diacetyl derivative with acetic
anhydride led to the colourless tetra-acetate (157), while reductive acetylation of xylerythrin furnished the dihydrotriacetate (158). Analogous
reactions take place with other members of this group and point to
the common quinone methide chromophore in these pigments.

Structure (155) for xylerythrin was confirmed by synthesis involving
a Perkin-type condensation between polyporic acid and (4-hydroxy-
phenyl)acetic acid (*329*). This reaction, performed in acetic anhydride

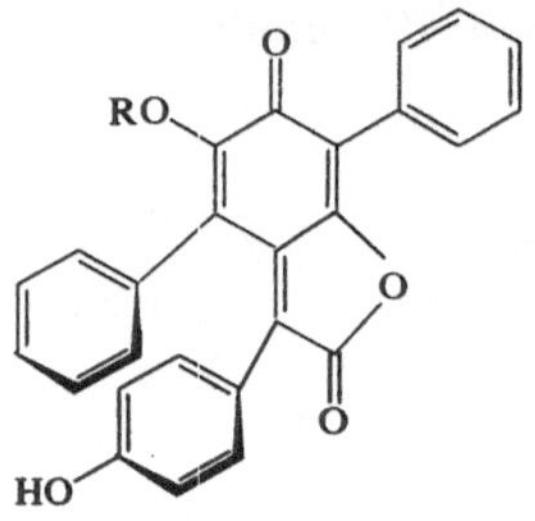

(155) R = H
(156) R = CH₃

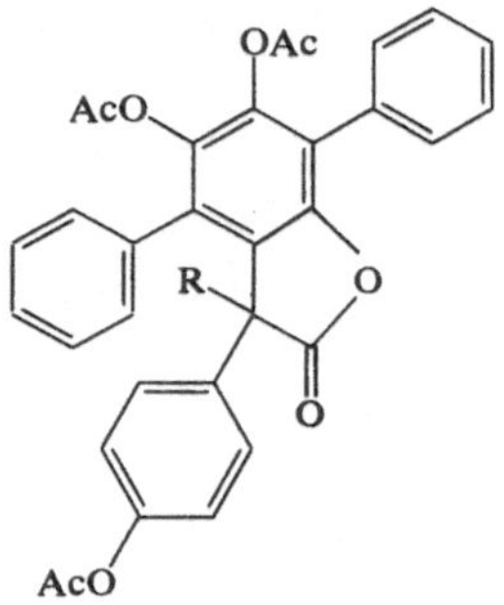

(157) R = OAc
(158) R = H

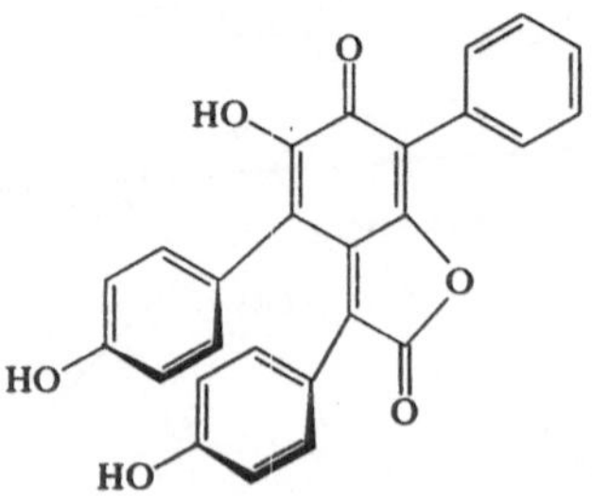

(159)

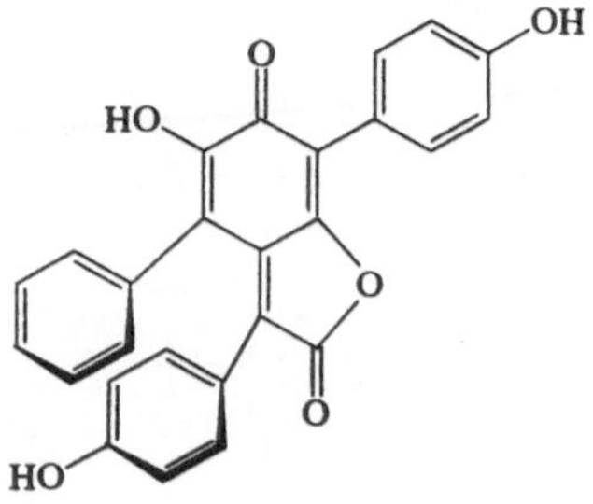

(160)

containing sodium acetate, led to the colourless tetra-acetate (157) from which xylerythrin was obtained in fair yield on acid hydrolysis and oxidation. Xylerythrin has also been synthesised by others (701). A single crystal X-ray analysis of the bis-bromoacetate derivative of xylerythrin revealed considerable steric strain in the molecule which causes the phenyl rings to twist out of the plane of the central chromophore (2, 3).

The structure of 5-O-methylxylerythrin (156) followed straightforwardly from its conversion to xylerythrin dimethyl ether with dimethyl sulphate and potassium carbonate and from its synthesis by condensation between polyporic acid monomethyl ether and (4-hydroxyphenyl) acetic acid (315, 329). In contrast, the unambiguous assignment of the positions of all hydroxy groups in peniophorin (159), a monohydroxy derivative of xylerythrin, proved impossible on chemical and spectroscopic grounds alone (330). Differentiation between the alternative structures (159) and (160) was ultimately achieved by X-ray analysis of peniophorin trimethyl ether (325). Similarly, the position of the *para*-hydroxylated phenyl ring in peniophorinin (161) was defined only after X-ray analysis of the derived dimethyl ether (317, 323). This crystallographic work served also to correct the structure originally proposed for peniophorinin in which the oxygen atom of the pyran ring

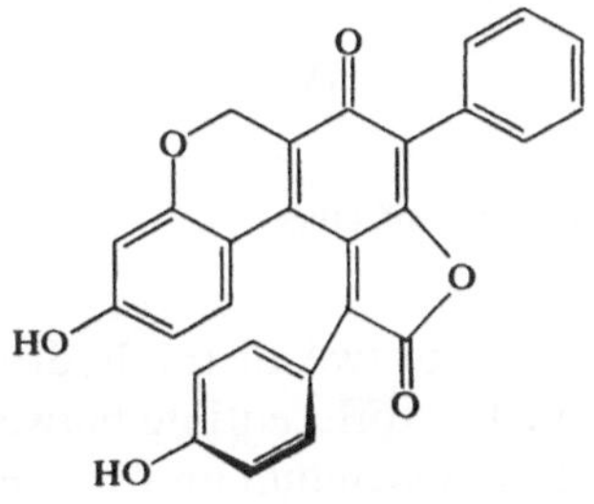

(161) 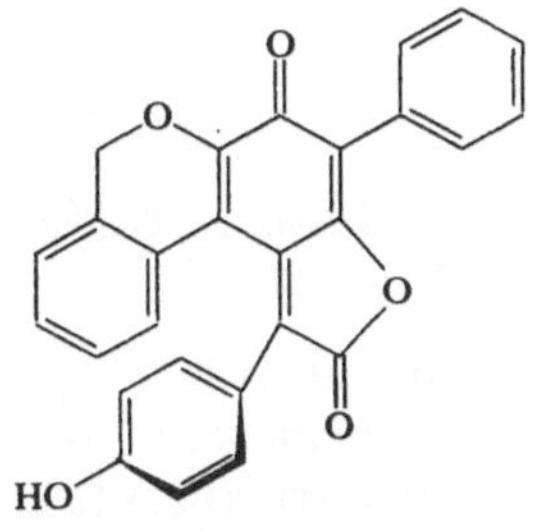   (162)

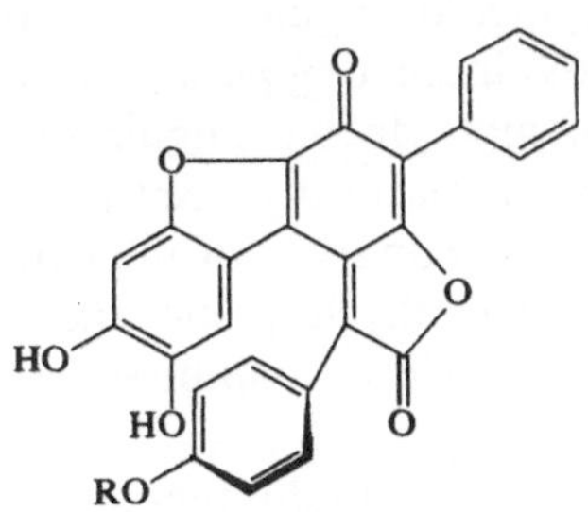

(163) R = H
(164) R = CH₃

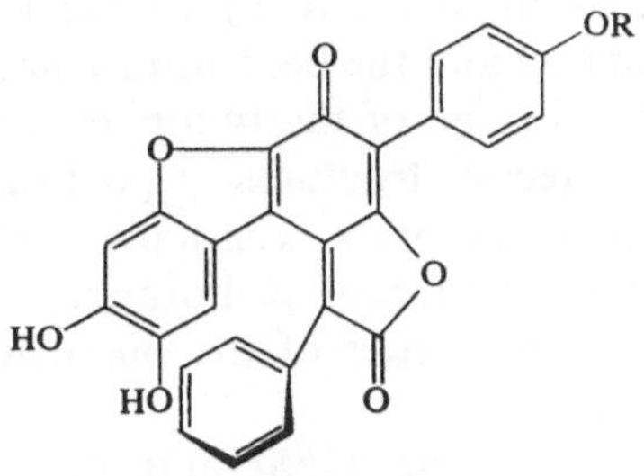

(165) R = H
(166) R = CH₃

adjoined the quinonoid nucleus (317) as is found in fact to be the case in xylerythrinin (162) (324).

In peniosanguin (163) and its O-methyl ether (164) the hydroxy group of the quinonoid ring has cyclised to form a dibenzofuran. Proof of the presence of the ether bridge was obtained on oxidation of peniosanguin trimethyl ether (= O-methylpeniosanguin dimethyl ether) (167) whereupon the methyl esters (168) and (170), together with esters of benzoic, p-anisic, and (4-methoxyphenyl)glyoxylic acids were isolated after treatment with diazomethane (Scheme 28) (318).

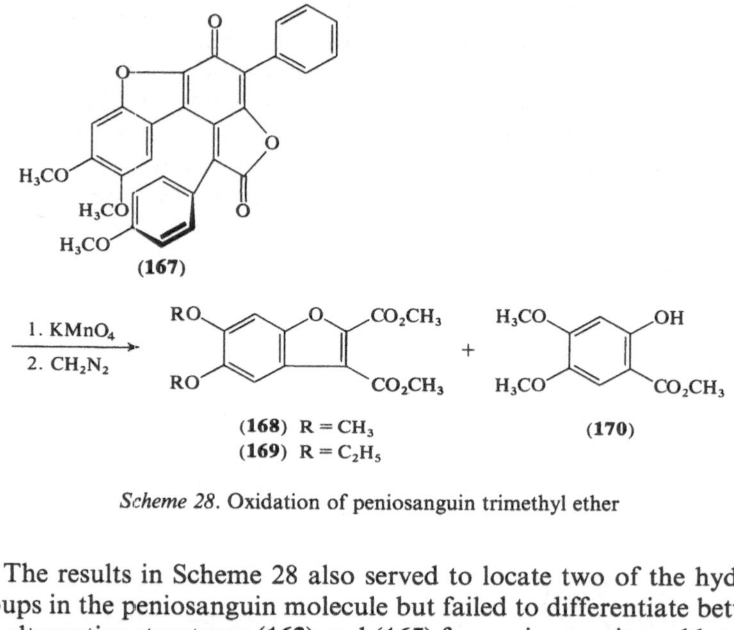

Scheme 28. Oxidation of peniosanguin trimethyl ether

The results in Scheme 28 also served to locate two of the hydroxy groups in the peniosanguin molecule but failed to differentiate between the alternative structures (163) and (165) for peniosanguin and between structures (164) and (166) for the methyl ether of this pigment. Similar oxidation of the diethyl ether of O-methylpeniosanguin gave methyl p-anisate and the benzofuran derivative (169) indicating para-methoxylation in one of the two phenyl rings in O-methylpeniosanguin (318). The precise structures of peniosanguin and its naturally occurring O-methyl ether were eventually settled by an X-ray analysis of the tribenzoyl derivative of peniosanguin (321a). Some physical and spectroscopic properties of the pigments of P. sanguinea are summarised in Table 15.

Xylerythrin (155) and the other quinone methide pigments of P. sanguinea clearly have their biogenetic origins in the assembly of three phenylpropanoid units and therefore pose a question concerning

Table 15. *Properties of Pigments from* Phanerochaete sanguinea

Pigment	Colour	M.p.	Infrared [ν_{CO}]	Ultraviolet/Vis. [λ_{max} (log ε), dioxan]	References
Xylerythrin (155)	Black	265–268°	1750, 1625, 1600 cm^{-1}	255 (4.37), 360 (3.94), 450 nm (4.18)	(329)
5-O-Methyl-xylerythrin (156)	Black	250–256°	1780, 1640, 1590 cm^{-1}	246 (4.37), 357 (4.03), 448 nm (4.24)	(329)
Peniophorin (159)	–	300–305° (dec.)	1775, 1630, 1600 cm^{-1}	265 (4.42), 395 (infl., 4.07), 453 nm (4.18)	(330)
Peniophorinin (161)	Dark brown	305–315° (dec.)	1760, 1620 cm^{-1}	278 (4.40), 292 (infl., 4.36), 380 (infl., 3.97), 447 nm (4.25)	(317)
Xylerythrinin (162)	Dark red	273–275° (dec.)	1780, 1615, 1595 cm^{-1}	273 (4.39), 365 (3.88), 437 nm (4.29)	(324)
Peniosanguin (163)	Black	310° (dec.)	1780, 1640, 1600 cm^{-1}	261 (4.37), 299 (4.15), 420 nm (4.12)	(318)
Peniosanguin-O-methyl ether (164)	–	290–295° (dec.)	1785, 1635, 1600 cm^{-1}	262 (4.47), 297 (4.36), 420 nm (4.33)	(318)
Penioflavin (171)[a]	Yellow	(i) 206° (ii) 215°	1630 cm^{-1}	266 (4.56), 298 (4.36), 314 (sh., 4.21), 372 nm (3.81)	(320)

[a] Electronic spectrum recorded in ethanol.

the nature of the 'dimer' to which the third C_6–C_3 moiety is appended. Despite the successful synthetic work of GRIPENBERG (329, 330), the obvious corollary that xylerythrin arises naturally by condensation between polyporic acid and (4-hydroxyphenyl)acetic acid (667) has so far found no support in labelling experiments. On the contrary, VON MASSOW and coworkers (693, 695) have suggested that the dimeric intermediate may be a pulvinic acid and have subsequently supported their hypothesis with incorporation experiments using ^{14}C-labelled phenylalanine and tyrosine (690, 691, 694), as well as [^{14}C]-4-hydroxypulvinic acid (694). Their results suggest the pathway depicted in Scheme 29.

According to this mode of biosynthesis, 4-hydroxypulvinic acid reacts further with phenylalanine ($\equiv$ phenylpyruvate) and with tyrosine ($\equiv$ 4-hydroxyphenylpyruvate) to give xylerythrin (155) and peniophorin (159), respectively, with loss of the carboxyl group of the amino acid. The reasonable assumption that this same carbon atom is retained ultimately as the methylene group of the pyran ring during the biosynthesis of peniophorinin (161) is, however, rendered impossible if VON MASSOW's observation that [1′-^{14}C]tyrosine is incorporated to the same

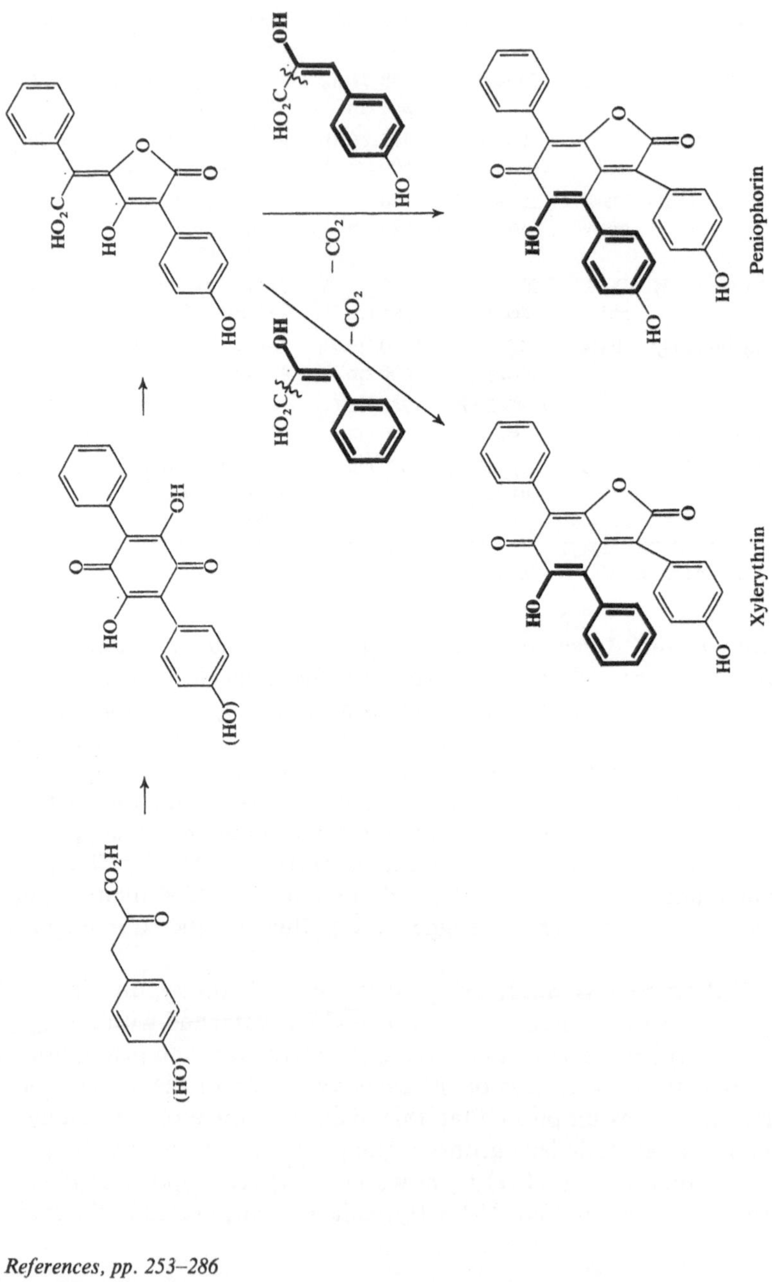

Scheme 29. Possible biosynthesis of quinone methide pigments in *Phanerochaete sanguinea*

degree both in peniophorin (**159**) and in peniophorinin (**161**) is correct (*691*). On his basis it appears that the pyran methylene carbon in both peniophorinin and xylerythrinin (**162**) must be introduced from an alternative C_1-source.

Recently, the isolation from *P. sanguinea* of an additional yellow pigment, penioflavin (**171**) (*320, 369*), has created further problems in considering the biogenesis of these pigments.

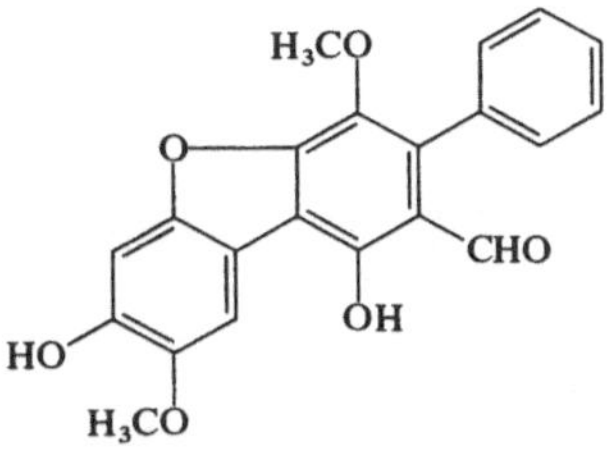

(**171**)

2.1.7. Summary

The structural and biosynthetic inter-relationships between the various arylpyruvate derived pigments described in Sections 2.1.1 to 2.1.6 are summarised in Scheme 30.

2.2. Compounds Derived from Phenylalanine and Tyrosine

L-DOPA, Betalaines and Muscaflavin

Several toadstools are able to convert tyrosine to L-3-(3,4-dihydroxyphenyl)alanine (L-DOPA; **172**). L-DOPA has been detected in the fruit bodies of *Agaricus bisporus* (*663*) and its presence in *Strobilomyces floccopus* (*605*), *Hygrocybe conica*, *H. ovina* (*619*), and *Rhodocybe mundula* (*599*) is responsible for the remarkable colour reaction ($\rightarrow$ red $\rightarrow$ black) due to melanin formation (*347*) (Scheme 31) which occurs on injury of these toadstools (*597*).

The orange, purple, and yellow compounds which in combination are responsible for the red pigmentation in the cap skin of that most distinctive of mushrooms, the 'fly agaric' *Amanita muscaria*, have been isolated and identified by Musso and coworkers. Historical aspects of the work, and details of the purification and structural elucidation of the pigments, have been reviewed by Musso (*502, 503*) and by others (*184*).

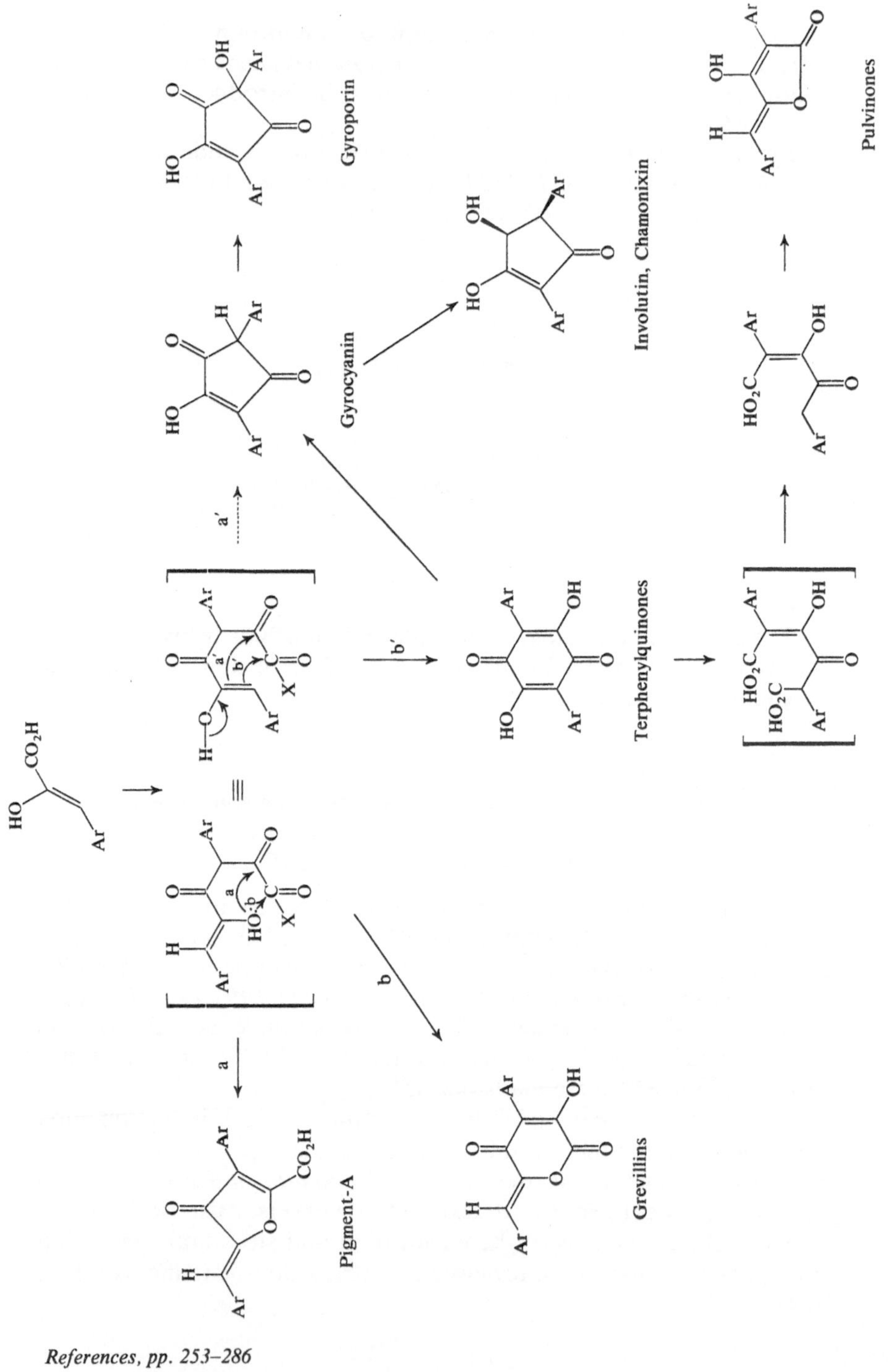
Gyroporin
Gyrocyanin
Involutin, Chamonixin
Pulvinones
Terphenylquinones
Pigment-A
Grevillins
a
b
a′
b′

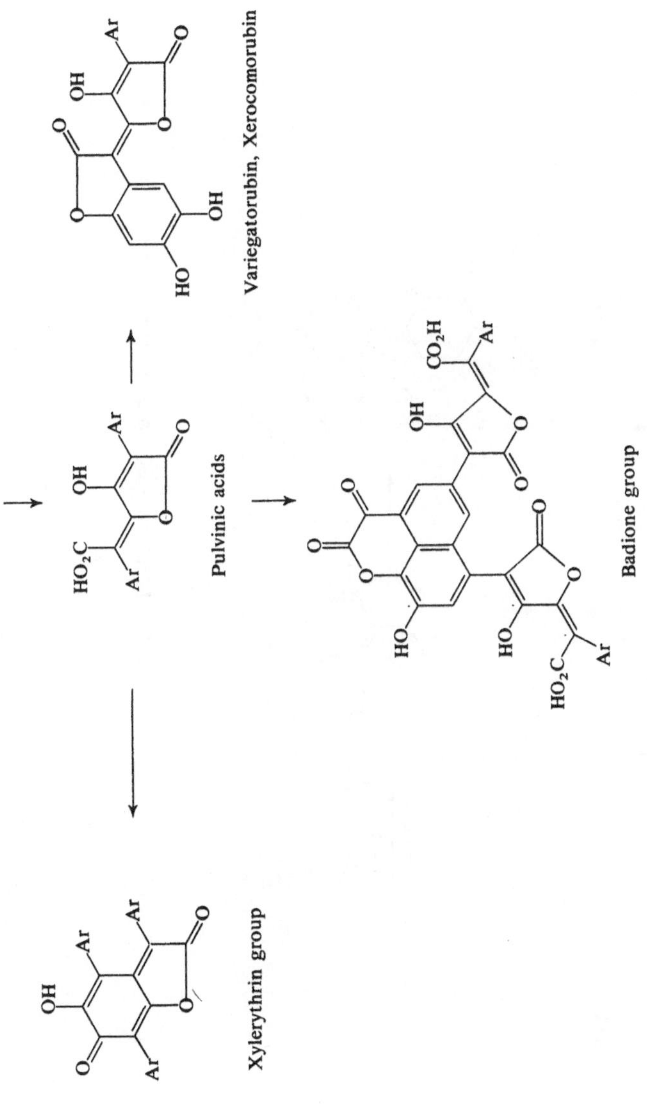

Scheme 30. Structural and biosynthetic inter-relationships between arylpyruvate derived pigments

Variegatorubin, Xerocomorubin

Pulvinic acids

Badione group

Xylerythrin group

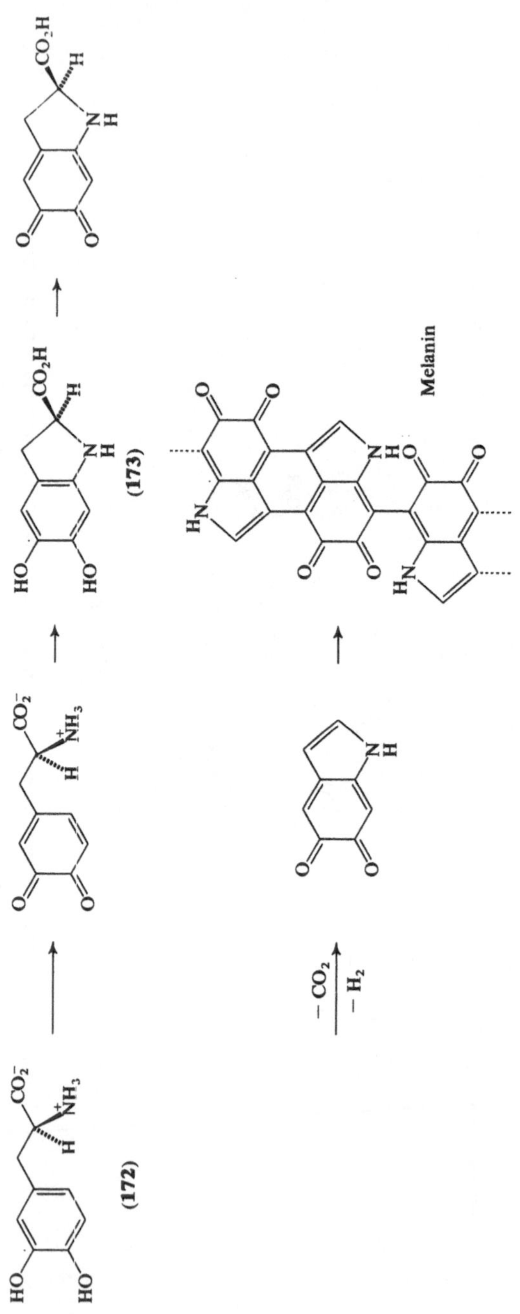

Scheme 31. Melanin formation from L-DOPA

The orange musca-aurins [UV/vis. (water): $\lambda_{max} = 260, 300, 375$ nm] exhibited electronic spectra which characterised them (190) as members of the betalaine family, a group of plant pigments derived by imine formation between betalamic acid (174) and a variety of amino acids. The musca-aurins-I (175), -II (176), and -VII (177) incorporating the amino acids ibotenic acid, stizolobic acid and histidine, respectively, have been isolated from *A. muscaria* in pure form (189, 192). Other musca-aurins correspond to mixtures of betalaines incorporating a variety of acidic (musca-aurins-III and -IV) and neutral (musca-aurins-V and -VI) amino acids (Table 16). Several of these betalaines occur in higher plants (486, 527, 528, 529).

(174) (175) (176) (177)

The isolation and identification of the light, acid, and base sensitive musca-aurins posed considerable separation problems which were overcome by careful, repeated chromatography on Sephadex (188, 192). Hydrolysis of the individual musca-aurins gave betalamic acid (174) together with the constituent amino acids which were detected by electrophoresis and paper chromatography and identified by linked gas chromatography-mass spectrometry of their *N*-trifluoroacetyl methyl esters (189, 192).

Table 16. *Pigments from* Amanita muscaria *which Constitute Mixtures of Betalaines*

Musca-aurin	Amino Acid Component[a]	Name of Betalaine
Musca-aurin-III[b]	α-Aminoadipic acid (trace)	
	Glutamic acid (major)	Vulgaxanthin-I (529)
	Aspartic acid	Miraxanthin-III (527, 528)
Musca-aurin-IV[b]	Glutamic acid	Vulgaxanthin-I
	Aspartic acid	Miraxanthin-III
Musca-aurin-V	Glutamine	Vulgaxanthin-II (529)
	Asparagine	
	Leucine	
	Valine	
	Proline	Indicaxanthin (486)
Musca-aurin-VI	Glutamine	
	Proline	Indicaxanthin

[a] All amino acids have the L-configuration.
[b] Musca-aurins-III and -IV also contain a novel dehydroglutamic acid component.

The composition of each of the musca-aurin betalaines was confirmed by partial synthesis from the appropriate amino acid and betalamic acid derived *in situ* from betanin (178), a pigment of the red beet, *Beta vulgaris*. The equilibrium depicted in Scheme 32 was monitored spectrophotometrically and the musca-aurins thus produced were purified by Sephadex chromatography (189).

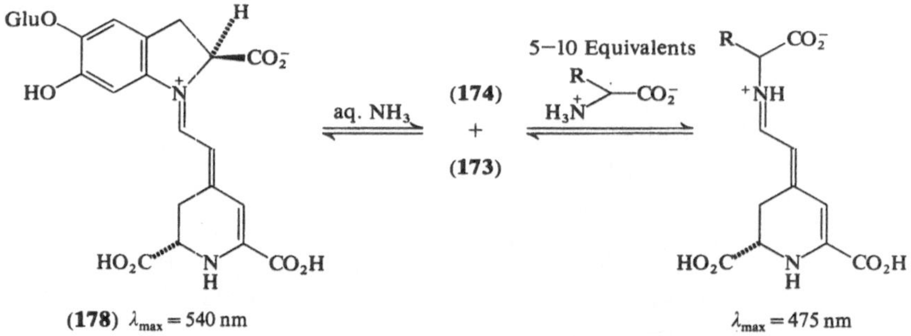

Scheme 32. Partial synthesis of musca-aurins

The purple pigment [UV/vis. (water): $\lambda_{max} = 303$, 540 nm] from *A. muscaria* has been named muscapurpurin and recently assigned the betalaine structure (179) (504) modifying an earlier, tentative, structural assignment (188, 190, 502).

The yellow pigment, muscaflavin [UV/vis. (water): $\lambda_{max} = 238$, 420 nm], is isomeric with betalamic acid (174) (190, 191) and was identi-

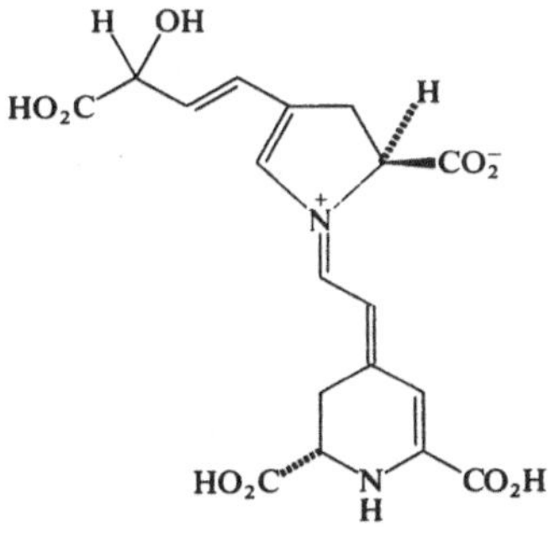

(179)

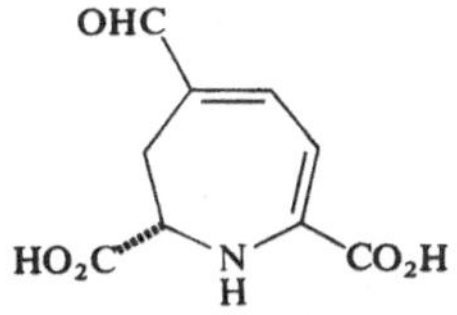

(180)

fied as the novel dihydroazepine (**180**) from its ^{1}H-n.m.r. spectrum together with those of its crystalline dimethyl ester and the product of sodium borohydride reduction (*65, 688*).

Muscaflavin (**180**) has also been isolated from the yellow, orange, and red fruit bodies of several *Hygrocybe* species [*H. chlorophana* $(2 \times 10^{-3} \%)$, *H. citrinovirens* (1.5×10^{-3}), *H. coccinea*, *H. intermedia* (5×10^{-4}), *H. punicea* (1.6×10^{-3}), *H. splendidissima* (2.9×10^{-3}), *H. quieta*] (*250*) and detected chromatographically in many others (*124, 443*) where it occurs in the form of Schiff bases $(\lambda_{max} = 450\text{–}470 \text{ nm})$ incorporating a variety of auxiliary amino acids (*124, 250, 443, 688*). The reconstitution of these labile conjugates (the 'hygro-aurins') from muscaflavin (**180**) and free amino acids in the presence of molecular sieves and *p*-toluenesulphonic acid has been achieved (*250*), and the distribution of the 'hygro-aurins' in *Hygrocybe* has been studied in an effort to use the pigment and amino acid patterns as an aid to the taxonomy of this genus (*124, 169, 443*).

The availability in quantity of muscaflavin (**180**) from *Hygrocybe* toadstools has permitted the measurement of its CD and ^{13}C-n.m.r. spectra for the first time. The assignment of ^{13}C-n.m.r. signals is shown in Table 17 (*250*).

Table 17. ^{13}C-*n.m.r. Data for Muscaflavin (180) (δ values, with TMS as internal standard)*

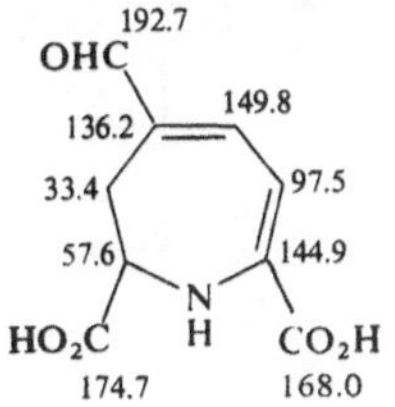

(180)   CD$_3$OD

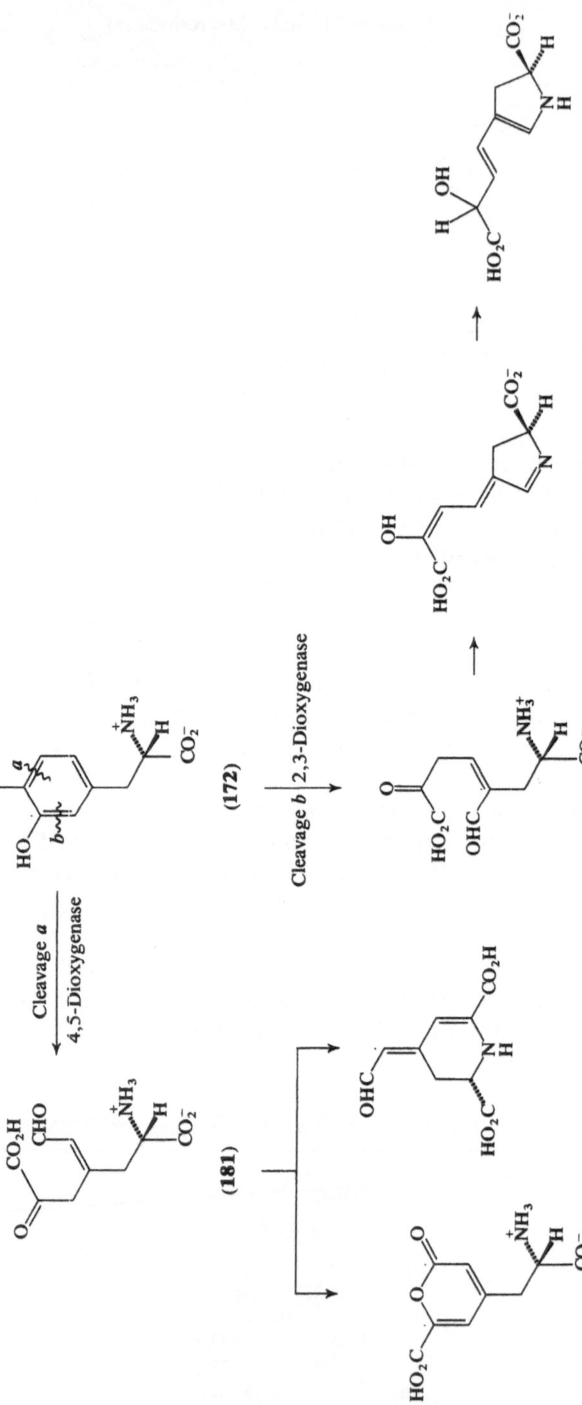

(172)

Cleavage *a*
4,5-Dioxygenase

Cleavage *b* 2,3-Dioxygenase

(181)

(183)

Betalamic acid

(182)

(184)

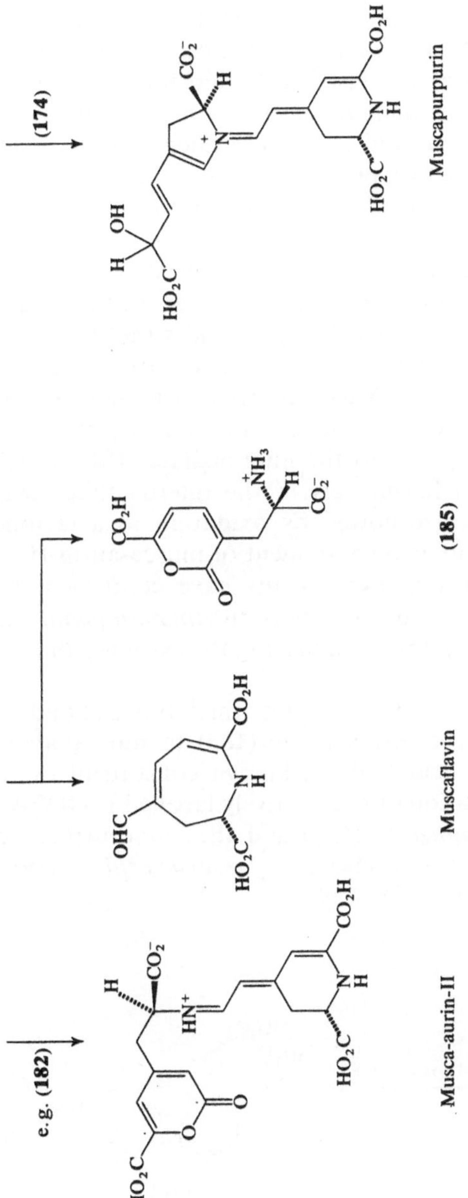

Scheme 33. Biosynthesis of *Amanita* and *Hygrocybe* pigments

The red caps of *Amanita caesarea* contain all of the pigments found in *A. muscaria* with the exception of musca-aurin-I (175) (*192*). This deficiency reflects the absence of the hallucinogenic amino acid ibotenic acid from this edible *Amanita* species. *Amanita citrina*, *A. pantherina*, *A. rubescens*, *A. spissa*, and *A. fulva* contain none of the *A. muscaria* pigments while only two pigments of the musca-aurin type occur in *A. flavoconia* (*192, 502*). Muscaflavin is a constituent of the cap skin of *A. phalloides* (*92*).

The biosynthesis of betalaine pigments in fungi almost certainly follows the same course as prevails in higher plants. It has been established using ^{14}C- and ^{3}H-labelled precursors that betalamic acid (174) is formed in plants belonging to the order Centrospermae from tyrosine and L-DOPA (172) (*242, 378*). The transformation involves extradiol (metapyrocatechase) cleavage of the catechol ring of L-DOPA at bond *a* (Scheme 33) giving rise to the intermediate (181) which can cyclise to betalamic acid. Alternatively, if the intermediate (181) undergoes hemiacetal formation followed by oxidation to a lactone, stizolobic acid (182), the amino acid component of musca-aurin-II (176), results. Successful incorporation experiments have confirmed that L-DOPA is a precursor of stizolobic acid both in *Amanita pantherina* (*553*) and in higher plants (*554*), thus establishing the existence of a 4,5-dioxygenase pathway in fungi.

The alternative extradiol cleavage which breaks bond *b* in L-DOPA (172) can lead *via* the intermediate (183) to muscaflavin (180) (*688*) and to stizolobinic acid (185), a known constituent of *A. pantherina* (*168*). The incorporation of radioactively labelled L-DOPA into stizolobinic acid in *A. pantherina* (*553*) and the co-occurrence of L-DOPA with muscaflavin in fruit bodies of *Hygrocybe conica* (*619*) add support to this 2,3-dioxygenase pathway.

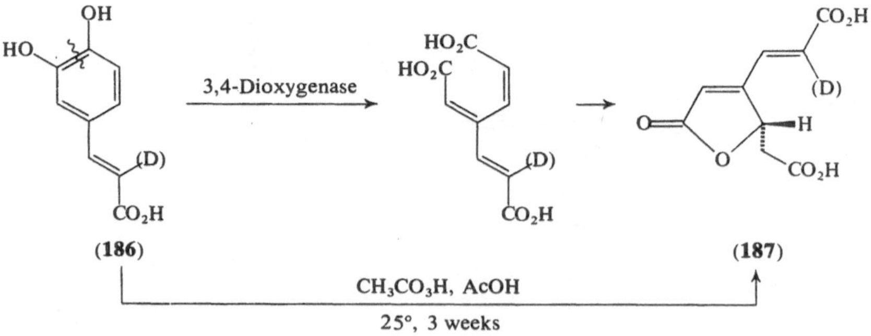

Scheme 34. Biosynthesis and synthesis of hygrophoric acid

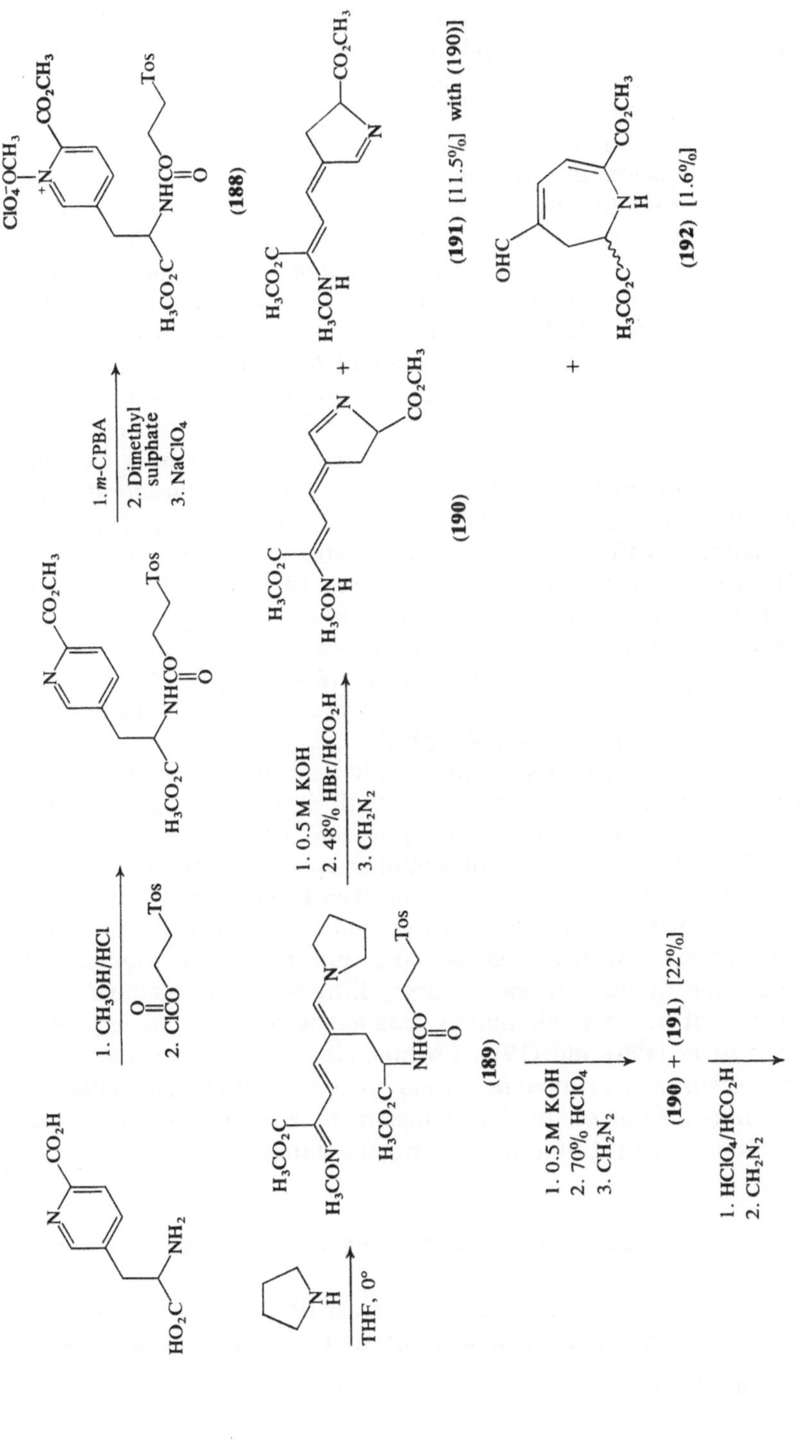

Scheme 35. Synthesis of muscaflavin

Muscapurpurin (179) could arise from the cleavage product (183) of L-DOPA by dehydrative cyclisation followed by hydrogenation to muscapurpurinic acid (184). Subsequent condensation between musca-purpurinic acid and betalamic acid (174) would then give rise to the purple pigment (179) of *A. muscaria* (*504*).

Toadstools belonging to *Hygrophorus,* section Discoidei, are able to cleave the aromatic ring of a 3,4-dihydroxylated precursor by intra-diol (pyrocatechase) cleavage, a reaction without parallel in Basidiomy-cetes. Thus, fruit bodies of *Hygrophorus aureus, H. chrysodon, H. hy-pothejus, H. lucorum, H. nemoreus,* and *H. speciosus* produce (+)-hy-grophoric acid (187) [UV (methanol): $\lambda_{max} = 215$, 256 nm] which was isolated in the form of its calcium salt (*250, 251*). The biosynthesis of (187) (Scheme 34) was established by feeding [α-^{2}H]caffeic acid (186) to fruit bodies of *H. lucorum.* The incorporation rate of deuterium labelled caffeic acid into hygrophoric acid was almost quantitative and, furthermore, administration of the precursor caused a considerable in-crease in the concentration of (187) in the fruit bodies (*250, 251*).

Oxidation of caffeic acid with peroxyacetic acid led to hygrophoric acid (187) in racemic form (*251;* see also ref. *573*). Ozonolysis of (+)-hygrophoric acid afforded (*S*)-malic acid after oxidative work up, thus establishing the absolute configuration shown.

Interestingly, *H. aureus, H. hypothejus, H. lucorum,* and *H. specio-sus* are also able to induce extradiol cleavage of L-DOPA leading to the formation of muscaflavin (180) (*250*).

The biogenetic hypothesis delineated in Scheme 33 for muscaflavin (180) provided MUSSO and coworkers with a synthetic route to this pigment (*65, 66*). The key step in this approach (Scheme 35) was genera-tion of the enamine (189) by nucleophilic opening of the pyridinium ion (188). It was hoped that (189), a masked form of the biosynthetic intermediate (183), would cyclise to give muscaflavin in the presence of acid. In the event, when the amino group in (189) was unmasked with base and cyclisation was induced, little or no muscaflavin was obtained. Rather, the predominant products were the isomeric pyrro-line derivatives (190) and (191). Fortunately, these heterocycles were jointly transformed in moderate yield to the dimethyl ester (192) of racemic muscaflavin on further exposure to acid. Finally, the ester was resolved by chromatography on potato starch.

2.3. Compounds Derived from Cinnamic Acids

The enzyme phenylalanine ammonia-lyase (PAL) catalyses non-oxi-dative deamination of L-phenylalanine to form *trans*-cinnamic acid in a host of basidiomycetous fungi (*702*). *p*-Coumaric acid is formed from cinnamate by enzymatic hydroxylation and by non-oxidative de-amination of L-tyrosine. Hydroxylation of *p*-coumaric acid gives caffeic acid. These cinnamate derivatives and their close relatives are precursors of the pigments discussed in this section.

2.3.1. Purpurogallin Derivatives

Sporophores of *Fomes fomentarius,* the earliest fungus known to have been used by man, contain the red-brown purpurogallin derivative fomentariol (**193**) [UV/vis. (methanol): λ_{max} (log ε) = 224 (4.21), 285 (4.29), 330 (4.58), 460 nm (3.55)] (*39, 631*) which is concentrated in the hard crust of the fruit bodies and is responsible for the intense blood red colour produced when this fungus is treated with alkali. The benzotropolone nucleus in fomentariol may be derived in the fungus by oxidative combination of two molecules of 2,3,4-trihydroxycin-namyl alcohol (**197**) in a manner analogous to the formation of purpu-rogallin from pyrogallol. The oxidation of alcohol (**197**) has been accomplished *in vitro* using as an oxidant potassium iodate or a cell free extract from *F. fomentarius* fruit bodies (Scheme 36).

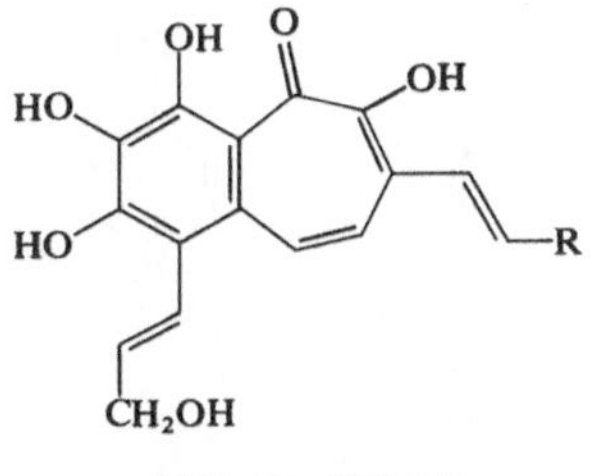

(**193**) R = CH$_2$OH
(**194**) R = CHO

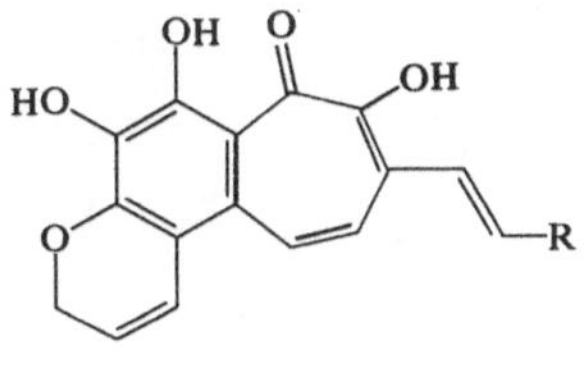

(**195**) R = CH$_2$OH
(**196**) R = CHO

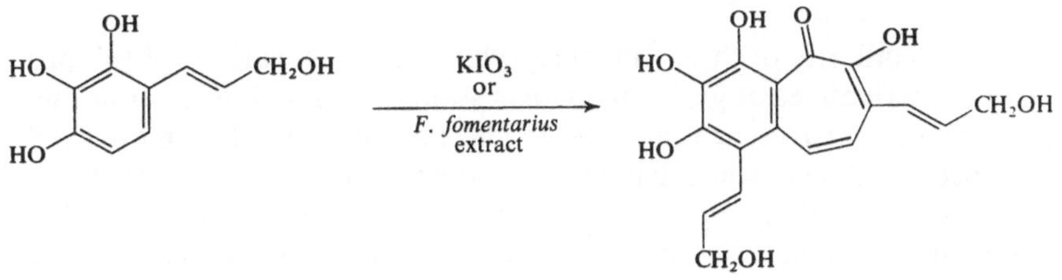

(**197**)

(**193**)

Scheme 36. Synthesis of fomentariol

Oxidation of fomentariol led to dehydrofomentariol (**194**) (*322*) which, together with the pyran derivatives anhydrofomentariol (**195**) (*229*) and anhydrodehydrofomentariol (**196**) (*228*), are minor constituents of *F. fomentarius*. The structures (**193**) to (**196**) are fully consistent with the ^{1}H-n.m.r. spectra of these pigments and the position of the aldehyde side chain in dehydrofomentariol (**194**) was confirmed by X-ray analysis. The metabolite (**194**) has been converted to the pyran (**196**) with *p*-toluenesulphonic acid in tetrahydrofuran (*228*).

The occurrence of these purpurogallin derivatives seems to be restricted to *Fomes fomentarius* (*38*).

2.3.2. Styrylpyrones

Some wood inhabiting fungi (Table 18) produce yellow styrylpyrone pigments such as bisnoryangonin (**198**) and hispidin (**199**). Hispidin was first isolated from *Inonotus hispidus* (= *Polyporus hispidus*) by ZOPF (*729*) and later by ZELLNER (*728*) but was first identified as 6-(3′,4′-dihydroxystyryl)-4-hydroxy-2-pyrone (**199**) by EDWARDS (*209*) and by BU'LOCK (*143*) and their coworkers. Some properties of hispidin and related styrylpyrone pigments are given in Table 20.

	R¹	R²
(**198**)	H	H
(**199**)	H	OH
(**200**)	OH	OH

(**201**)

The similarity of these fungal styrylpyrones with various phenylpropanoid derivatives of green plants has stimulated much interest in their biogenesis and role in fungi (*141*). The biosynthesis of hispidin (**199**) has been studied in some detail both at the tracer and at the enzymological level. Labelling experiments with cultures of *I. hispidus* (*522, 523, 662*) and *Phaeolus schweinitzii* (*351*) have proved that the styryl moiety derives from phenylalanine *via* cinnamic acid, *p*-coumaric acid, and caffeoylcoenzyme-A (**202**) and that the pyrone ring is acetate derived (Scheme 37).

Table 18. *Occurrence of Bisnoryangonin and Hispidin*[a]

Bisnoryangonin (**198**)	*Gymnopilus aeruginosus, G. aurantiophyllus (350), G. bellulus (297), G. braendlei (350), G. chrysopellus (297), G. decurrens [0.55%] (348), G. luteofolius, G. obscurus (350), G. penetrans (179, 297), G. punctifolius (350, 544), G. spectabilis (297, 349), Hypholoma radicosum, H. subviride (297), Inonotus hispidus [cultures] (523), Pholiota alnicola (= Flammula alnicola = F. apicrea), Ph. aurivella, Ph. connissans (= Flammula connissans), Ph. lucifera (297), Ph. squarroso-adiposa (116), Ph. tuberculosa (297)*
Hispidin (**199**)	*Gymnopilus aeruginosus, G. aurantiophyllus (350), G. bellulus (297), G. braendlei (350), G. chrysopellus (297), G. luteofolius, G. obscurus (350), G. penetrans (179, 297), G. punctifolius (350, 544), G. spectabilis (297), Hymenochaete luteobadia [trace] (232), H. mougeotii (232, 235), H. pinnatifida, H. rubiginosa, H. salei [trace] (232), Hypholoma marginatum (= H. dispersum), H. radicosum, H. subviride (297), Inonotus dryophilus [trace] (232), I. hispidus (143, 145, 209, 232, 235, 728), [c] (141, 662), I. nidus-pici [trace], I. nodulosus, I. radiatus, I. rheades, I. tamaricis, Onnia tomentosa [trace] (232), Phaeolus schweinitzii [1.2%] (670), (232), [c] (351), Phellinus chrysoloma, Ph. conchatus, Ph. contiguus [trace], Ph. erectus, Ph. ferreus [trace], Ph. ferrugineo-fuscus [trace], Ph. ferruginosus, Ph. hartigii, Ph. igniarius (232), [c] (412), Ph. laevigatus, Ph. lundellii [trace], Ph. nigrolimitatus, Ph. pini, Ph. pseudopunctatus [trace], Ph. punctatus [trace], Ph. ribis (= Phylloporia ribis) [trace] (232), Ph. rimosus (= Ph. robiniae) (232, 235), Ph. robustus, Ph. torulosus, Ph. tremulae, Ph. trivialis (232), Ph. tuberculosus (= Ph. pomaceus) [0.32%] (414), (232), Ph. viticola [trace] (232), Pholiota alnicola, Ph. aurivella, Ph. connissans, Ph. lucifera, Ph. tuberculosa (297)*

[a] Chemical yields of styrylpyrones vary depending on the age of the fruit body and the method of purification; those quoted refer to dried fungus. In many other cases the yields of bisnoryangonin and hispidin have been determined by spectroscopic assay and are cited in the original literature.

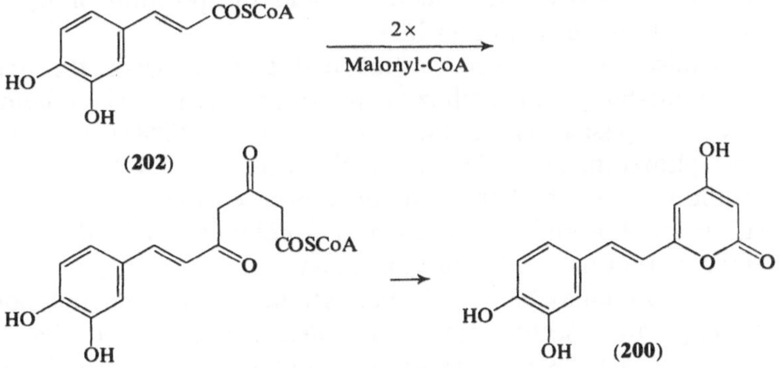

Scheme 37. Biosynthesis of hispidin

The induction by light of several enzymes in this pathway has been established (*507, 662, 679, 680*), and one of these, responsible for the hydroxylation of *p*-coumaric acid to caffeic acid in *I. hispidus*, is also capable of hydroxylating bisnoryangonin (**198**) to produce hispidin (**199**) (*508*). Further hydroxylation of hispidin leads to leucohymenoquinone (**200**) which occurs with the red hymenoquinone (**201**) [UV/vis. (methanol): λ_{max} (log ε) = 233 (4.41), 277 (4.10), 315 (4.18), 345 (sh., 4.09), 390 nm (sh., 3.89)] in the ox blood coloured fruit bodies of *Hymenochaete mougeotii* (*232, 415*). Hymenoquinone is also present in *H. pinnatifida* (*232*).

Hispidin (**199**) is oxidised enzymatically during the ripening of sporophores of *I. hispidus* and the resulting brown-black polymer becomes firmly bound to the cell wall material. This polymer has been identified with 'fungal lignin' and is believed to be responsible for the darkening and significant toughening of ageing sporophores of *I. hispidus* and other characteristically 'woody' fungi which produce styrylpyrones (*141, 143, 145, 412*). The natural occurrence of 'dimeric' styrylpyrones lends credence to this idea (see below).

There are reports of the hallucinogenic properties of *Gymnopilus spectabilis* (*138, 699*) but no clinical evidence exists to link these effects with the presence in this toadstool of bisnoryangonin (*349*). Hispidin (**199**) and bisnoryangonin (**198**) exhibit good *in vitro* antimicrobial activity against Gram-positive test organisms including the acid-fast *Mycobacterium smegmatis* (*76*).

Styrylpyrones have been synthesised by condensing aromatic aldehydes with 4-methoxy-6-methyl-2-pyrone (**203**) in the presence of magnesium methoxide (*144*). The process is exemplified (Scheme 38) by the synthesis of hispidin in which methoxymethyl ethers serve to protect the catechol hydroxy groups (*212*). A range of hispidin analogues prepared in the same way showed an interesting correlation between the infrared carbonyl stretching frequency and the position of hydroxy substitution in the phenyl ring (*211*).

Tri-*O*-methylhispidin and several analogues in which the phenyl ring bears methoxy and methoxymethoxy groups produce colourless dimers when exposed to sunlight. Structures of type (**204**) were deduced for these photodimers by detailed analysis of their ^{1}H-n.m.r. spectra (*67*). Although dimers of this type have been encountered during the derivatisation of hispidin from *Inonotus hispidus* (*143, 209*) it is considered unlikely that they are natural products.

Hispidin (**199**) may be transformed into more elaborate metabolites on coupling with a second pyrone. *Phellinus tuberculosus* (= *Phellinus pomaceus*), a fungus causing white rot on plum trees, produces hispidin together with its dehydro dimer, 3,14′-bihispidinyl (**205**) (*414*). The

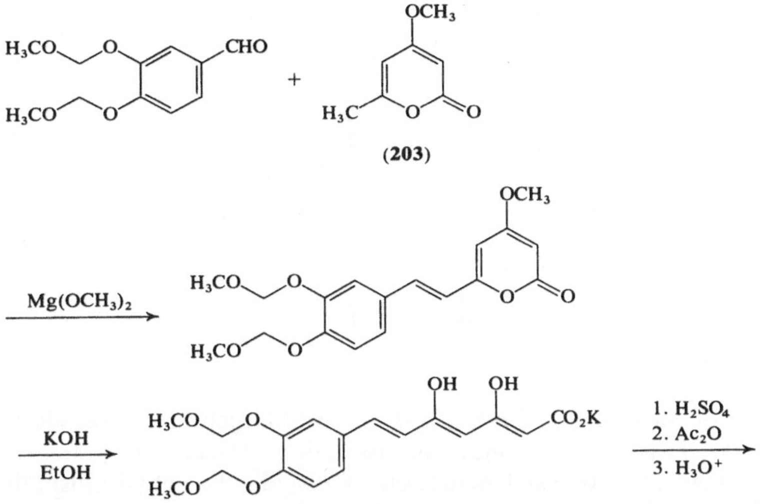

Scheme 38. Synthesis of hispidin

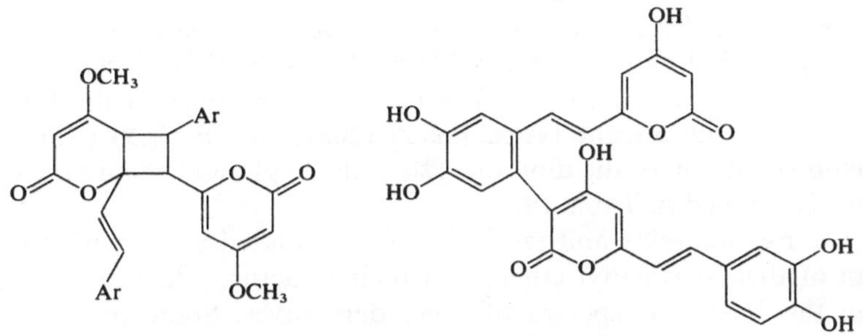

(204)   (205)

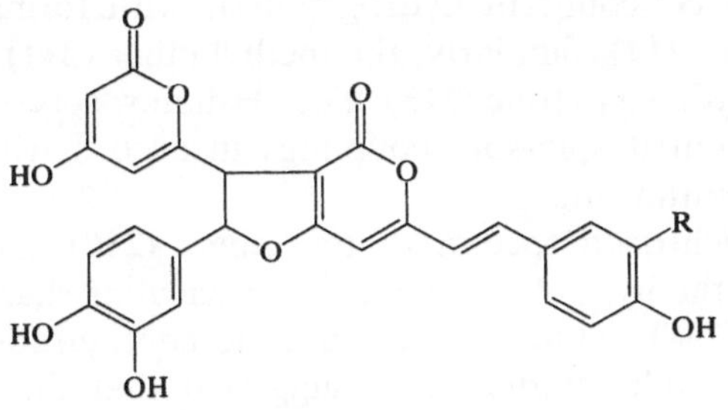

(206) R = H
(207) R = OH

(208) R = H
(209) R = OH

structure of the dehydro dimer (205) followed largely from considera-
tion of its ¹H-n.m.r. spectrum. The formation of 3,14′-bihispidinyl is
readily explained by initial oxidation of hispidin to dehydrohispidin
(214) followed by conjugative attack at the *ortho*-quinone by the nucleophilic (C-3) centre of the pyrone ring in a second hispidin molecule
(Scheme 39, route *a*). The compound (205) is of obvious interest with
regard to the oxidative polymerisation of hispidin.

The bright yellow pigmentation of the common 'sulphur tuft', *Hypholoma fasciculare*, is due to the more elaborate hispidin condensation
products hypholomin-A (206) and hypholomin-B (207). These pigments
are accompanied in *H. fasciculare* by other dipyrones in the form of
the strongly fluorescent fasciculins-A (208) and -B (209) (*238*). The
wider distribution of the dipyrone (205), the hypholomins and the fasciculins is detailed in Table 19.

The hypholomins and fasciculins were isolated and purified in the
form of their permethyl ethers and their structures deduced *inter alia*
from the ¹H-n.m.r. spectra of these derivatives. Some properties of
these dipyrones and their ethers are summarised in Table 20.

Catalytic hydrogenation of the methyl ether (210) of hypholomin-A
brought about concomitant hydrogenolysis with formation of the 4-
hydroxypyrone (212). Similarly, the methyl ether (211) of fasciculin-B
afforded the hydroxypyrone (213). The ¹H-n.m.r. spectra of both (212)
and (213) exhibited spin-spin couplings in accord with the cleavage
of the dihydrofuran ring.

Circular dichroism spectra of the ethers (210) and (211) clearly
indicated that the hypholomins and the fasciculins share the same stereochemistry (*238*). Although the absolute configuration has not yet
been defined, model studies have suggested that the 3,4-dihydroxy-
phenyl and 4-hydroxypyrone rings in the natural products are mutually
trans disposed (*566*).

Table 19. *Occurrence of Dipyrones*[a]

3,14′-Bihispidinyl **(205)**	*Inonotus hispidus*, *Phellinus chrysoloma*, *Ph. ferrugineo-fuscus* [trace], *Ph. hartigii*, *Ph. igniarius*, *Ph. laevigatus*, *Ph. lundellii* [trace], *Ph. nigrolimitatus*, *Ph. pini*, *Ph. rimosus*, *Ph. robustus*, *Ph. torulosus*, *Ph. tremulae*, *Ph. trivialis (232)*, *Ph. tuberculosus* (= *Ph. pomaceus* = *Inonotus pomaceus*) [5×10^{-2}%] † *(414)*, *(232)*
Hypholomin-A **(206)** + Hypholomin-B **(207)**	*Hypholoma elongatipes* (= *H. elongatum*), *H. ericaeoides (297)*, *H. fasciculare* [HA: 1.4×10^{-3}%; HB: 6×10^{-4}%]* *(238)*, *(297)*, *H. marginatum* (= *H. dispersum*), *H. subla-teritium*, *H. subviride*, *Pholiota alnicola*[b] (= *Ph. apicrea*[b]), *Ph. as-tragalina*[b], *Ph. connissans*[b], *Ph. decussata*[b] *(297)*, *Ph. flammans (295, 297)*, *Ph. henningsii*[b], *Ph. lenta*[b], *Ph. scamba*[b], *Ph. squarrosa* *(297)*
Hypholomin-A **(206)**	*Pholiota gummosa*[b] *(297)*
Hypholomin-B **(207)**	*Hypholoma capnoides (297)*, *Inonotus dryophilus (232)*, *I. hispidus (232, 235)*, *I. nidus-pici* [trace], *I. nodulosus*, *I. radiatus*, *I. rheades*, *I. tamaricis*, *Onnia tomentosa*, *O. triqueter*, *Phellinus chrysoloma*, *Ph. conchatus*, *Ph. contiguus* [trace], *Ph. erectus*, *Ph. ferreus* [trace], *Ph. ferruginosus*, *Ph. hartigii*, *Ph. pini*, *Ph. pseudopuncta-tus* [trace], *Ph. punctatus*, *Ph. ribis* (= *Phylloporia ribis*) *(232)*, *Ph. rimosus* (= *Ph. robiniae*) *(232, 235)*, *Ph. robustus*, *Ph. torulo-sus*, *Ph. viticola* [trace] *(232)*
Fasciculin-A **(208)** + Fasciculin-B **(209)**	*Hypholoma elongatipes*, *H. ericaeoides (297)*, *H. fasciculare* [FA: 2×10^{-4}%; FB: 1.8×10^{-3}%] * *(238)*, *(297)*, *H. radicosum*, *H. sublateritium*, *H. subviride*, *Pholiota alnicola*[b], *Ph. astraga-lina*[b], *Ph. connissans*[b], *Ph. decussata*[b] *(297)*, *Ph. flammans (295, 297)*, *Ph. henningsii*[b], *Ph. lenta*[b], *Ph. scamba*[b], *Ph. squarrosa* *(297)*
Fasciculin-A **(208)**	*Hypholoma marginatum*, *Pholiota alnicola*[b] *(297)*
Fasciculin-B **(209)**	*Hypholoma capnoides (297)*

[a] Chemical yields refer to the dried (†) or the fresh (*) fungus. In many of the other cases the yields of dipyrones have been determined by spectroscopic assay and are cited in the original literature.

[b] Referred to *Flammula* in the literature cited.

	n	R
(210)	1	H
(211)	0	OCH₃

Table 20. *Properties of Some Fungal Styrylpyrones*

Pigment	Colour M.p.	TLC R_f	Ultraviolet/Vis. [λ_{max} (log ε)]	Infrared [ν_{CO}]	Permethyl Derivative				References
					Colour M.p.	TLC R_f	Ultraviolet/Vis. [λ_{max} (log ε)]	Infrared [ν_{CO}]	
Bisnoryangonin (198)	Yellow 240–244°	0.45[a]	217 (4.36), 364 nm (4.44) [ethanol]	1647 cm^{-1}	Yellow-green 157–158°	0.67[a]	218 (4.31), 358 nm (4.45) [ethanol]	1709 cm^{-1}	(348, 350)
Hispidin (199)	Yellow 259° (dec.)	0.32[a] 0.54[b]	257 (4.07), 373 nm (4.28) [ethanol]	1684, 1661 cm^{-1}	Pale yellow 153–158°	0.59[a] 0.66[c]	251 (4.17), 366 nm (4.39) [ethanol]	1701 cm^{-1}	(209, 235, 350)
Leucohymeno-quinone (200)	Green-yellow 228–230°	0.35[b]	220 (4.33), 340 (3.98), 395 nm (4.23) [methanol]	1670 cm^{-1}	—	—	—	—	(235, 415)
Hypholomin-A (206)	Bright yellow —	0.42[b]	276, 288, 375 nm [methanol]	—	Pale yellow 155–158°	0.59[c]	275, 289 (sh.), 375 nm [methanol]	1729 cm^{-1}	(238)
Hypholomin-B (207)	Golden yellow —	0.32[b]	259, 275, 382.5 nm [methanol]	—	Yellow 140–143°	0.52[c]	255, 271, 378 nm [methanol]	1721 cm^{-1}	(238)
Fasciculin-A (208)	Colourless[d] —	0.38[b]	285, 349 nm [methanol]	—	Pale yellow 110–112°	0.55[c]	254.5, 278.5, 342.5 nm [methanol]	1722 cm^{-1}	(238)
Fasciculin-B (209)	Colourless[d] —	0.28[b]	286, 357 nm [methanol]	—	Ivory 140–145°	0.49[c]	280, 344 nm [methanol]	1723 cm^{-1}	(238)

[a] Silica gel (methyl formate:hexane:formic acid = 100:50:1).
[b] Merck Kieselgel F_{254} (benzene:ethyl formate:formic acid = 5:4:2).
[c] Merck Kieselgel F_{254} (ethyl acetate:chloroform = 1:1).
[d] Blue fluorescence on TLC under ultraviolet light.

	n	R
(212)	1	H
(213)	0	OCH₃

Biogenetically the hypholomins and fasciculins, like the dehydro-dimer (205), may be regarded as condensation products of dehydrohis-pidin (214). Thus, the hypholomins-A and -B might result from attack at the electrophilic double bond in (214) by the nucleophilic (C-3) centre of the pyrone ring in bisnoryangonin and hispidin, respectively (Scheme 39, route b). Subsequent conjugate addition to the intermediate quinone methide (215) would furnish the dihydrofuran ring system of the hypholomins. The fasciculins-A and -B presumably arise from (214) in analogous fashion by condensation with 6-(4'-hydroxyphenyl)- and with 6-(3',4'-dihydroxyphenyl)-4-hydroxy-2-pyrone, respectively. Thus, the mechanism for dipyrone formation is not unlike the mechanisms described for lignin production in higher plants. It is noteworthy that the arylpyrones implicated in the biosynthesis of the fasciculins have not yet been isolated from fungi.

During a reinvestigation of the pigments of *Inonotus hispidus*, extraction of freeze-dried fruit bodies with methanol gave, besides hispidin (199) (0.8% dry weight), three other hispidin related pigments (569). The 3,3'-linked dehydro dimer (216) (0.36%) [UV/vis. (methanol): $\lambda_{max}(\log \varepsilon) = 222$ (4.51), 251 (4.35), 379 nm (4.66)] was identified from spectroscopic data, especially the [1]H-n.m.r. spectrum in which the 3-H signal from the pyrone ring of hispidin is absent. Cleavage of (216) with aqueous sodium dithionite yielded hispidin.

The structure of the ethylidene bridged derivative (217) (0.3%) [UV/vis. (methanol): $\lambda_{max}(\log \varepsilon) = 230$ (3.98), 250 (3.93), 380 nm (4.16)] followed from the spectroscopic data and from its formation in high yield by reaction of hispidin (199) with acetaldehyde. The pigments (216) and (217) are also present in the fruit bodies of *Phaeolus schweinitzii*.

The isolation of the keto-ester (218) (0.01%) [UV/vis. (methanol): $\lambda_{max}(\log \varepsilon) = 215$ (3.55), 260 (3.63), 375 nm (4.19)] from *I. hispidus*

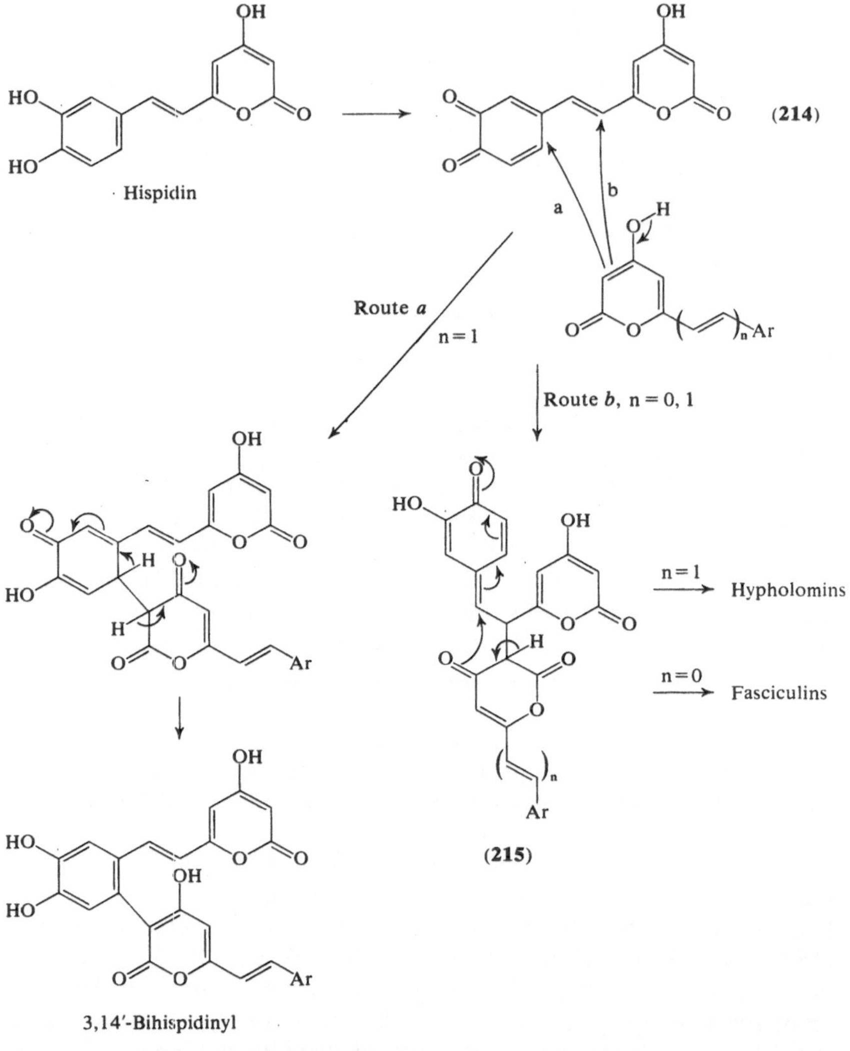

Scheme 39. Biogenesis of fungal dipyrones

might represent interception by the solvent of a biogenetic precursor to hispidin (Scheme 37).

Following the pioneering work of GABRIEL on the chromatography of pigments from *Gymnopilus* (*259*) and *Hypholoma* (*260*) more recent systematic studies by GLUCHOFF-FIASSON and FIASSON on the distribution of styrylpyrone pigments in Agaricales and Aphyllophorales have highlighted the value of this class of metabolites as chemotaxonomic

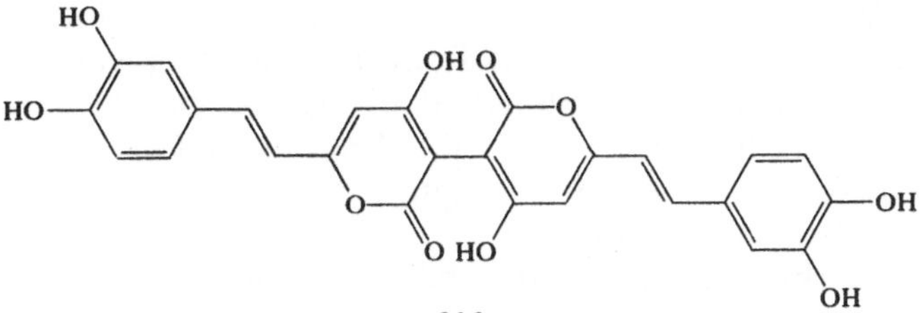

(216)

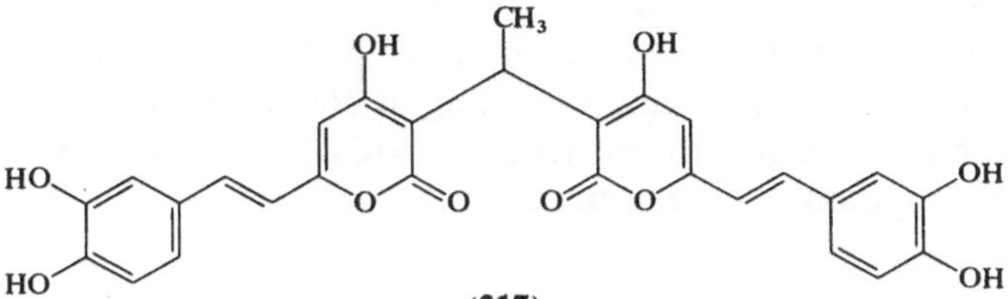

(217)

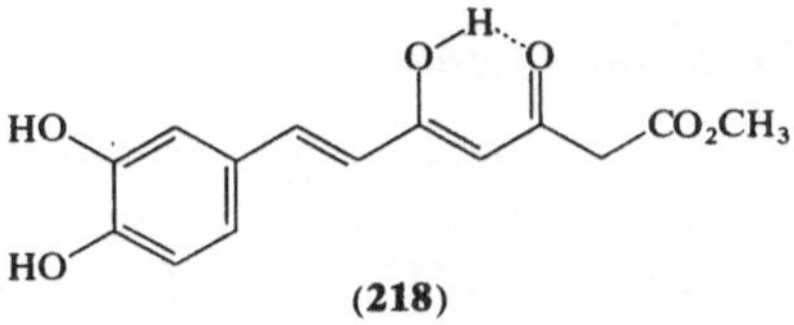

(218)

markers for *Hypholoma* (*294, 297*) and revealed their restriction, among the order Aphyllophorales, to the family Hymenochaetaceae (*232, 233*).

In addition to the styrylpyrones detailed above two higher vinylogues are known which arise naturally by condensation between a cinnamic acid starter unit and five and six acetate units, respectively. Geogenin (**219**) [UV/vis. (acetonitrile): $\lambda_{max} = 217$, 254, 260 (sh.), 301 (sh.), 313, 382 (sh.), 398, 420 nm (sh.)] has been isolated from the

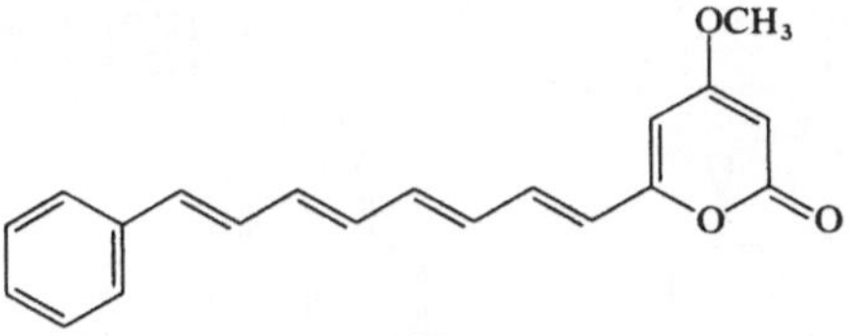

(219)

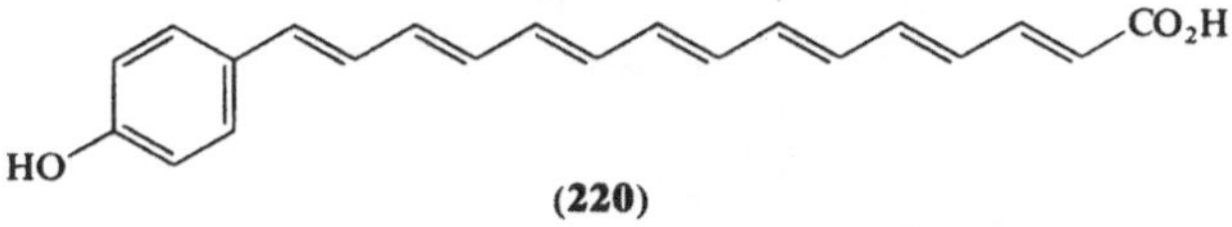

(220)

bright yellow mycelial mats of *Hydnellum geogenium* and its structure
has been confirmed by synthesis from 4-methoxy-6-methyl-2-pyrone
(**203**) and 7-phenyl-2,4,6-heptatrienal in the presence of magnesium
methoxide (*611*). The red fruit bodies of *Cytidia salicina* (= *Corticium
salicinum*) contain the violet-red pigment cortisalin (**220**) (*306*) which
is mentioned again in Section 3.7.

It is remarkable in view of the ability of fungi to produce pigments
by condensation of cinnamic, *p*-coumaric and caffeic acid derivatives
with one, two, five, and six acetate units that they appear essentially
devoid of metabolites involving side chain extension of the same aromatic starter units by three and by four acetate units. This situation
is in marked contrast to that prevailing in higher plants where secondary metabolites derived from triketide extension of cinnamoyl derivatives are abundant and widespread.

2.4. Compounds Derived from *p*-Hydroxybenzoic Acid

The Boviquinone Group

Fruit bodies of *Suillus bovinus* and *Chroogomphus rutilus* (= *Gomphidius rutilus*) develop a bright pink stain when moistened with ethanol, and produce purple colours with alkali (*57*). The pigments responsible have been characterised by Beaumont and Edwards (*70*, *72*)

(**221**) n = 3
(**222**) n = 4

	n	m
(**223**)	3	4
(**224**)	4	4

	n	m
(**225**)	3	3
(**226**)	3	4
(**227**)	4	4

as being dihydroxybenzoquinones bearing polyprenyl side chains. The principal pigment ('bovinone') of the fresh fruit bodies of *S. bovinus* was identified as the geranylgeranyl substituted quinone (**222**) from the spectroscopic data and degradative studies. The mass spectrum proved particularly informative in revealing the loss from an abundant molecular ion of three consecutive isoprene units; cleavage of a fourth isoprene unit then produced a cluster of diagnostic hydroxybenzylium ions. The 2,5-substitution pattern in the benzoquinone ring followed from the isolation of (2,3,5,6-tetramethoxyphenyl)acetic acid after oxidation of 'bovinone leucotetramethyl ether' with potassium permanganate.

From air-dried sporophores of *S. bovinus* Japanese workers isolated the methylene linked quinone 'amitenone' (**227**) (*487*). BEAUMONT and EDWARDS (*70*) were able to show, however, that the 'amitenone' content of the fresh fungus was minimal and consequently suggested that this diquinone was present in dried material largely as an artefact formed from 'bovinone' during the drying process. 'Amitenone' (**227**) has been synthesised by reaction of 'bovinone' (**222**) with formaldehyde in acetic acid solution (*70*).

The flesh of *Chroogomphus helveticus* contains considerable quantities of a second isoprenoid quinone, 'helveticone' (**221**) (*606*). In the mass spectrum 'helveticone' exhibited characteristic ions arising by fragmentation of a farnesyl side chain and the structure (**221**) was corroborated by comparison of this and other spectroscopic data with those of the isoprenologue (**222**).

Table 21. *Occurrence of Simple Boviquinone Derivatives*[a]

Boviquinone-3 (**221**) (Helveticone)	*Chroogomphus helveticus* [0.5%] (*606*), *Ch. rutilus* (= *Gomphidius rutilus*) [0.15%](*606*), (*72*)
Boviquinone-4 (**222**) (Bovinone)	*Chroogomphus rutilus* [trace: MS] (*72*), *Suillus bovinus* [3×10^{-2}%] (*70*), (*72*), *S. pictus* [2.5×10^{-2}%] (*448*)
Diboviquinone-3,4 (**223**)	*Chroogomphus rutilus* [1.6×10^{-3}%] (*72*)
Diboviquinone-4,4 (**224**)	*Chroogomphus rutilus* [trace: MS] (*72*), *Suillus bovinus* [6×10^{-3}%] (*72*)
Methylenediboviquinone-3,3 (**225**)	*Chroogomphus rutilus* [1.4×10^{-3}%] (*72*)
Methylenediboviquinone-3,4 (**226**)	*Chroogomphus rutilus* [MS] (*72*)
Methylenedibovi-quinone-4,4 (**227**) (Amitenone)	*Chroogomphus rutilus* [trace: MS] (*72*), *Suillus bovinus* (dried sporophores: *487*)

[a] Yields refer to percentage of fresh weight of fungus.

In view of the occurrence of several of these closely related isoprenoid quinones in fungi (Table 21), the trivial nomenclature has been simplified (72) so that 'helveticone' and 'bovinone' become boviquinone-3 and boviquinone-4, respectively.

The boviquinones-3 and -4 have been synthesised in moderate yields from 2,5-dihydroxy-1,4-benzoquinone and the corresponding allylic bromide in the presence of a tertiary amine (Scheme 40) (389). It is interesting that the success of these reactions was crucially dependent on the direct use of farnesyl bromide and of geranylgeranyl bromide without special purification. The preparation and properties of the chromenols and chromanols of boviquinone-4 have been reported (71).

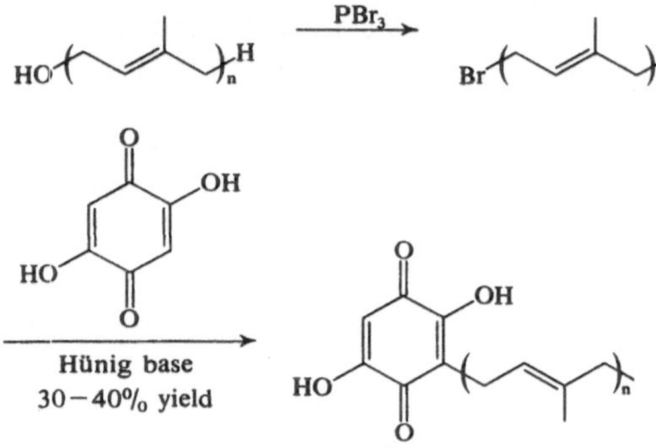

Scheme 40. Synthesis of boviquinones

Boviquinone-3 (221) is also the predominant pigment of *Chroogomphus rutilus* where it occurs accompanied by minor quantities of the diboviquinones (223) and (224), and the methylenediboviquinones (225), (226) and (227) (72).

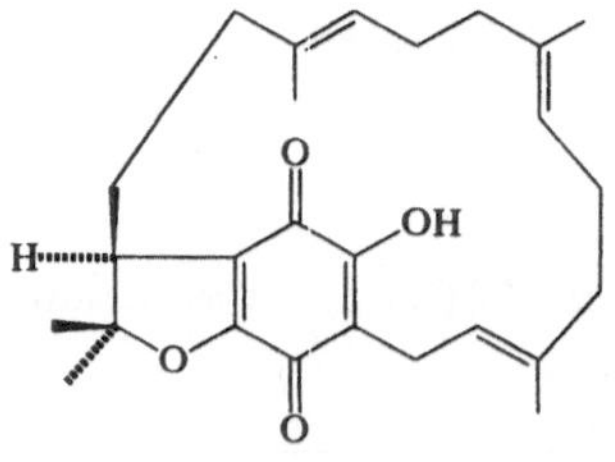

(228) 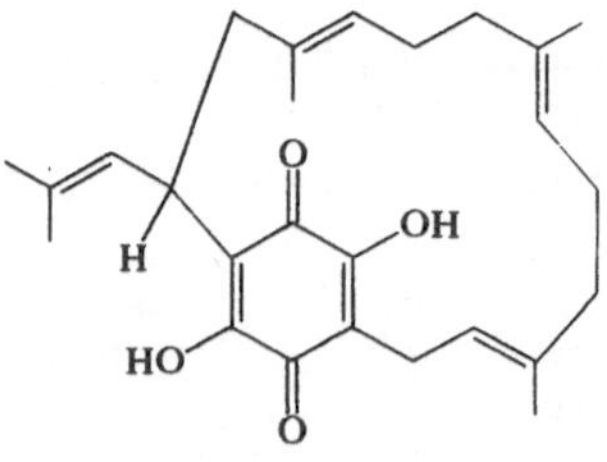   (229)

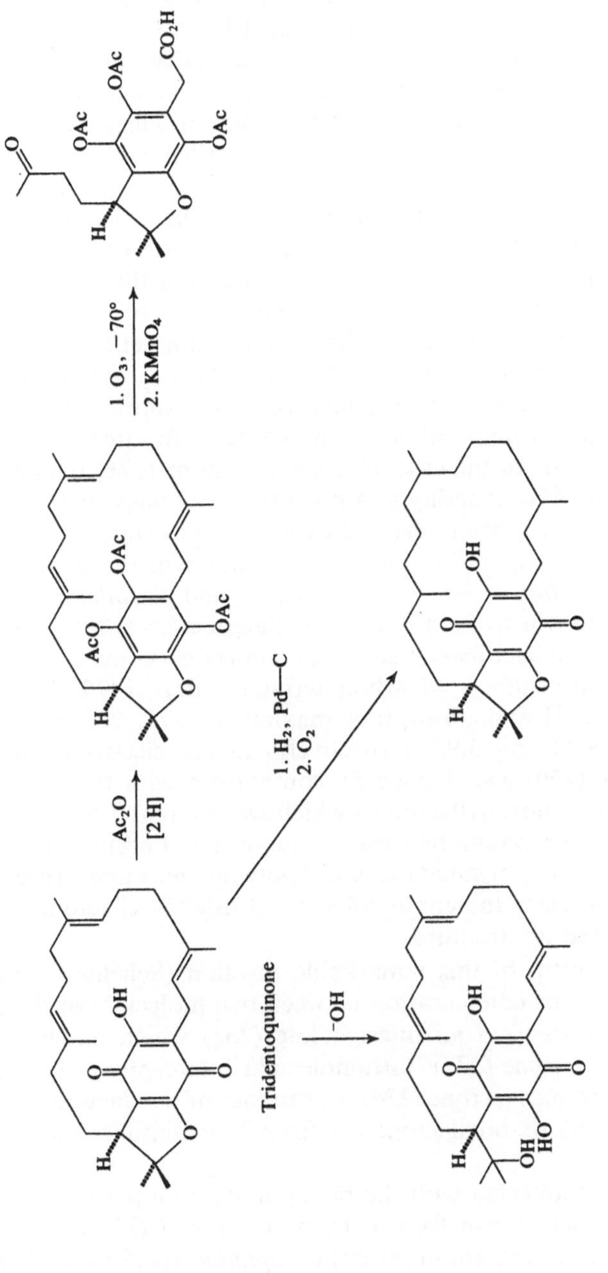

Scheme 41. Reactions of tridentoquinone

Several fungi contain more elaborate pigments which are closely related to boviquinone-4. In tridentoquinone (**228**), which is present to the extent of 0.05% of the fresh weight of *Suillus tridentinus*, the geranylgeranyl side chain of boviquinone-4 has taken part in ansa ring formation. The structure of the ansaquinone (**228**) was deduced from spectroscopic analysis and chemical investigations (Scheme 41) and was confirmed by an X-ray structure determination which also suggested the absolute configuration shown (*97*). A detailed analysis of the mass spectra of tridentoquinone and its derivatives has been published (*571*).

A second ansaquinone, rhizopogone (**229**), has been isolated from an unidentified species of *Rhizopogon* (*389*). The ^{1}H-n.m.r. and mass spectral data pointed to structure (**229**) which is further consistent with the formation of a diacetyl derivative. Rhizopogone is optically active ([α]$_{578}$+ 28°) but the absolute configuration is not yet known.

The ansa rings of pigments (**228**) and (**229**) are obviously formed by intramolecular electrophilic attack of an appropriately functionalised side chain terminus at the quinone or hydroquinone ring in a suitable precursor. In the case of tridentoquinone (**228**) the possibility of boviquinone-4 itself acting as a precursor was suggested earlier (*97*) but has subsequently been ruled out by feeding experiments (see below).

The pigment responsible for the pink colour of the stem base and mycelium of *Suillus americanus*, *S. bovinus*, and *S. collinitus* has been isolated and shown to be a dimer, bovilactone-4,4 (**230**), of boviquinone-4 (*389*). Bovilactone-4,4 showed complex carbonyl absorption in the infrared spectrum [ν_{CO}(carbon tetrachloride) = 1795, 1760, 1730, 1645, 1620 cm^{-1}] and absorption maxima at 275, 382 and 460 nm (log ε 4.06, 3.92 and 3.95, respectively) in the electronic spectrum. The structure (**230**) was derived by comparison with the model compound (**231**) ('bisnorbovilactone') which was prepared by heating 2,5-dihydroxy-1,4-benzoquinone under reflux in ethyl acetate in the presence of acetylated polyamide (*392*). Applying the same procedure to boviquinone-4 (using toluene as solvent) afforded bovilactone-4,4 (**230**) which confirmed its structure.

The mechanism of this remarkable reaction (Scheme 42) may involve an aldol type condensation between two molecules of the appropriate quinone to yield an intermediate (**232**) which undergoes ring opening to the ketene (**233**). Intramolecular interception of the ketene (**233**) would produce lactone (**234**), a tautomer of bovilactone-4,4 (**230**). The biosynthesis of bovilactone-4,4 from boviquinone-4 may follow similar lines.

Intimately connected with the boviquinone group are some fungal metabolites which derive from benzene-1,2,4-triol (**235**). This simple chromogen is itself present in several *Gomphidius* species (*G. glutinosus*,

(230)

(231)

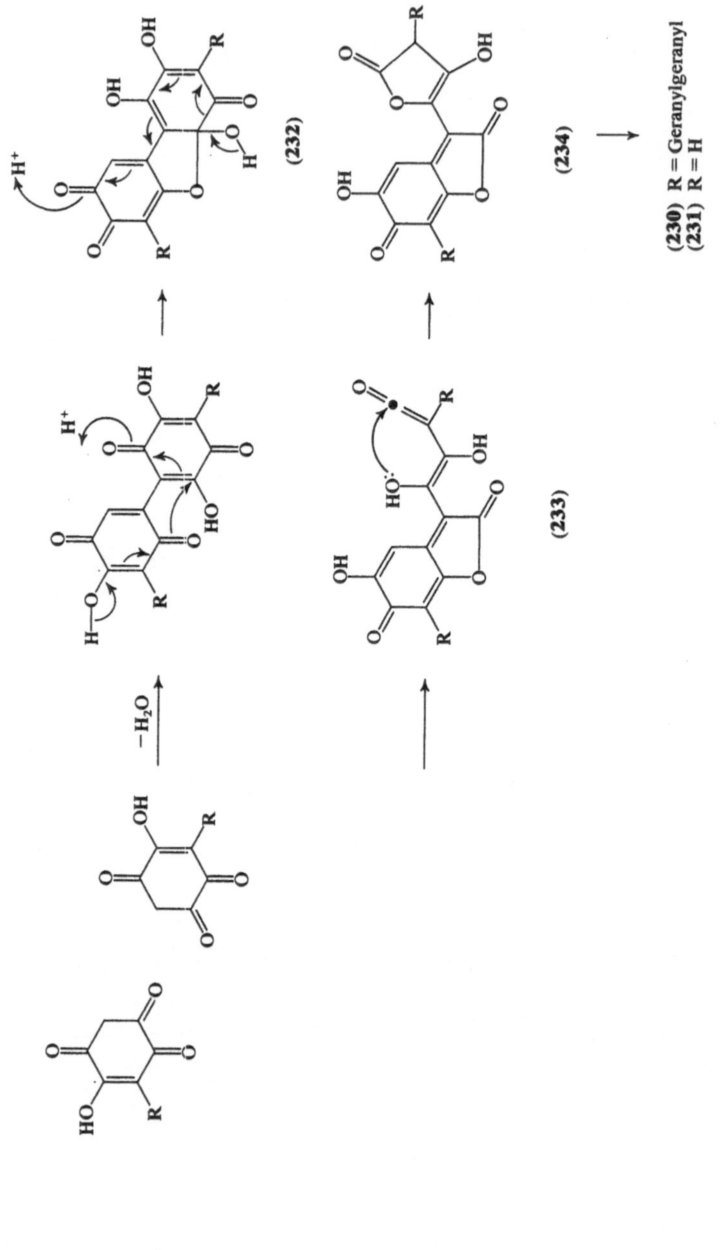

Scheme 42. Synthesis of bovilactones

G. maculatus and *G. roseus*) and is responsible for the appearance of wine red stains which soon turn black when fruit bodies of *G. maculatus* are injured (*689*). Further investigation of the constituents of *G. maculatus* and *G. glutinosus* led to the isolation of the colourless biphenyl (**236**), and the red gomphilactone (**237**) [UV/vis. (acetonitrile): λ_{max} (log ε) = 273 (3.56), 295 (sh., 3.40), 355 (3.39), 465 nm (3.44); infrared (disc): ν_{CO} = 1785, 1760 cm^{-1}] (*391*). The stereochemistry of gomphilactone is consistent with the ^{1}H-n.m.r. spectrum in which no nuclear Overhauser enhancement of the signal due to the proton at C-3 in the 4-ylidenebutenolide ring (δ 8.48, J = 6 Hz) was observed upon irradiation of the aromatic protons in (**237**).

(235)   (236)   (237)

A mechanism similar to that involved in the formation of bovilactone-4,4 (**230**) can be assumed to be operating in the biosynthesis of gomphilactone (**237**). In this case, however, the first step must involve oxidative phenolic coupling of (**235**) (Scheme 43).

The cap skins of *Suillus americanus*, *S. collinitus*, *S. granulatus*, *S. luteus*, and *S. variegatus* contain the colourless hydroquinone derivative suillin (**238**), which is closely related to boviquinone-4 (*390*). Its structure followed from the spectroscopic data and its conversion into the cyclic carbonate (**239**) (Scheme 44) which proved the position of the acetoxy group. Hydrolysis of suillin (**238**) with dilute aqueous acid led to 3-geranylgeranyl-1,2,4-trihydroxybenzene (**240**) which has been synthesized by the method outlined in Scheme 44 (*390*).

Structurally the boviquinones bear a great resemblance to the ubiquinones whose biosynthesis is known to originate from *p*-hydroxybenzoate. Accordingly, preliminary feeding experiments with ^{14}C-labelled *p*-hydroxybenzoic acid and mevalonic acid have shown good incorporation of both substrates into boviquinone-3 (**221**) by fruit bodies of *Chroogomphus helveticus* (*32*). Recently, it has also been established that [1-^{13}C]-*p*-hydroxybenzoic acid is incorporated into tridentoquinone (**228**) when administered to young fruit bodies of *Suillus tridentinus*. Furthermore, it was evident from the ^{13}C-n.m.r. spectrum of the

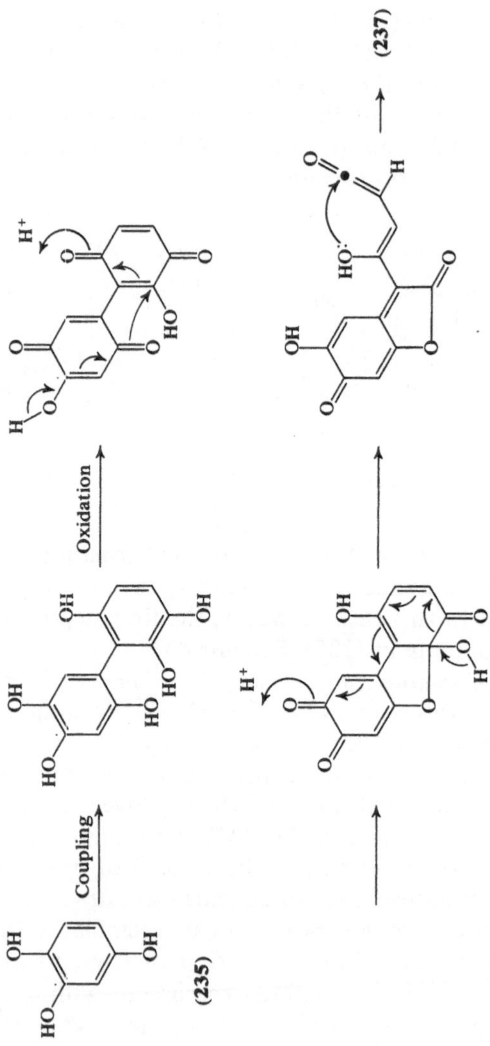

Scheme 43. Possible biosynthesis of gomphilactone

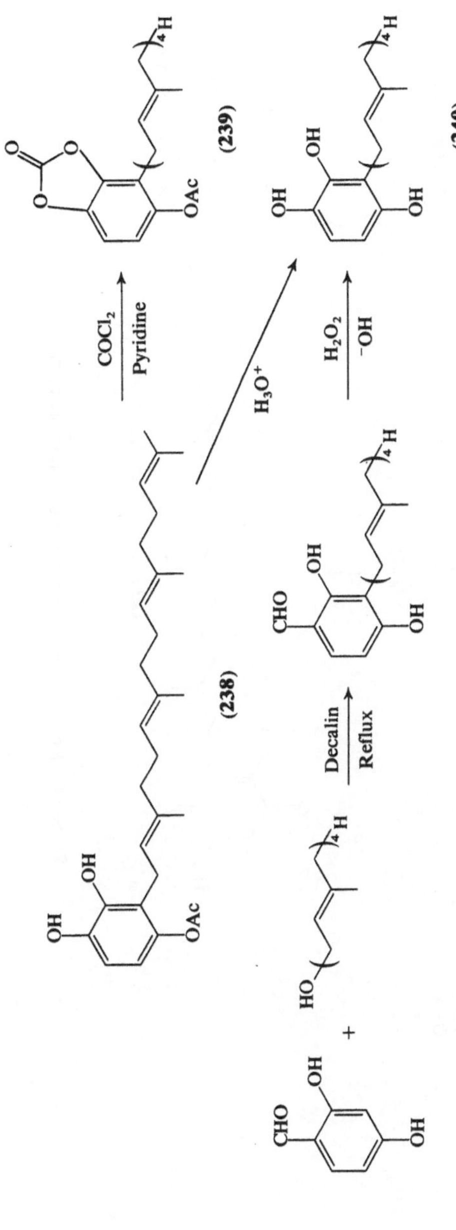

Scheme 44. Chemistry of suillin

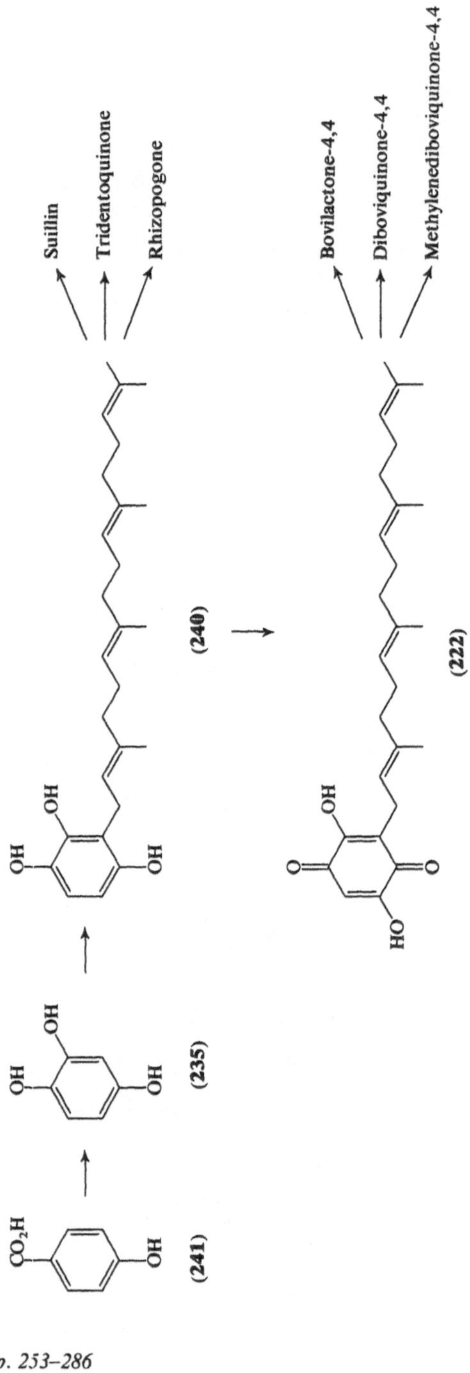

Scheme 45. Relationships between the boviquinone pigments

pigment (228) that the label had been incorporated only into that car-bonyl group adjacent to the free hydroxy function. This important result proved that *p*-hydroxybenzoic acid is the source of the quinonoid ring in tridentoquinone and at the same time precluded the involvement of a symmetrical intermediate such as boviquinone-4 or its terminal epoxide in the closure of the ansa ring. This conclusion was further supported by the lack of incorporation of both boviquinone-4 and its epoxide when they were fed in labelled form to *S. tridentinus* (*448*). The following relationships between the boviquinone pigments may be suggested (Scheme 45).

The isolation of biogenetically related polyprenylquinones from species of *Suillus*, *Gomphidius*, and *Chroogomphus* reinforces the taxonomic link which exists between the Boletaceae and the Gomphidiaceae. Similarly, the presence of the ansaquinone (229) in *Rhizopogon* adds further support to the proposed close taxonomic relationship of this genus with Boletales.

2.5. Miscellaneous Compounds Which May be Derived from the Shikimate-Chorismate Pathway

2,5-Dimethoxy-1,4-benzoquinone (thermophillin) (242) has been isolated from cultures of *Bjerkandera fumosa* (= *Polyporus fumosus*) (*140*), *Gloeophyllum trabeum* (= *Lenzites thermophila*) (*449*), and *Trametes lilacino-gilva* (*530*), and from the wood-destroying fungus *Gloeophyllum sepiarium* (*506*). The biosynthesis of thermophillin is not known but its similarity to benzene-1,2,4-triol suggests an origin from shikimic acid.

(242)   (243)   (244)

(245)   (246)

A shikimate-chorismate biogenesis can be more confidently predicted for 6-nitro-*iso*-vanillic acid (243), a constituent of an as yet unidentified Australian toadstool belonging to *Cortinarius* (*280*). The compound (243), the only example of a nitroaromatic metabolite from *Cortinarius*, produces an intensely yellow dianion [UV/vis. (ethanol + alkali): λ_{max} (log ε) = 270.5 (4.53), 429 nm (4.76)] which is responsible for the bright yellow colour of the cut flesh of this fungus. A yellow nitro compound, 3,5,6-trichloro-1,4-dimethoxy-2-nitrobenzene, has been isolated from *Phellinus rimosus* (= *Fomes robiniae*) (*147*).

Successful incorporation experiments involving [U-^{14}C]tyrosine suggested that the quinone ring and C-methyl group of coprinin (244), a yellow pigment from cultures of *Lentinus cyathiformis* (= *L. degener*), are derived from shikimate (*517*). However, there have been contradictory results (mentioned later) which point towards a tetraketide origin for coprinin (244) and its 6-hydroxy derivative.

Polyporus tumulosus, the cause of brown rot in fallen jarrah (*Eucalyptus marginata*), produces the orange-red (2,4,5-trihydroxyphenyl)-glyoxylic acid (245) when grown on a synthetic medium containing 'Marmite' (*535*). This pigment, which has also been isolated from fruit bodies of *Serpula lacrimans* (*90*) and *Paxillus statuum* (*268*), must owe its colour to a quinone methide tautomer such as the one shown. A similar rationalisation may also explain the yellow colour of the *o*-hydroxyphenylglyoxylic acid derivative (246), a pigment which has been isolated from cultures of *Antrodia sinuosa* (= *Poria sinuosa*) (*151*).

In *Polyporus tumulosus*, the pigment (245) co-occurs with hydroxylated phenylacetic acids (*173, 535*) from which it may be derived biosynthetically. In *Serpula lacrimans* and in *Paxillus statuum* the formation of (245) by degradation of a pulvinic acid may be suggested.

Two antibacterial agents, 2-methoxy-6-(1-propyl)-1,4-benzoquinone and its 3-methoxy derivative, have been isolated from the culture fluids of the fungus *Camarops microspora* (*686*). The biosynthesis of these compounds is not known.

3. Pigments from the Acetate-Malonate Pathway

In contrast to its widespread importance in the biosynthesis of secondary metabolites in Lower Fungi (*667, 668*) the acetate-malonate pathway is only of restricted importance to pigment production in Macromycetes. It is confined to a few basic systems which are categorised below in line with TURNER's classification (*667*) which is based on the size of the common 'polyketide' progenitor involved in each case.

Most of the pigments in this section, with the notable exception of the very large group of anthraquinones and pre-anthraquinones, are produced among Ascomycetes (e.g. members of such genera as *Daldinia, Bulgaria, Hypoxylon, Chlorosplenium*) or Aphyllophorales (e.g. *Gloeophyllum, Phlebia, Sarcodontia*) living on wood.

3.1. Tetraketides

Two meroterpenoids with chromogenic properties have been isolated from fruit bodies of *Albatrellus* species. Cristatic acid (**247**), a farnesyl phenol modified by incorporation of a furan ring in the side chain, is present in high concentrations in *Albatrellus cristatus* (*725*). This compound is closely related to the pigments responsible for the greenish colour of the fruit bodies. A North American *Albatrellus* species closely aligned with *A. cristatus* contains grifolic acid (**248**) instead of (**247**) (*725*). Clearly, grifolic acid (**248**) is derived by farnesylation of orsellinic acid and serves, in turn, as the precursor of cristatic acid (**247**).

(**247**)

(**248**)   (**249**)

Farnesylation has occurred at the alternate site in orsellinic acid during the formation of scutigeral (**249**) by *Albatrellus ovinus* and *A. subrubescens* (*98*). This compound is responsible for the yellow stain often seen on the sporophores and for the yellow colour reaction of the flesh of these fungi produced with aqueous alkali. The distribution of the substituents about the benzene ring in (249) was deduced using ^{13}C-n.m.r. spectroscopy and confirmed by synthesis starting from 2,3,4-trihydroxy-6-methylbenzaldehyde (250) (Scheme 46).

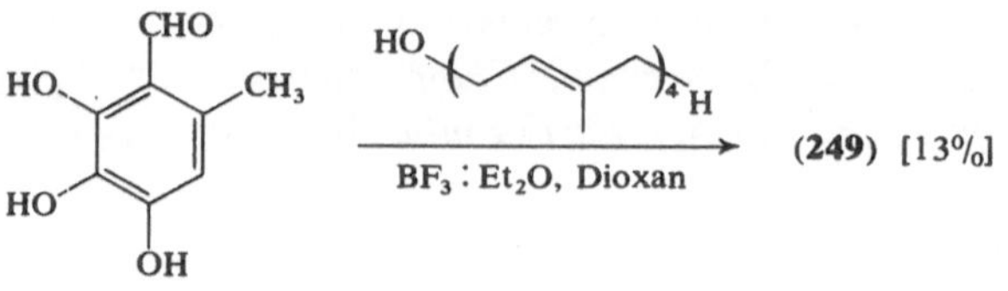

Scheme 46. Synthesis of scutigeral

Albatrellus species are a rich source of further meroterpenoids (*98, 696*) and biological screening of the group has revealed remarkable cytotoxic activities (*725*).

Extraction of the sporophores of the wood-destroying fungi *Gloeophyllum odoratum* (= *Osmoporus odoratus*) (*221, 341*) and *G. sepiarium* (*221*) with methanol produces the red-brown colouring matter, trametin [UV/vis. (ethanol): λ_{max} = 226, 277, 432 nm]. The fluorone structure (251) was established for this substance by detailed spectroscopic studies and by extensive chemical derivatisation and degradation (Scheme 47) (*221*).

A closer inspection of the extraction procedure revealed compound (251) to be an artefact formed from the genuine pigment of these *Gloeophyllum* species as a result of methanolysis, the two methoxy groups in (251) being derived from the solvent (*221*). Accordingly, extraction of *G. sepiarium* with ethanol yielded the homologue (252), but the failure of experiments to extract any pigment from fruit bodies of *G. odoratum* using polar solvents such as dimethylformamide suggested that the genuine colouring matter is present in the fungus as a polymer or is bound to a polymeric substance. The presence of 'bound trametin' in fungal sporophores may be ascertained by a simple test. Thus, on addition of aluminium oxide to ethanolic extracts of the fruit body acetal (252) is adsorbed with formation of an olive green colour. Treatment with dilute hydrochloric acid then yields an orange solution containing the xanthylium salt of (252) (*92, 221*).

The taxonomic significance of the occurrence of trametin in *Gloeophyllum* has been assessed chromatographically (*92, 180*).

The structures (251) and (252) suggest a biosynthesis of the *Gloeophyllum* pigment from two molecules of the tetraketide precursor (253) (Scheme 48) in a fashion analogous to the *in vitro* fluorone synthesis of Hatsuda (*353*).

It is interesting that an isomer (**254**) (fomecin-A) of the tetraketide (**253**) has been isolated along with the yellow dialdehyde (**255**) (fomecin-B) from cultures of *Antrodia juniperina* (=*Fomes juniperinus*) (*479*). Both fomecins-A and -B have been synthesised (*356*) and found to exhibit antibiotic properties.

The benzoquinone coprinin (**244**) has been isolated from cultures of *Coprinus radians* (=*C. similis*) (*18*) and is produced, together with its 6-hydroxy derivative, by cultures of *Lentinus cyathiformis* (=*L. degener*) (*524*) and *Gloeophyllum trabeum* (=*Lenzites thermophila*) (*668*). The biogenesis of these simple quinones has been the subject of much controversy. Experimental evidence supporting their formation *via* the corresponding quinols by decarboxylation and subsequent oxidation of 6-methylsalicylic acid (*525*) is in conflict with results indicating the involvement of tyrosine (*517*). The respective arguments have been evaluated by TURNER (*667*).

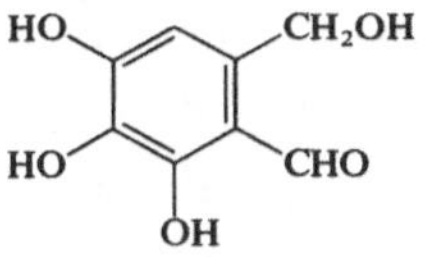

(254) 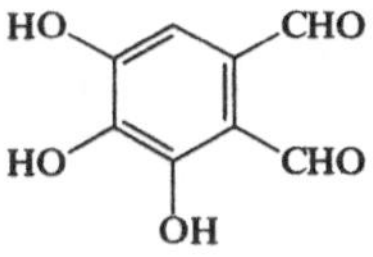   (255)

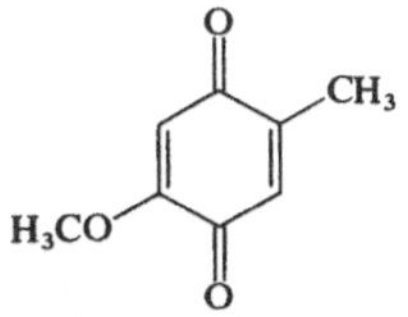

(244)

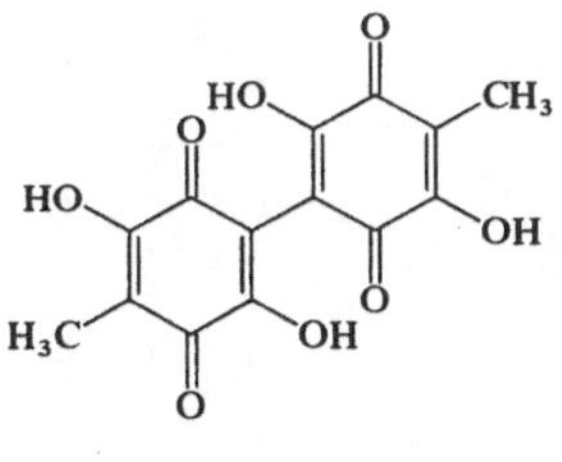

(256)

Biogenetically closely related to these simple benzoquinones is oosporein (**256**), a dibenzoquinone known from Ascomycetes and Fungi Imperfecti, which has been isolated along with its leuco derivative from cultures of *Phlebia mellea* and *Ph. albida* (*649*).

(251) R = CH₃
(252) R = C₂H₅

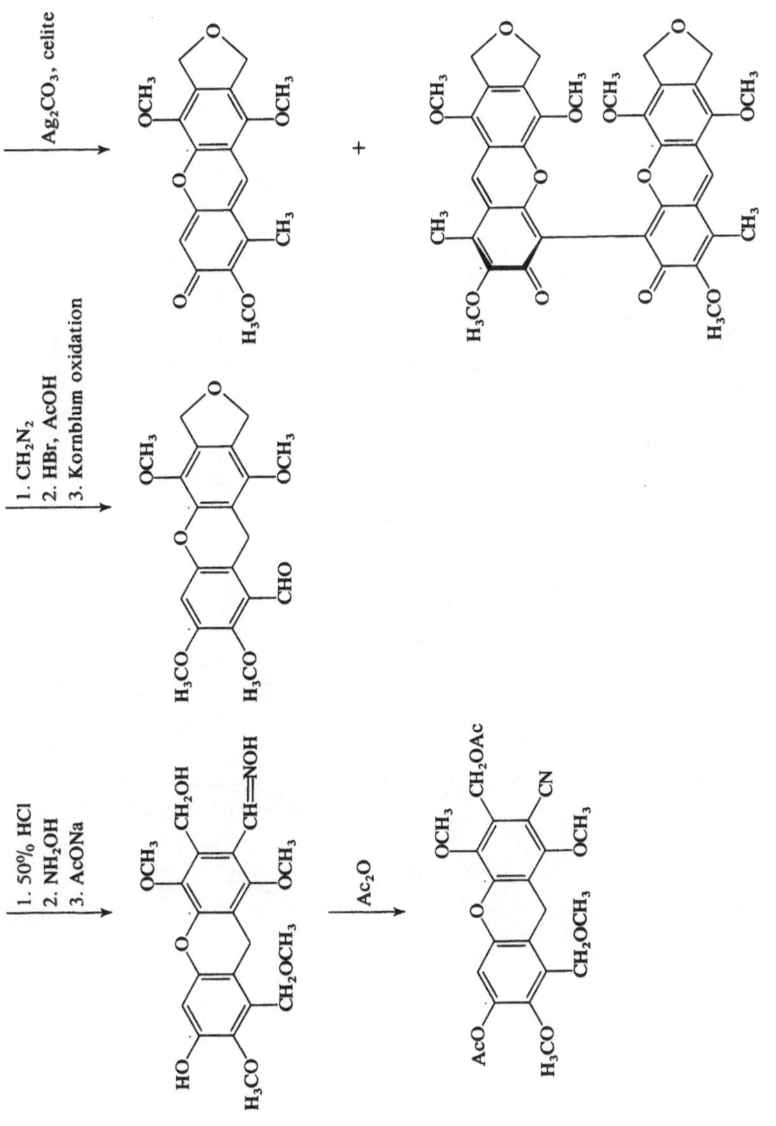

Scheme 47. Chemistry of trametin

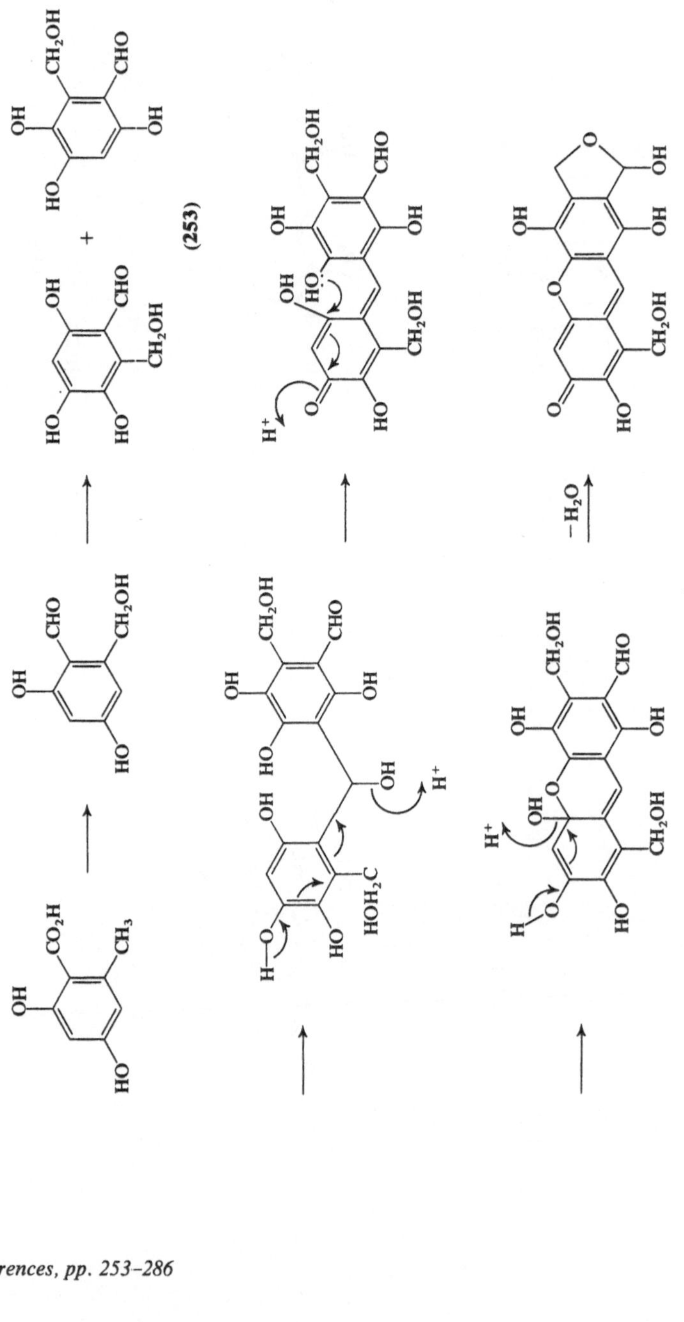

(253)

Scheme 48. Possible biogenesis of a trametin precursor

Veracruzalone (**257**), reported as a constituent of *Onnia tomentosa* (= *Polyporus tomentosus*) (*193*), is presumably derived from a polyketide precursor in a fashion similar to other fungal tropolones, e.g. sepedonin, a metabolite of *Sepedonium chrysospermum* (*477, 723*).

(**257**)

3.2. Pentaketides

Investigation of the brown-black sporophores of *Daldinia concentrica* by Bu'LOCK (*14, 15, 16*) and by others (*19*) has led to the discovery of a number of metabolites produced by oxidative coupling of naphthalene-1,8-diol (**258**) (Table 22). The main tractable pigment in *D. concentrica* fruit bodies is the dark red perylenequinone (**260**) which is accompanied by large quantities of black polymeric material (*14, 15, 19*). Careful extraction of the sporophores of *D. concentrica* gave the sensitive dinaphthyl (**259**) which could be further coupled, *in vivo* and *in vitro*, to afford the perylenequinone (**260**) together with polymers such as (**261**) and (**262**) (*15*).

From the mycelium of colourless strains of *D. concentrica,* Bu'LOCK (*16*) was able to isolate the ethers (**263**) and (**264**) of naphthalene-1,8-diol (**258**). Although the diol (**258**) itself has not been detected in fungi, the isolation of its ethers and its facile oxidation to the dinaphthyl (**259**) plus polymers strongly supports its involvement in the biosynthetic relationships depicted in Scheme 49.

Table 22. *Occurrence of Naphthalene-1,8-diol Derivatives*

4,4′,5,5′-Tetrahydroxy-1,1′-binaphthyl (**259**)	*Daldinia concentrica* [3.5 × 10^{-5}% fresh fungus; 1% by spectroscopic assay] (*15*), (*16*), *Hypoxylon fuscum* [0.5%, dried fungus] (*339*)
4,9-Dihydroxyperylene-3,10-quinone (**260**)	*Bulgaria inquinans* (*210*), *Daldinia concentrica* (*14, 15, 16, 19*), *Hypoxylon fuscum* (*339*), *H. sclerophaeum* var. *microspora* (*713*)
Bulgarein (**265**)	*Bulgaria inquinans* (*210*)
Bulgarhodin (**266**)	*Bulgaria inquinans* (*210*)
Hypoxylone (**269**)	*Hypoxylon sclerophaeum* (*107*)

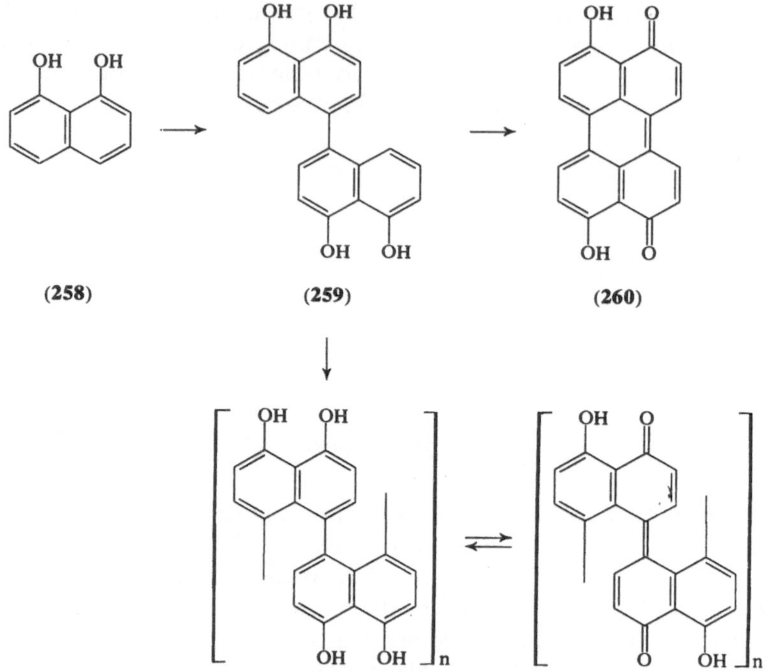

(258)   (259)   (260)

(261)   (262)

Scheme 49. Inter-relationship of *Daldinia concentrica* metabolites

In vivo, the oxidative polymerisation process shown in Scheme 49 occurs within the cellular matrix during development of the fruit bodies and thus exerts a cross-linking action. These events are analogous to melanisation, e.g. of insect cuticles, and the black *Daldinia* polymer is a type of fungal melanin (*141*).

The perylenequinone (260) is also present in the black sporophores of *Bulgaria inquinans* where it occurs along with the purple pigments bulgarein (265) and bulgarhodin (266) (*210*). The highly insoluble pigments (260), (265), and (266) were most conveniently separated in the form of the corresponding leuco-peracetates (Table 23) from which the quinones themselves could be regenerated by hydrolysis followed by aerial oxidation. The unique benzo[*j*]fluoranthene nucleus in the compounds (265) and (266) was deduced from the electronic spectra of the leuco-peracetates (Table 23) and is in accord with a biogenesis *via* oxidative coupling involving (258) or, possibly, naphthalene-1,2,8-triol.

Bulgarhodin (266) formed a tetra-acetate (268) which is believed to be a derivative of the *ortho*-quinonoid tautomer (267) of the pigment.

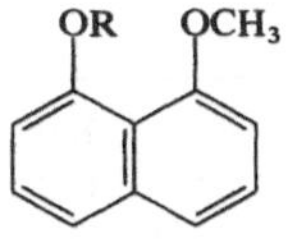

OR OCH₃

(263) R = H
(264) R = CH₃

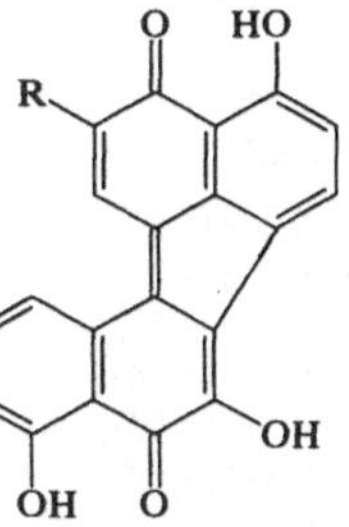

(265) R = H
(266) R = OH

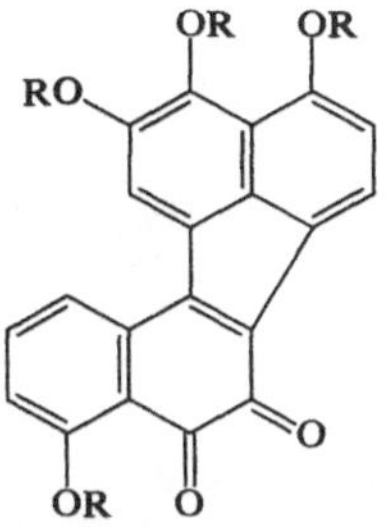

(267) R = H
(268) R = Ac

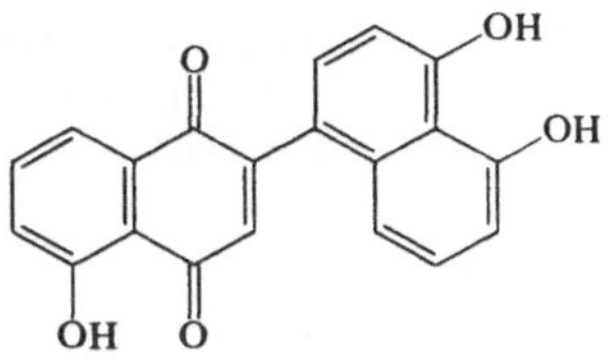

(269)

Neither bulgarhodin nor bulgarein is the same as the crystalline pigment 'bulgariin' isolated from *B. inquinans* by ZOPF (*733*).

The close relationship of *Bulgaria* and *Daldinia* with *Hypoxylon* in the Xylariaceae is illustrated by the occurrence in some *Hypoxylon* species of perylenequinones and their relatives (Table 22). The structure of hypoxylone (269), the main pigment of the dark purple carpophores of *Hypoxylon sclerophaeum* collected in French Guyana, followed from analysis of the ^{1}H- and ^{13}C-n.m.r. spectra and comparison with model systems (*107*). The naphthyl-naphthoquinone structure of hypoxylone suggests oxidative coupling between naphthalene-1,8-diol (258) and naphthalene-1,4,5-triol.

There exists no direct evidence that naphthalene-1,8-diol (258) and consequently the quinones (260), (265), (266), and (269) are of polyketide origin. However, this hypothesis is supported indirectly by the presence in colourless mutants of *Daldinia concentrica* of several resorcinol derivatives formed by an alternative folding of a pentaketide chain (*16*).

6-Methyl-1,4-naphthoquinone (270) has been isolated from cultures of *Marasmius graminum* (*74*). The related naphthoquinone (271) occurs in fruit bodies of *Heterobasidion annosum* (= *Fomes annosus*) (*194*). The structure of (271) was evident from the spectroscopic data and the precise position of the aromatic methyl group was established by synthesis. The possibility that (271) is an artefact formed from the corresponding 2-hydroxy-3-isopentenylnaphthoquinone during isola-

Table 23. *Properties of Naphthalene-1,8-diol Derivatives*

Pigment	Colour	Ultraviolet/Vis. $[\lambda_{max}(\log \varepsilon)]$	Leuco-peracetate Derivative			References
			Colour	M.p.	Ultraviolet/Vis. $[\lambda_{max}(\log \varepsilon)]$	
(260)	Dark red	265(4.33), 340(3.73), 419(4.32), 444(4.50), 493(3.77), 526(4.00), 567 nm(4.17) [tetrachloroethane]	Yellow	>300°	251(4.48), 259(4.48), 295(3.47), 403(4.15), 425(4.46), 452 nm(4.54) [chloroform]	(19, 210)
(265)	Purple	253, 300(sh.), 372, 565(sh.), 660 nm [ethanol]	Yellow	234–237°	248(4.80), 274(3.96), 284(4.15), 303(4.32), 315(4.47), 323(4.54), 328(4.52), 336(4.24), 358(3.75), 374(4.02), 385(3.89), 393(4.14), 420(3.66), 444 nm(3.53) [chloroform]	(210)
(266)	Purple	258(4.45), 324(3.97), 348(3.91), 410(4.03), 450(sh., 3.80), 670 nm(4.00) [sulphuric acid]	Yellow	274–276°	248(4.76), 274(3.96), 295(4.16), 309(4.41), 322(4.60), 341(4.04), 376(3.99), 395(4.11), 420(3.45), 450 nm(3.15) [chloroform]	(210)
(269)	Dark violet	228(4.4), 250(3.9), 319(3.7), 334(3.7), 418 nm [ethanol]	–	–	–	(107)

tion and purification cannot be excluded. The biosynthesis of these pigments has not yet been elucidated.

(270)   (271)

3.3. Hexaketides

The fruit bodies of *Hypoxylon fragiforme* contain a mixture of the azaphilones (+)-mitorubrin (272) (0.43%), (+)-mitorubrinol (273)

(0.12%), (+)-mitorubrinol acetate **(274)** (0.23%) and mitorubrinic acid **(275)** *(612)*. (+)-Mitorubrinol **(273)** becomes the principal pigment in the brick red stromata of *H. rubiginosum* where it occurs together with smaller quantities of **(272)**, **(274)**, and **(275)** *(155)*. Azaphilones have also been detected in *H. howeianum* *(713)*.

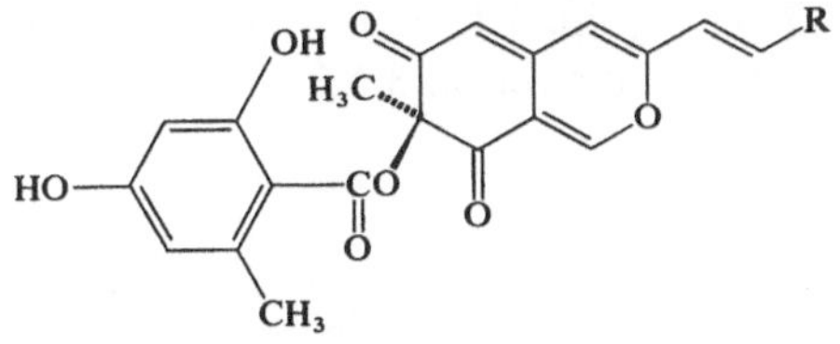

(272) R = CH₃
(273) R = CH₂OH
(274) R = CH₂OAc
(275) R = CO₂H

The absolute configuration of the *Hypoxylon* pigments followed X-ray crystallographic analysis of related sclerotiorin (= azaphilone) mould metabolites *(714)* and comparison of chiroptical data *(632)*. The production of dextrorotatory azaphilones by *Hypoxylon* is noteworthy since the laevorotatory enantiomers of **(272)**, **(273)**, and **(275)** have been isolated from *Penicillium* species *(139)*.

Surveys using thin layer chromatography of the distribution of pigments among *Hypoxylon* species have revealed that chemical similarities appear to parallel morphological resemblance, suggesting a valuable role for these pigments as taxonomic markers in this and related genera *(303, 712, 713)*.

3.4. Heptaketides

From the wine red alcohol extracts of mature fruit bodies of *Cortinarius rufoolivaceus* the naphthoquinones **(276)** and **(277)** have been isolated *(250)*. Young sporophores of this fungus are not highly pigmented and appear to contain these and other quinones (see Section 3.5.3) as their leuco derivatives.

The naphthoquinones **(276)** and **(277)** are closely related to torachrysone **(278)** and to torachrysone-8-*O*-methyl ether **(279)**. The ketone

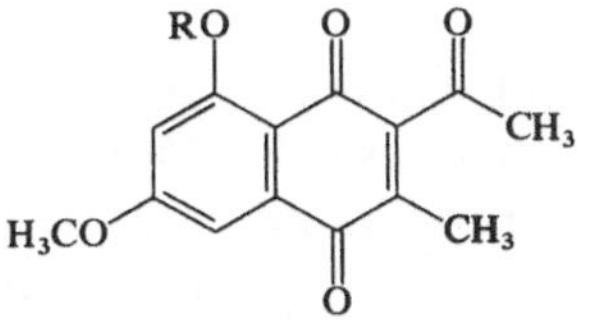

(276) R = H
(277) R = CH₃

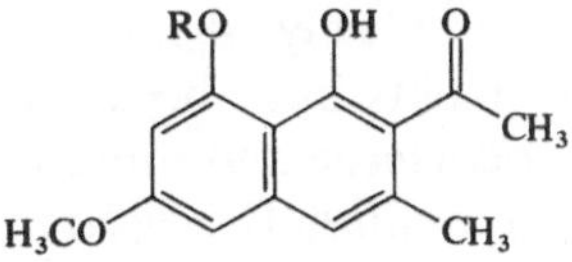

(278) R = H
(279) R = CH₃

(**279**) is also present in extracts of *C. rufoolivaceus* (*252*) and in cultures of *C. orichalceus* (*384*, *388*) and both (**278**) and (**279**) are known as constituents of higher plants (*577*). Torachrysone-8-*O*-methyl ether has been synthesised using a novel photochemical elimination of sulphinic acid (Scheme 50) (*634*).

The red-violet principal pigments of *C. rufoolivaceus* possess structures in which a unit of (**279**) is joined to a 1,4-anthraquinone; these compounds are discussed further in Section 3.5.3.

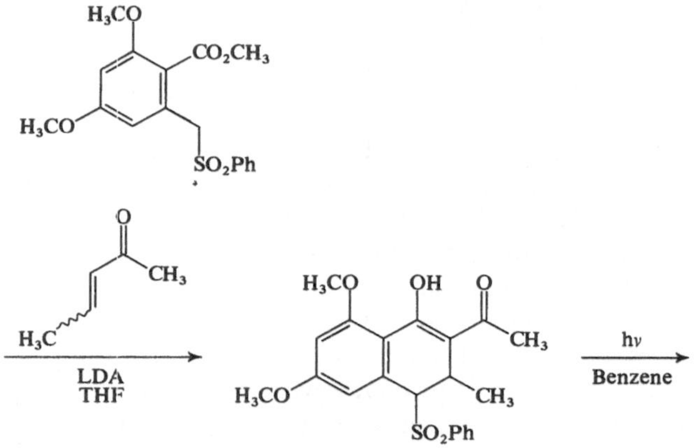

Scheme 50. Synthesis of torachrysone-8-*O*-methyl ether

Trichione (**280**) is the red pigment of fruiting bodies of the slime mould *Trichia floriformis* (*438*). The substitution pattern in the naphthoquinone nucleus and the nature of the unsaturated side chain and terminal hemimalonate moiety followed from the spectroscopic data.

Recently, a minor yellow-orange pigment from *Trichia floriformis* has been identified as 2,3,5-trihydroxy-1,4-naphthoquinone (**281**) by comparison with synthetic material (*156*). This simple naphthoquinone is most likely derived by oxidative degradation from a trichione type precursor. Further pigments are produced by *T. floriformis* as a result of intermolecular condensation between molecules of the trichione type. The structures of these more complex naphthoquinone pigments remain to be completely elucidated (*156*, *438*).

Pigments have been recognised as useful taxonomic characters in the Trichiaceae and other families of Myxomycetes (*105*, *586*).

The naphtho[1,2-*b*]furan derivative (**282**) occurs in the dark green velvety mycelial mat produced upon cultured growth of the ascomycete *Roesleria pallida*. This compound which exhibits strong blue fluorescence was isolated from the green extracts of the fungus and the struc-

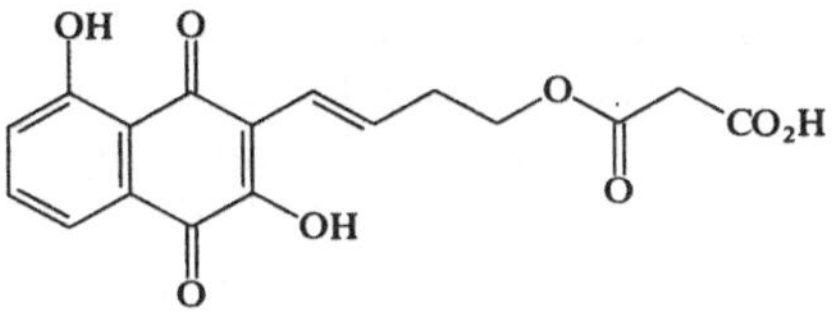

(280)

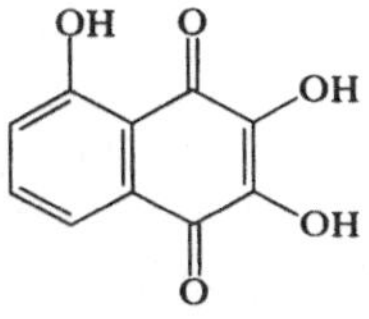

(281)

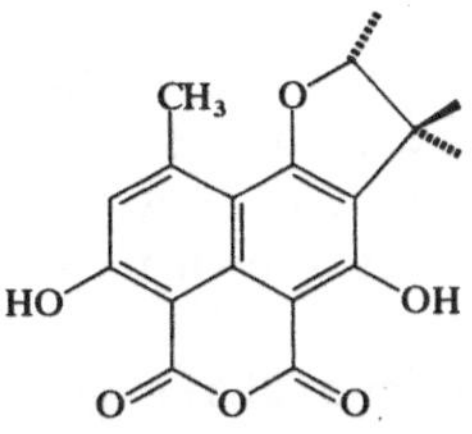

(282)

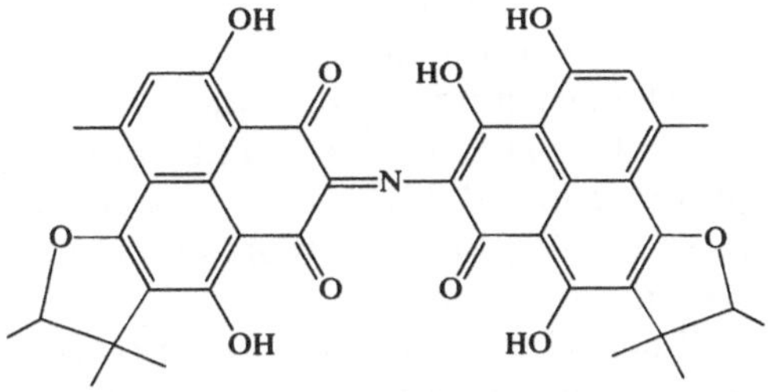

(283)

ture was established by synthesis involving degradation of the well known heptaketide, atrovenetin (*681*).

The anhydride (**282**) has also been described from *R. hypogea* (*59*) which is synonymous with *R. pallida* (*183*). The green colouring matter present in this fungus has been isolated (*59*) and proved to be identical with the green pigment (**283**) previously characterised from the virulently pathogenic ascomycete *Gremmeniella abietina* (*55*). The pigment (**283**) forms a stable 2:1 chelate complex with zinc ions in which form it may in part be present in *Roesleria* (*59*).

The cultured mycelium of the ascomycete *Cordyceps ophioglossoides* rapidly turns greenish black on exposure to alkali. The pigments responsible for this colour change have been isolated and characterised,

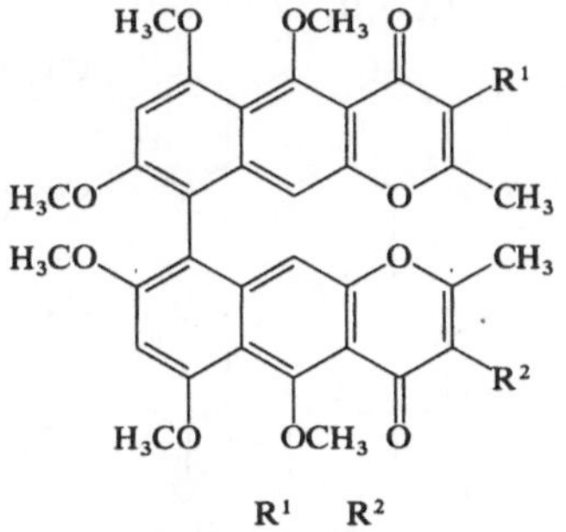

	R^1	R^2
(**284**)	H	H
(**285**)	CH$_3$	H
(**286**)	CH$_3$	CH$_3$

after extensive chromatography, in the form of their permethyl derivatives produced on treatment of the crude mycelial extract with diazomethane (100). The principal pigment obtained in this way was the hexamethyl ether (285) of 3-methylustilaginoidin-A [UV/vis. (methanol): λ_{max} (log ε) = 205 (3.96), 228 (4.08), 252 (sh., 4.08), 262 (4.20), 288 (4.24), 333 (3.31), 396 nm (3.50)] which was accompanied by minor amounts of the hexamethyl ethers (284) and (286) of ustilaginoidin-A and 3,3′-dimethylustilaginoidin-A, respectively. Ustilaginoidin-A is a metabolite of *Ustilaginoidea virens* (667).

The structures of the ethers (284), (285), and (286) were deduced from the mass and ¹H-n.m.r. spectra and have been confirmed by proton coupled ¹³C-n.m.r. experiments. The ¹H-n.m.r. data for the ether (285), which are representative of the group, are shown in Table 24.

Each of the ustilaginoidin derivatives (284), (285), and (286) is optically active by virtue of restricted rotation about the biaryl linkage. By comparing their CD spectra, which all show a negative Cotton effect at longer wavelength, with those of atrovirin-B (336) and skyrin (339) (see Section 3.5.3) the (R)-absolute configuration has been deduced for the chiral axis in the ustilaginoidin pigments described here.

The mixture of pigments extracted from *Cordyceps ophioglossoides* exhibited no methoxyl resonance in the ¹H-n.m.r. spectrum and must therefore contain a series of ustilaginoidins in which all phenolic hydroxy groups are free. Analogy may therefore be drawn between the rapid darkening produced in the mycelium and in the fruit bodies of *C. ophioglossoides* with base and similar processes which take place

Table 24. ¹H-N.m.r. Data for the Hexamethyl Ether (285) of 3-Methylustilaginoidin-A (δ values, with TMS as internal standard)

CDCl₃

as a result of extended quinone formation from hydroxydinaphthyls in *Daldinia concentrica* (see Scheme 49, Section 3.2). It is significant in this regard that during preliminary work on the constituents of *Cordyceps ophioglossoides* a pigment corresponding to the monomer unit of 3,3′-dimethylustilaginoidin-A was isolated and characterised as its acetyl derivative (*364*).

In contrast to the pigments of *C. ophioglossoides*, the colouring matter present in the bright orange-red fruit bodies of *C. militaris* appears to contain a polyolefinic chromophore (*249*).

Hypocrellin (**287**) [UV/vis. (ethanol): $\lambda_{max}(\log \varepsilon) = 215$ (4.78), 268 (4.60), 283 (4.59), 342 (3.50), 367 (3.40), 465 (4.48), 539 (4.04), 581 (4.13), 608 nm (3.80)], a dark red pigment from the acetone extract of the fungus *Hypocrella bambusae* (*167*), exhibits photodynamic activity towards microorganisms. The perylenequinone structure of hypocrellin was deduced from the ^{1}H- and ^{13}C-n.m.r. spectra and was confirmed by an X-ray analysis which also established the relative stereochemistry shown in (**287**).

(**287**)

3.5. Octaketides

The largest and most important chemical class within this category comprises the anthraquinone and pre-anthraquinone pigments found in great variety in toadstools belonging to the genera *Dermocybe*, *Cortinarius*, *Tricholoma* and *Leucopaxillus*.

3.5.1. Neutral Anthraquinones and Anthraquinone Carboxylic Acids

Chromatographic analysis of the distribution of neutral anthraquinones and anthraquinone carboxylic acids particularly in *Dermocybe* and *Cortinarius* has proved a valuable aid to the taxonomy of these genera. Thus, chemotaxonomic studies on *Dermocybe* species pioneered by GABRIEL (*261, 262, 263, 265*) and continued by GRUBER (*332, 334*), using paper chromatography, and by KELLER (*400, 401, 402*), KELLER and AMMIRATI (*403*), and HøILAND (*374*), using thin layer chromatography, showed that many species possess characteristic anthraquinonoid pigment patterns which can be conveniently applied to the differentiation of infrageneric taxa. Species of *Dermocybe* collected in South America (*335, 400, 402*), North America (*400, 403*), Scandinavia (*373, 374*), and New Zealand (*400, 402*), as well as Europe (*261, 262, 263, 265, 266, 332, 334, 400, 401, 652*), and *Cortinarius* species from Europe (*264, 333, 617*) and Nordic countries (*118, 119, 372, 375*) have been studied in this way. The data collected on European dermocybes support the systematic arrangement proposed by MOSER (*497*). The demonstration by WATLING (*410*) that chromatographic techniques based on anthraquinones can be applied effectively to the reclassification of herbarium material, including specimens of *Dermocybe cinnabarina* over one hundred years old, complements the earlier observations regarding the endurance of these pigments which were made by MOSER (*496*). The results of the many chromatographic surveys are collected in Tables 25 and 26 along with references to chemical work.

Although both ZOPF (*732*) and BACHMANN (*57*) were undoubtedly dealing with anthraquinones during their studies of the colouring matters of *Cortinarius bulliardi* and *C. armillatus*, respectively, the first anthraquinones to be chemically characterised from fungi emerged from the investigation of the pigments of the blood red fruit bodies of *Dermocybe sanguinea* by KÖGL and POSTOWSKY (*432*). Since that time the number of anthraquinonoid pigments isolated from macromycetes has risen considerably and Tables 25 and 26 represent, respectively, the known distribution of neutral anthraquinones and of anthraquinone carboxylic acids.

	R^1	R^2	R^3
(288)	H	H	H
(289)	H	H	Glu
(290)	H	CH_3	H
(291)	CH_3	H	H
(292)	CH_3	CH_3	H

	R^1	R^2
(293)	H	OH
(294)	H	OAc
(295)	OH	H
(296)	NH_2	H

	R¹	R²
(297)	H	H
(298)	OH	H
(299)	OH	Glu

	R¹	R²
(300)	H	H
(301)	OH	H
(302)	OH	OH

Table 25. *Occurrence of Neutral Anthraquinones*[a]

Emodin (**288**)	*Cortinarius bulliardi, C. mallochii (405), Dermocybe anthracina (542), D. aspenensis (403), D. egmontiana (402), D. flavotomentosa (400), D. malicoria (334, 401), D. punicea (334), D. sanguinea* [0.15%] *(616)*, *(102, 334, 401, 403, 432, 602), D. sanguinea* var. *sierraensis (403), D. semisanguinea* [5×10^{-3}%] *(616), D. sommerfeltii (=D. cinnamomeobadia) (401), D. vinicolor (402), Hypomyces aurantius (87)*
Emodin-1-β-D-glucopyranoside (**289**)	*Dermocybe aspenensis (403), D. egmontiana (402), D. flavotomentosa (400), D. malicoria (334, 401), D. punicea (334), D. sanguinea* [0.12%] *(615), (334, 401, 403), D. sanguinea* var. *sierraensis (403), D. sommerfeltii (401), D. vinicolor (402)*
Physcion (**290**)	*Cortinarius bulliardi (543), C. elegantior (387, 617), C. flavescentium (371), C. fulmineus (515, 617), C. mallochii (405), C. orichalceus* [cultures] *(338, 388, 617), C. rufoolivaceus (252, 617), C. subfulgens (515, 617), C. xanthophyllus (371), Dermocybe californica (403), D. canaria (404), D. cinnabarina* [3×10^{-2}%] *(621), (334, 401, 403), D. cramesina, D. egmontiana (402), D. hesleri (403), D. olivaceonigra (402), D. sanguinea (401, 616), D. sanguinea* var. *vitiosa (401), D. semisanguinea* [trace] *(400, 616) D. sommerfeltii (401), D. vinicolor (402), Tricholoma auratum (515)*
Questin (**291**)	*Dermocybe crocea (=D. cinnamomeolutea)* [1.4×10^{-3}%] *(657), (629), D. semisanguinea (87)*
Emodin-6,8-di-*O*-methyl ether (**292**)	*Cortinarius armillatus (96), C. elegantior (387, 617), C. flavescentium (371), C. mallochii (405), C. orichalceus* [c] *(338, 388), C. rufoolivaceus (252, 617), C. xanthophyllus (371), Dermocybe semisanguinea (87)*
Fallacinol (**293**)	*Cortinarius bulliardi (405), Dermocybe californica (403), D. cinnabarina* [0.12%] *(621), (334, 401, 403), D. hesleri* [trace] *(403)*
ω-*O*-Acetylfallacinol (**294**)	*Dermocybe cinnabarina* [1.5×10^{-2}%] *(621)*
Erythroglaucin (**295**)	*Cortinarius elegantior (387, 617), Dermocybe canaria (404), D. cinnabarina* [1×10^{-2}%] *(621), D. semisanguinea* [trace] *(616)*

Table 25 *(continued)*

4-Aminophyscion (**296**)	*Dermocybe canaria (404)*
Dermoglaucin (**297**)	*Dermocybe aspenensis (403), D. egmontiana (402), D. flavoto-mentosa (400), D. marylandensis (403), D. phoenicea (334, 401), D. phoenicea* var. *occidentalis (403), D. punicea (334), D. sanguinea* [1×10^{-2}%] *(616), (334, 401, 403, 602), D. sanguinea* var. *sierraensis (403), D. sanguinea* var. *vitiosa (401), D. semisanguinea* [1×10^{-2}%] *(616), (334, 401, 403), D. sommerfeltii (401), D. sphagnogena (400), D. vinicolor (402)*
Dermocybin (**298**)	*Dermocybe aspenensis (403), D. egmontiana (402), D. marylandensis (403), D. phoenicea (334, 401), D. phoenicea* var. *occidentalis (403), D. punicea (334), D. sanguinea* [6×10^{-2}%] *(616), (102, 334, 401, 403, 432, 602), D. sanguinea* var. *sierraensis (403), D. sanguinea* var. *vitiosa (401), D. semisanguinea* [4×10^{-2}%] *(616), (334, 401, 403), D. vinicolor (402)*
Dermocybin-1-β-D-glucopyranoside (**299**)	*Dermocybe aspenensis (403), D. egmontiana (402), D. marylandensis (403), D. phoenicea (334, 401), D. phoenicea* var. *occidentalis (403), D. punicea (334), D. sanguinea* [4×10^{-2}%] *(615), (334, 401, 403), D. sanguinea* var. *sierraensis (403), D. sanguinea* var. *vitiosa (401), D. semisanguinea (334, 401, 403), D. vinicolor (402)*
6-Methylxanthopurpurin-3-O-methyl ether (**300**)	*Dermocybe* sp. *(28), D. splendida, D. umbonata (402)*
Austrocortinin (**301**)	*Dermocybe* sp. *(28), D. splendida, D. umbonata (402)*
Xanthorin (**302**)	*Dermocybe* sp. *(28)*

[a] In the following references the pigments have been identified by chromatographic comparison with authentic substances: *87, 334, 371, 400, 401, 402, 403*. Other citations are of work in which the pigments were isolated and characterised chemically and/or spectroscopically. Yields refer to percentage of fresh weight of fungus; in references *616* and *621* they correspond to yields of aglycone after acid hydrolysis of the corresponding glycosides.

During their work, KÖGL and POSTOWSKY *(432)* isolated emodin (**288**) together with a second pigment which they named dermocybin. Structural studies by BIRKINSHAW *(102)* established dermocybin as either the 6- or 7-O-methyl ether of 1,5,6,7,8-pentahydroxy-3-methyl-9,10-anthraquinone ('nordermocybin') which was synthesised, thus establishing the oxygenation pattern in dermocybin itself.

In a reinvestigation of the constituents of *D. sanguinea* STEGLICH *et al. (602, 616)* working with 14 kg of fresh material, were able simultaneously to confirm emodin as the major pigment, to establish the structure (**298**) for dermocybin, and to identify physcion (**290**) and a new anthraquinone, dermoglaucin (**297**), in the neutral pigment fraction. The structure of dermoglaucin was deduced principally by comparison of the ^{1}H-n.m.r. spectrum of its pertrimethylsilyl ether with that of physcion (290), and the recognition that only one of the quinonoid carbonyl groups is hydrogen bonded to an adjacent *peri*-hydroxy group. Oxidation of dermoglaucin with lead tetra-acetate gave a violet diquinone which furnished dermocybin (298) when treated with sodium hydroxide. Interestingly, with concentrated aqueous ammonia, dermo-cybin (298) was transformed into the aminoanthraquinone (318) (p. 132) in a reaction characteristic of a 1,2,4-trihydroxylation pattern in the anthraquinone nucleus.

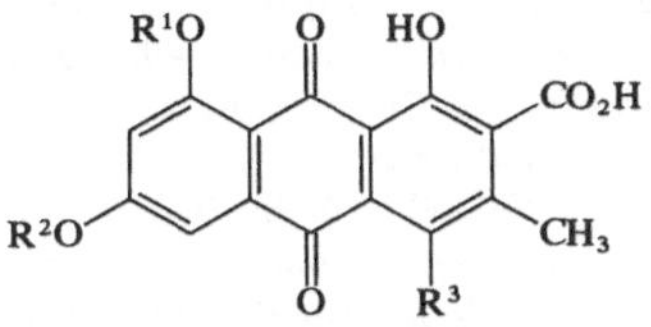

	R^1	R^2	R^3
(303)	H	H	H
(304)	H	H	OH
(305)	CH$_3$	H	H
(306)	CH$_3$	H	OH
(307)	H	CH$_3$	H
(308)	H	CH$_3$	OH
(309)	CH$_3$	CH$_3$	H
(310)	CH$_3$	CH$_3$	OH

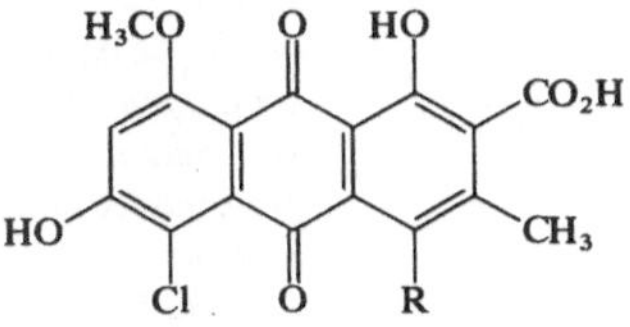

(311) R = H
(312) R = OH

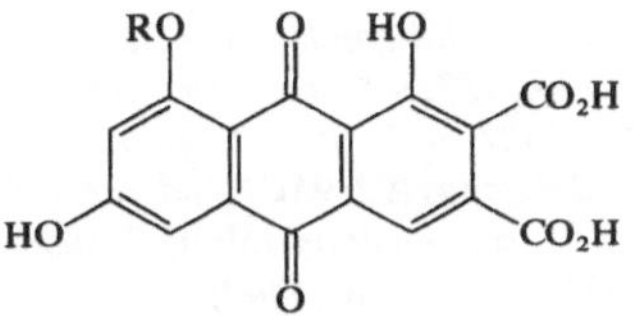

(313) R = H
(314) R = CH$_3$

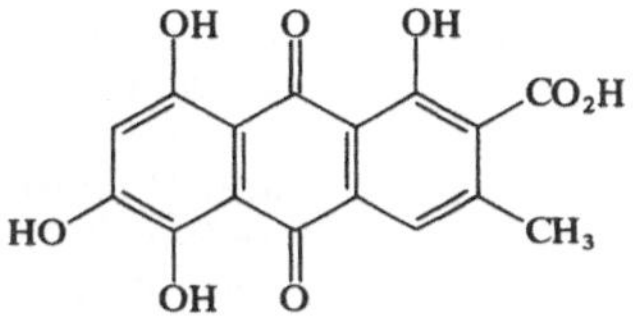

(315)

Careful chromatographic separation of the acidic pigments of *D. sanguinea* (616) led to isolation of the known endocrocin (303) and the novel anthraquinone carboxylic acids dermolutein (305) and der-morubin (306). The quinones (305) and (306) are accompanied in *D. sanguinea* by their 5-chloro derivatives (311) and (312), respectively.

Dermolutein (305) was chemically correlated with questin (291) and with endocrocin (303) by decarboxylation and demethylation, respec-tively, while demethylation of dermorubin (306) afforded the quinone (304) and thence catenarin after decarboxylation. Differentiation be-tween the alternative 1- and 8-*O*-methyl derivatives of structure (304) for dermorubin was made on the basis of acylation shifts. Thus, the

Table 26. *Occurrence of Anthraquinone Carboxylic acids*[a]

Endocrocin (**303**)	*Claviceps purpurea (246, 247), Cortinarius armillatus* [trace] *(543), C. aureofulvus (371), C. auroturbinatus (515, 617), C. bulliardi (543), C. citrinolilacinus (371), C. elegantior* [1×10^{-2}%]* *(384), (387, 617, 657), C. eufulmineus (371), C. fulmineus (515, 617), C. incognitus (400), C. odoratus (515, 617), C. odorifer (617), C. olivellus (515, 617), C. personatus* (= *C.* spec. 1) *(384, 617), C. pseudocolus (543), C.* spec. 2 *(617), C. splendens (371), C. subcroceofolius (400), C. subfulgens, C. sulfurinus, C. superbus (371), C. tubarius (400), C. vitellinus (515, 617), C. xanthochlorus (371), Dermocybe alcalisensibilis, D. alienata (402), D. alnophila (401), D. amoena (402), D. anthracina (401, 542), D. austronanceiensis (402), D. californica (403), D. canaria, D. cardinalis, D. castaneodisca (402), D. chrysophtalma (400), D. cinnabarina* [2×10^{-3}%] *(621), (334, 403), D. cinnamomea (334), D. cramesina (402), D. crocea* (= *D. cinnamomeolutea* = *D. saligna) (334, 400, 401, 629), D. hesleri (403), D. holoxantha (401), D. icterina, D. leptospermarum, D. luteostriatula (402), D. malicoria (334, 401), D. marylandensis (403), D. obscuroolivea* var. *brunnea (400), D. olivaceonigra, D. oliveoicterina, D. olivipes (402), D. palustris* var. *sphagneti (334), D. parientalis (402), D. phoenicea (334), D. phoenicea* var. *occidentalis (403), D. punicea (334), D. sanguinea (334, 401, 616), D. schaefferi* (= *D. carpineti) (334, 401, 405), D. semisanguinea (334, 401, 403), D. sphagnogena (400), Leucopaxillus tricolor (86, 94, 515), Tricholoma auratum* (= *T. equestre)* [1×10^{-4}%] *(515), (294), T. flavobrunneum (294), T. sulphureum (293, 294, 515)*
Nordermorubin (**304**)	*Cortinarius subfulgens (515, 617)*
Dermolutein (**305**)	*Cortinarius armillatus (96, 543), C. aureofulvus (371), C. auroturbinatus (515, 617), C. cedretorum (371), C. cotoneus (590), C. elegantior* [1.8×10^{-2}%]* *(384), (387, 617, 657), C. fulmineus (515, 617), C. incognitus (400), C. miniatopus (92, 96), C. odoratus, C. orichalceus (371), C. psittacinus, C. subannulatus (590), C. subcroceofolius (400), C. subfulgens (371), C. tubarius (400), C. venetus* var. *montanus (590), Dermocybe alienata (402), D. alnophila (401), D. amoena (402), D. anthracina (334, 401, 542), D. aspensis (403), D. aurea, D. austronanceiensis (402), D. cardinalis (402), D. cinnamomea (334, 401), D. crocea* [1.7×10^{-2}%] *(657), (334, 400, 401, 629), D. holoxantha (401), D. icterina (402), D. malicoria (334, 401), D. marylandensis (403), D. obscuroolivea* var. *brunnea, D. oliveoicterina, D. olivipes (402), D. palustris (334), D. palustris* var. *sphagneti (334, 629), D. phoenicea (334, 401), D. phoenicea* var. *occidentalis (403), D. punicea (334), D. sanguinea (334, 401, 403, 616), D. sanguinea* var. *sierraensis (403), D. sanguinea* var. *vitiosa (401), D. semisanguinea (334, 401, 403, 616), D. sommerfeltii* (= *D. cinnamomeobadia) (334, 401), D. sphagnogena (400), D. uliginosa (334, 401), D. vinicolor (402), Tricholoma auratum* [1×10^{-4}%] *(515), (294), T. sulphureum (515)*
Dermorubin (**306**)	*Cortinarius armillatus (96, 543), C. incognitus (400), C. miniatopus (92, 96), C. subcroceofolius (400), C. subfulgens (118, 515, 617), C. tubarius (400), Dermocybe anthracina (334, 401, 542), D. cinna-*

Table 26 (continued)

	momea (334, 401), D. crocea [3.5 × 10⁻³%] *(657), (334, 400, 401, 629), D. flavotomentosa (400), D. malicoria (334, 401), D. marylandensis (403), D. palustris* var. *sphagneti (629), D. phoenicea (334, 401), D. phoenicea* var. *occidentalis (403), D. punicea (334), D. sanguinea (334, 401, 403, 616), D. sanguinea* var. *sierraensis (403), D. sanguinea* var. *vitiosa (401), D. semisanguinea (334, 401, 403), D. sommerfeltii, D. uliginosa (334, 401)*
Cinnalutein (307)	*Cortinarius bulliardi (405, 543), C. pseudocolus (543), Dermocybe californica (403), D. cinnabarina* [3 × 10⁻²%] *(621), (334, 401, 403), D. hesleri (403)*
Cinnarubin (308)	*Cortinarius bulliardi, C. pseudocolus (543), Dermocybe californica (403), D. cinnabarina* [2 × 10⁻²%] *(621), (334, 401, 403), D. cramesina (400), D. hesleri (403)*
Endocrocin-6,8-di-*O*-methyl ether (309)	*Cortinarius armillatus, C. miniatopus (96)*
Dermorubin-6-*O*-methyl ether (310)	*Cortinarius armillatus, C. miniatopus (96)*
5-Chlorodermolutein (311)	*Dermocybe alnophila, D. malicoria, D. phoenicea (401), D. sanguinea (401, 616), D. sanguinea* var. *vitiosa, D. sommerfeltii (401)*
5-Chlorodermorubin (312)	*Cortinarius subcroceofolius (400), Dermocybe crocea (401), D. marylandensis (403), D. phoenicea (334), D. phoenicea* var. *occidentalis (403), D. punicea (334), D. sanguinea (334, 401, 403, 616), D. sanguinea* var. *sierraensis (403), D. sanguinea* var. *vitiosa (401), D. semisanguinea (403), D. sommerfeltii, D. uliginosa (401)*
Cortinaric Acid (313)	*Cortinarius elegantior* [2 × 10⁻²%]* *(384), (387, 617)*
Cortinaric Acid-8-*O*-methyl ether (314)	*Cortinarius elegantior* [1.6 × 10⁻²%]* *(384), (387, 617)*
Clavorubin (315)	*Claviceps purpurea (246, 247)*

ᵃ In the following references the pigments have been identified by chromatographic comparison with authentic substances: *86, 92, 118, 334, 371, 400, 401, 402, 403*. Other citations are of work in which the pigments were isolated and characterised chemically and/or spectroscopically. Yields designated by an asterisk refer to dried fungal material; other yields are based on fresh fungus.

differences in chemical shifts for the corresponding aromatic protons in the ¹H-n.m.r. spectra of the pertrimethylsilylated and peracetylated derivatives of 1,8-dihydroxyanthraquinones serve as useful parameters for ascertaining the positions of *O*-alkyl and *O*-glycosyl groups (*614*); in the case of dermorubin (306) these shifts clearly indicated an 8-*O*-methyl ether (*616*). 5-Chlorodermolutein (311) and 5-chlorodermorubin (312) were separated by extensive thin layer and paper chromatography and their structures followed largely from ¹H-n.m.r. considerations.

Most of the emodin and dermocybin in *D. sanguinea* is present in the form of the 1-β-D-glucopyranosides (**289**) and (**299**), respectively (*615*). The glucosides were obtained by aqueous extraction of fresh fruit bodies and separated by chromatography on columns of poly-amide. Emodin-1-β-D-glucopyranoside (**289**) formed a hexa-acetyl de-rivative and with mineral acid or β-glucosidase (but not α-glucosidase) it was cleaved to emodin and D-glucose. The position of the sugar residue was established by methylation of (**289**) with diazomethane and subsequent hydrolysis, whereupon emodin-6,8-di-*O*-methyl ether (**292**) was obtained which was identical with synthetic material (see below).

Dermocybin-1-β-D-glucopyranoside (**299**) was similarly cleaved with β-glucosidase and, with aqueous ammonia, the deoxyaminogluco-side (**317**) was produced indicating a 'free' 5,7,8-trihydroxylation pat-tern in the parent glucoside. Hydrolysis of (**317**) with mineral acid (Scheme 51) afforded the aminoquinone (**318**) which had earlier been isolated upon ammonolysis of dermocybin itself (*616*). The position of the glucose residue in (**299**) was further corroborated by the acylation shifts observed in the ^{1}H-n.m.r. spectra (*614, 615*).

The pigments of *D. sanguinea* fall rationally into the biosynthetic pattern shown in Scheme 52. It has been established by feeding experi-ments with young sporophores of *D. sanguinea* that endocrocin, radio-labelled both at C-9 and in the carboxyl group, is effectively incorpo-rated into the pigments dermolutein (**305**) and dermorubin (**306**) but that this carboxylic acid does not serve as progenitor of the neutral

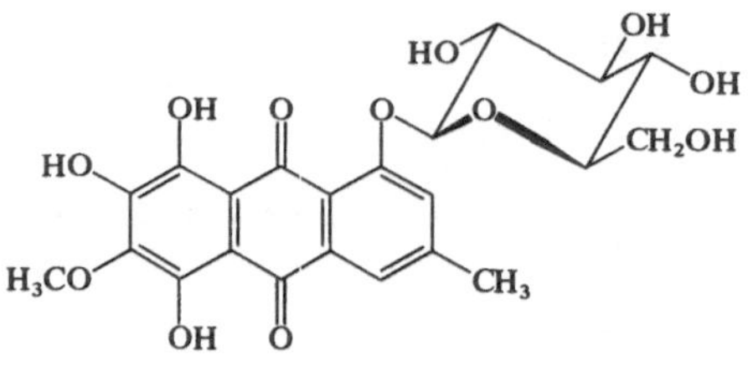

(**299**)

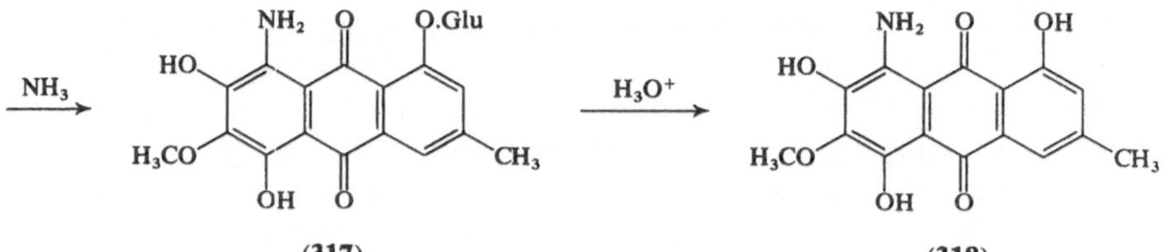

(**317**)   (**318**)

Scheme 51. Transformation of dermocybin glucoside

anthraquinones (*601*). Labelled (**303**) was prepared for these experiments from methyl (3,5-dimethoxyphenyl)acetate according to the reactions shown in Scheme 53 (*542, 620*). Decarboxylation of the intermediate (**309**) gave emodin-6,8-di-*O*-methyl ether (**292**) (*615*).

That emodin (**288**) is the precursor of both dermoglaucin (**297**) and dermocybin (**298**) has been established by the successful incorporation of [2,4-^{3}H$_2$]emodin-6-β-D-glucopyranoside into these neutral pigments in young fruit bodies of *Dermocybe semisanguinea* (*601*), a species related to *D. sanguinea*[1].

Tritium labelled emodin glucoside was not incorporated by *D. semisanguinea* into the acidic pigments (**303**), (**305**) or (**306**), a result one might have expected, and it is apparent that both the neutral and the acidic anthraquinones in *D. sanguinea* and closely related species share a common polyacetate biogenesis which diverges at a stage prior to complete formation of the anthraquinone nucleus (*271, 338, 596, 598, 601*). A common precursor to both groups could exist at the anthrone level or at a stage at which the carboxylated ring is not yet aromatic (see Scheme 60, Section 3.5.2).

Dermoglaucin (**297**) has been synthesised by ROBERGE and BRASSARD (*546*) by a route which nicely illustrates the power of cycloaddition chemistry for the construction of specifically substituted anthraquinones. These workers exploited the regioselectivity inherent in the reaction of highly oxygenated butadienes such as (**319**) and (**320**) with chloroquinones and one of their routes to dermoglaucin (**297**) is summarised in Scheme 54.

A contrasting mode of anthraquinone elaboration occurs in *Dermocybe cinnabarina* (Scheme 55) (*621*). In the neutral pigments produced in fruit bodies of this fungus hydroxylation has occurred both at the C-3 methyl group and within the A-ring itself, while the carboxylic acids cinnalutein (**307**) and cinnarubin (**308**) are etherified at the 6- rather than the 8-hydroxy group. The major pigment present is fallacinol (**293**) and all of the anthraquinones occur almost entirely as their glycosidic derivatives.

The structures of the corresponding aglycones followed chemical correlation with known anthraquinones. Thus, cinnalutein (**307**) on decarboxylation with copper powder in quinoline afforded physcion (**290**), while with boron tribromide endocrocin (**303**) was produced. A synthesis of cinnalutein is illustrated above in Scheme 53 (*620*). Decarboxylation and demethylation of cinnarubin (**308**) afforded erythroglaucin (**295**) and nordermorubin (**304**), respectively.

---

[1] In ref. *596* it is wrongly reported that this experiment was performed using *D. sanguinea* itself.

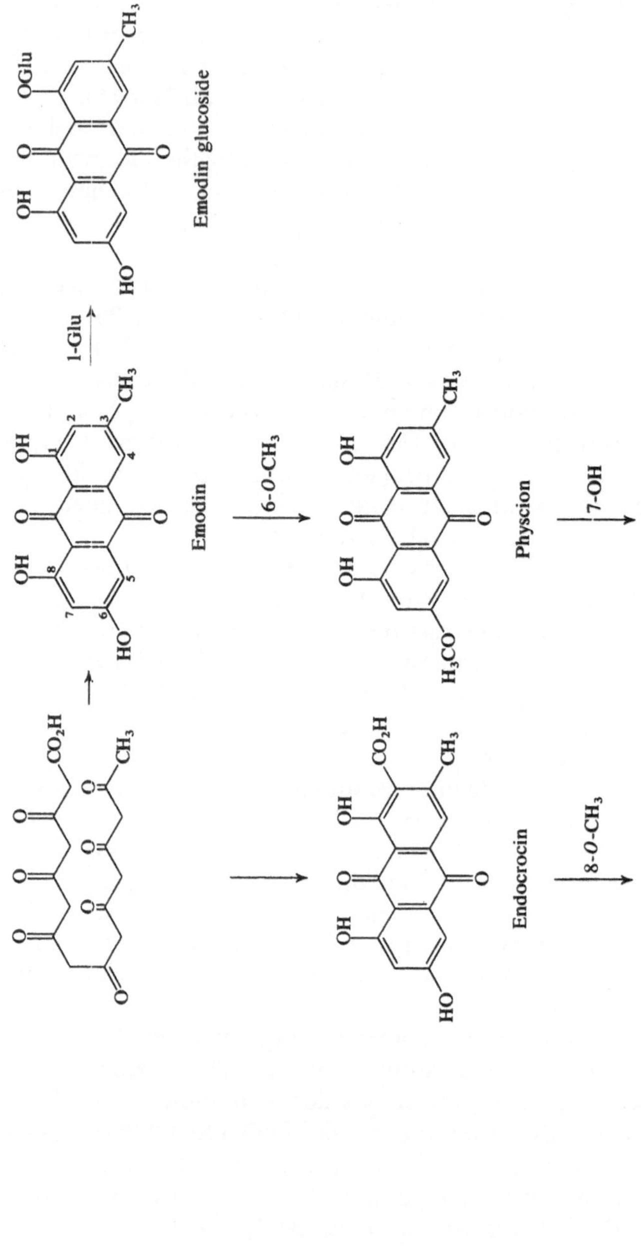
Emodin glucoside
1-Glu
Emodin
6-O-CH₃
Physcion
7-OH
Endocrocin
8-O-CH₃

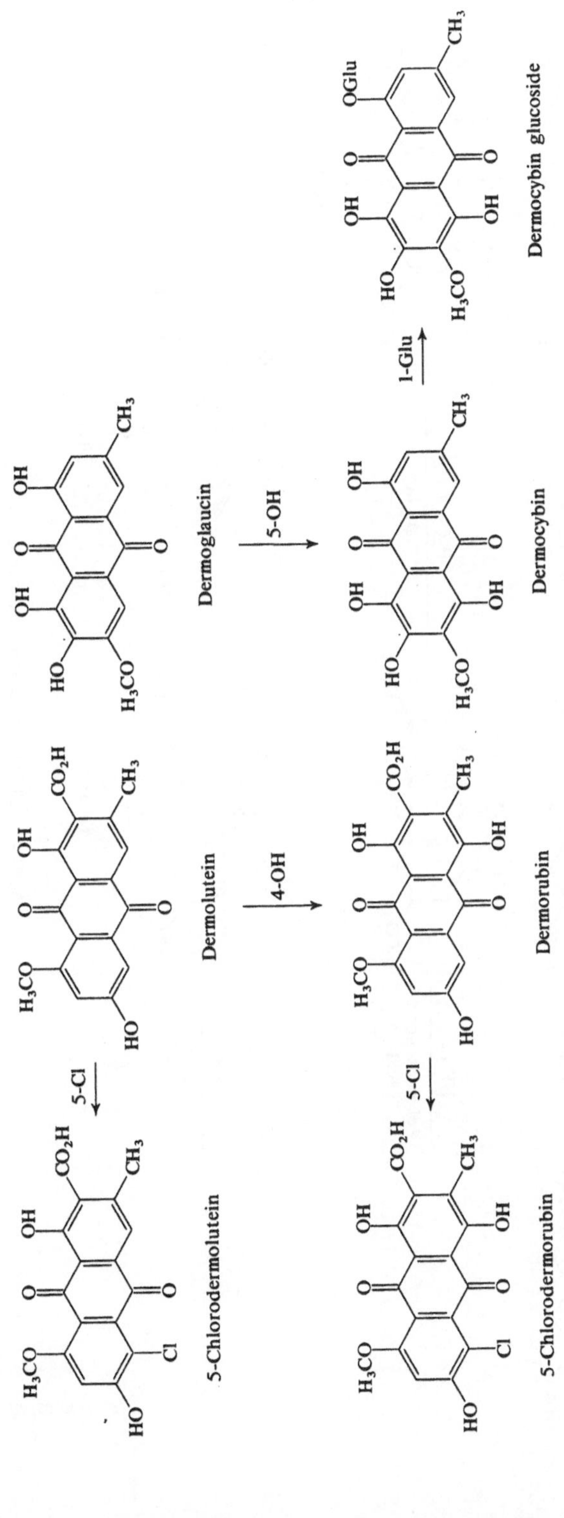

Scheme 52. Biosynthetic relationships between pigments of *Dermocybe sanguinea*

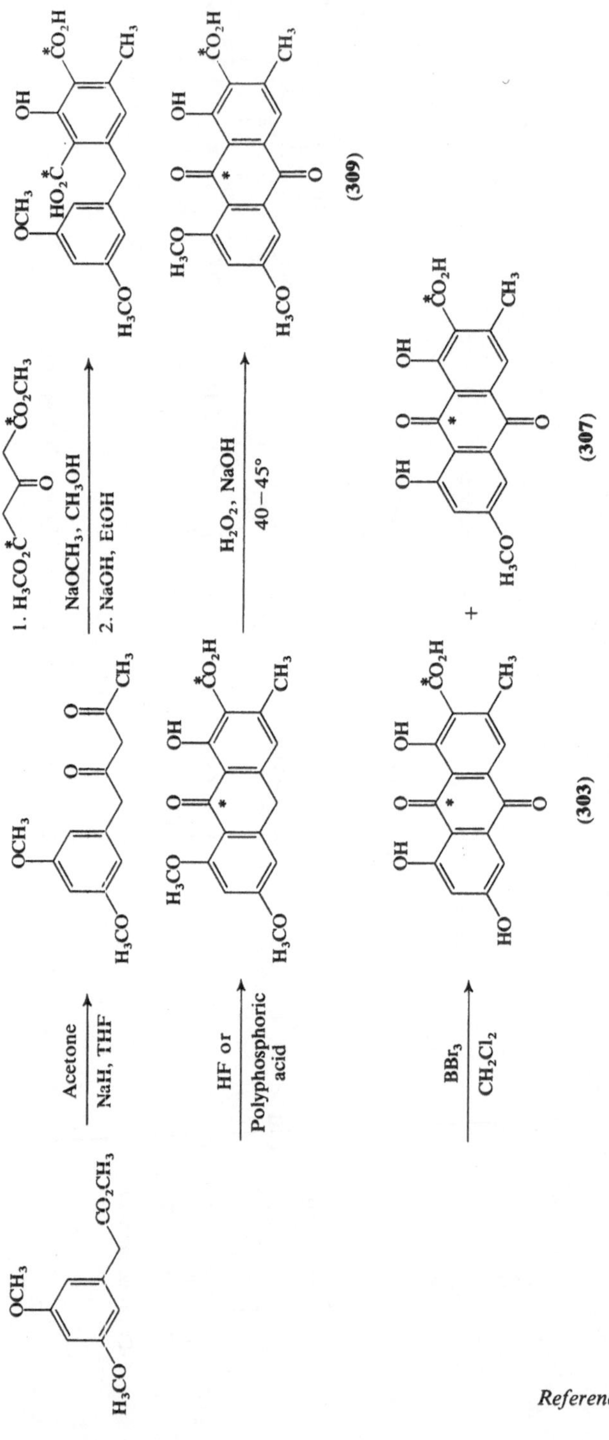

Scheme 53. Synthesis of labelled endocrocin

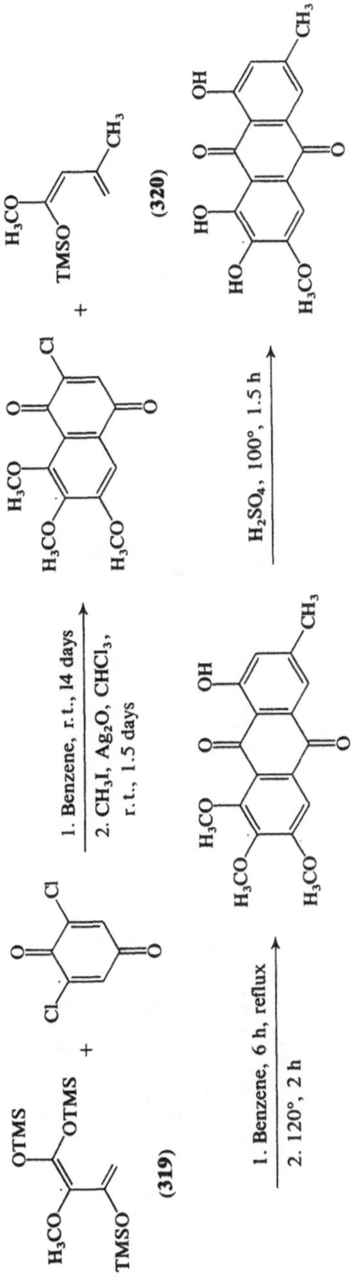

TMS = Trimethylsilyl-

Scheme 54. Synthesis of dermoglaucin

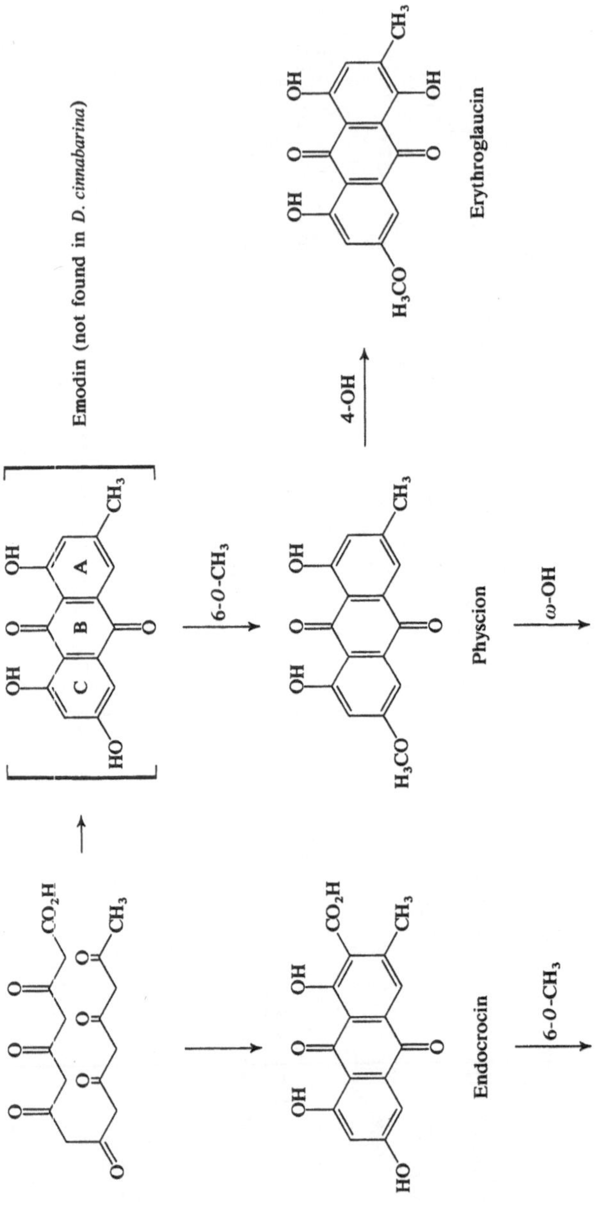

References, pp. 253–286

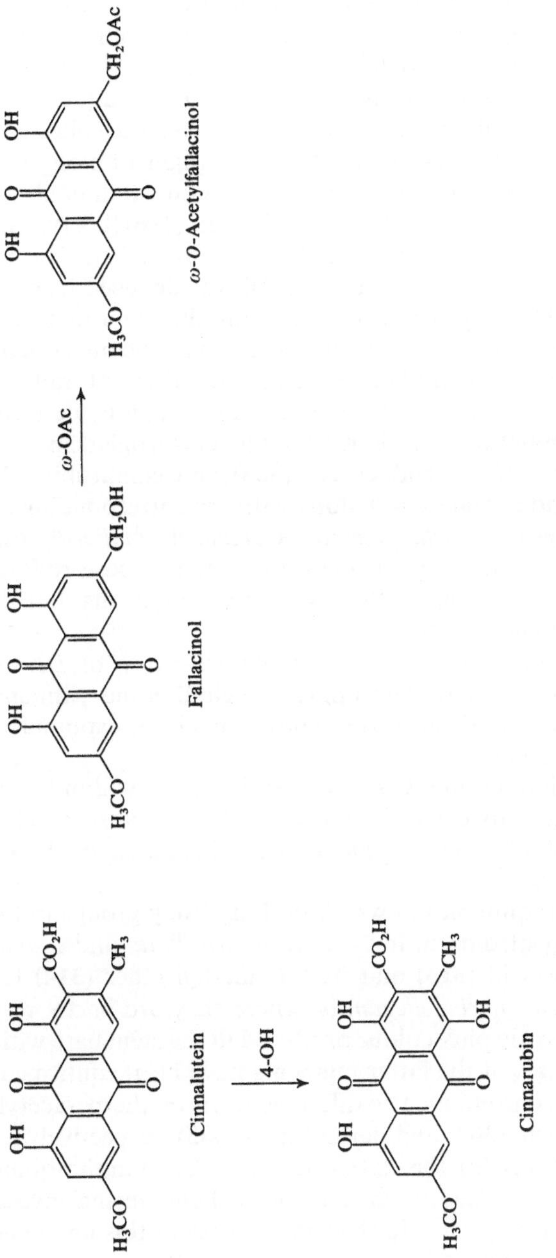

Scheme 55. Biosynthetic relationships between pigments of *Dermocybe cinnabarina*

The likely biosynthetic pathway for pigment production in *D. cinna-barina* is presented in Scheme 55. This pathway is supported by the results of feeding experiments using young fruit bodies of *D. cinnaba-rina* during which a 40% rate of incorporation of [2,4-^{3}H$_2$]emodin-6-β-D-glucopyranoside into fallacinol (**293**) was observed (*542, 621*).

In the bright yellow fruit bodies of *D. canaria* collected in New Zealand physcion (**290**) occurs as the major pigment along with minor amounts of erythroglaucin (**295**) and the unique aminoanthraquinone (**296**) (*404*). The structure of the novel pigment (**296**) [UV/vis. (ethanol): λ_{max} (log ε) = 231 (4.16), 266 (3.87), 311 (sh., 3.54), 398 (3.35), 418 (3.32), 495 (sh., 3.64), 519 (3.79), 566 nm (3.74)] was deduced from the mass, infrared and ^{1}H-n.m.r. spectra, and from the smooth formation of physcion (**290**) on deamination using nitrous acid in aqueous tetra-hydrofuran. The pigment (**296**) represents the first naturally occurring anthraquinone bearing an amino group and it may be derived biosyn-thetically by amination of its cometabolite, erythroglaucin.

The D-glucosides of endocrocin (**303**) and cinnarubin (**308**) have been isolated and characterised along with the corresponding aglycones from fruit bodies of *D. schaefferi* and *Cortinarius bulliardi*, respectively (*405*). There is spectroscopic evidence that the glucose residue resides at the C-8 hydroxy group in these very polar pigments, but this detail requires further confirmation.

The significance to the systematic classification of *Dermocybe* of various modes of pigment development, including the 'sanguinea type' and 'cinnabarina type' illustrated above, has been expounded further by KELLER (*401, 403*).

Hydroxylation in the C-ring of endocrocin without subsequent methylation leads to clavorubin (**315**). The structure of clavorubin, a metabolite of *Claviceps purpurea*, was elucidated by FRANCK and RESCHKE (*246, 247*).

Acidic anthraquinones in which both hydroxy groups in the C-ring have been methylated occur in *Cortinarius armillatus* and *C. miniatopus*, while cortinaric acid (**313**) and its 8-*O*-methyl ether (**314**) have been isolated from *Cortinarius elegantior* where they are partly responsible for the characteristic pink colour produced in the stem base with ammo-nia. The structures of the latter pigments have been confirmed by syn-thesis involving chromium trioxide oxidation of the peracetyl deriva-tives of endocrocin (**303**) and dermolutein (**305**), respectively (*387*).

A third pathway for the elaboration of neutral anthraquinone pig-ments in *Dermocybe* has recently come to light during investigations of a red Australian species. In the sporophores of this unnamed taxon, 6-methylxanthopurpurin-3-*O*-methyl ether (**300**), austrocortinin (**301**), and xanthorin (**302**) co-occur as minor constituents alongside the major

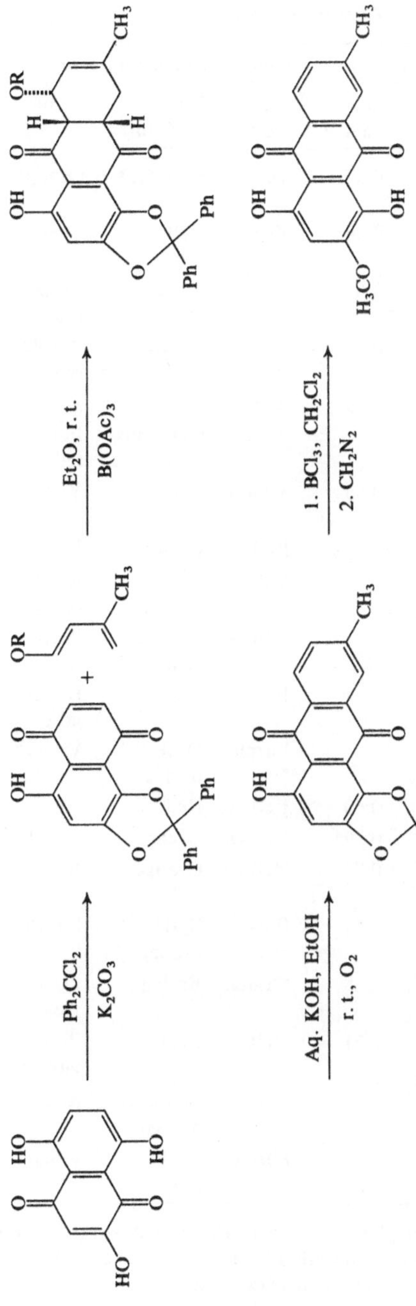

Scheme 56. Synthesis of austrocortinin

R = Trimethylsilyl or acetyl

Table 27. *Chromatographic Properties of Some Fungal Anthraquinones and Anthraquinone Carboxylic Acids*

Pigment	TLC R_f[a]		Colour of Pigment on TLC			Refer-ence
	Solvent A	Solvent B	Daylight	UV (366 nm)	+ Base	
Physcion (**290**)	0.70	0.67	Yellow	Ochreous yellow	Orange[b]	(*403*)
6-Methylxanthopurpurin-3-O-methyl ether (**300**)	0.70	0.66	Yellow	Yellow	Orange-yellow[c]	(*29*)
Austrocortinin (**301**)	0.68	0.61	Orange	Yellow	Purple-red[c]	(*29*)
Xanthorin (**302**)	0.68	0.64	Pink-red	Yellow	Purple-red[c]	(*29*)
Erythroglaucin (**295**)	–	0.68	Red	Violet	Violet[c]	(*617*)
Emodin (**288**)	0.61	–	Yellow	Ochreous orange	Purple-red[b]	(*403*)
Emodin-6,8-di-O-methyl ether (**292**)	–	0.55	Yellow	Orange	Yellow[c]	(*403*)
Cinnarubin (**308**)	0.50	–	Red	Orange	Violet[b]	(*403*)
Cinnalutein (**307**)	0.49	–	Yellow	Orange	Red-orange[b]	(*403*)
Fallacinol (**293**)	0.48	–	Yellow	Orange-yellow	Purple-red[b]	(*403*)
Dermoglaucin (**297**)	0.40	–	Brown	Dark	Brown-grey[b]	(*403*)
Dermocybin (**298**)	0.40	–	Purple	Dark purple	Violet[b]	(*403*)
Nordermorubin (**304**)	–	0.40	Red	Red	–	(*617*)
Endocrocin (**303**)	0.39	0.39	Yellow	Orange	Lilac[c]	(*403, 617*)
Dermolutein (**305**)	0.33	0.29	Yellow	Orange-red	Red[c]	(*403, 617*)
Dermorubin (**306**)	0.32	0.31	Pink-red	Red-orange	Purple[c]	(*403, 617*)
Cortinaric Acid (**313**)	–	0.28	Orange	Brown	Dark violet[c]	(*387*)
Cortinaric Acid-8-O-methyl ether (**314**)	–	0.21	Yellow	Brown	Pink-violet[c]	(*387*)
Emodin-1-β-D-glucopyranoside (**289**)	0.05	–	Yellow	Ochreous orange	Red[b]	(*403*)
Dermocybin-1-β-D-glucopyranoside (**299**)	0.03	–	Purple	Dark purple	Violet[b]	(*403*)

[a] Silica gel 60, Merck precoated plates (0.25 mm) (Solvent A – toluene:ethyl formate:formic acid = 5:4:1; Solvent B – benzene:ethyl formate:formic acid = 10:5:3).
[b] Plate sprayed with aqueous potassium hydroxide solution.
[c] Plate exposed to ammonia vapour.

tetrahydroanthraquinones (330) and (331) (Section 3.5.2) from which they almost certainly derive (28). The structures of the quinones (300) and (302) followed from comparison with authentic materials isolated from moulds and from lichens, respectively. Austrocortinin (301), a new natural product, proved identical with synthetic material (Scheme 56) (408).

Several hydroxylated anthraquinones have been separated and identified by linked gas chromatography-mass spectrometry of their pertrimethylsilyl ethers (682), but this technique appears limited in its general applicability to the identification of fungal anthraquinones. On the other hand, thin layer chromatography has found extensive use in this area, as has been fully discussed earlier. The thin layer chromatographic properties and colour reactions of most of the anthraquinones discussed in this section are summarised in Table 27.

3.5.2. Monomeric Pre-anthraquinones

The larger and structurally more complex group of pre-anthraquinones includes octaketides in which the anthraquinone moieties are incompletely developed. Two groups are known: the so-called monomeric pre-anthraquinones which have structures based on either a 3,4-dihydroanthracen-1($2H$)-one or a 1,2,3,4-tetrahydroanthraquinone nucleus and 'dimeric' pre-anthraquinones which may be formed by phenolic coupling between two dihydroanthracenone sub-units. The latter group is discussed in detail in Section 3.5.3.

Among toadstools belonging to *Cortinarius* subgenus *Phlegmacium* the principal pigments are derivatives of 3,4-dihydroanthracen-1($2H$)-one and, in contrast to the situation in *Dermocybe*, anthraquinones themselves play only a subordinate role (617). The pivotal biosynthetic intermediate atrochrysone (321) is present in *Cortinarius atrovirens* and *C. odoratus* (Table 28) where it is accompanied by the 4-hydroxy derivatives (322) and (323) of varying stereochemistry. The dimethyl ether

	R¹	R²
(321)	H	H
(322)*	OH	H
(323)*	H	OH

(324) R = H
(325) R = OH

* Relative configuration only.

Table 28. *Occurrence of 3,4-Dihydroanthracen-1(2H)-one Derivatives*[a]

(−)-(3R)-Atrochrysone (**321**)	*Cortinarius atrovirens (384, 515), C. odoratus* [1.2×10^{-3}%] *(515)*
(−)-*cis*-4-Hydroxyatrochrysone (**322**)	*Cortinarius odoratus* [1×10^{-3}%] *(515), (617)*
(+)-*trans*-4-Hydroxyatrochrysone (**323**)	*Cortinarius atrovirens* [3×10^{-4}%] *(515), (384, 617)*
(−)-(3S)-Torosachrysone-8-*O*-methyl ether (**324**)	*Cortinarius fulmineus, C. pseudosulfureus, C. splendens (516, 617), Tricholoma auratum* [2×10^{-4}%], *T. sulphureum (515)*
(+)-(3R,4R)-4-Hydroxytorosachrysone-8-*O*-methyl ether (**325**)	*Cortinarius splendens, C. vitellinus (515, 617)*

[a] Yields refer to percentage of fresh weight of fungus.

(**324**) of atrochrysone and its *trans*-4-hydroxy derivative (**325**) occur in only small quantities in fungi.

The structure of atrochrysone (**321**) followed from its ¹H-n.m.r. spectrum (Table 32) and the relative stereochemistry of its 4-hydroxy derivatives (**322**), (**323**), and (**325**) was established after the formation of acetonide derivatives and their investigation by ¹H-n.m.r. spectroscopy (*515*). The absolute configuration of atrochrysone (**321**) (from *C. odoratus*), torosachrysone-8-*O*-methyl ether (**324**) (from *T. auratum*), and *trans*-4-hydroxyatrochrysone-8-*O*-methyl ether (**325**) (from *C. splendens*) have been established, as shown, by correlation of their respective CD spectra with that of (3S)-ω-hydroxytorosachrysone-8-*O*-methyl ether (**327**). The reference compound (**327**) was itself derived from (−)-quinic acid (**326**) according to the reactions shown in Scheme 57 (*558, 568, 634*).

Closely related to the fungal dihydroanthracenones is torosachrysone (**328**), a yellow pigment from the seeds and seedlings of the higher plant *Cassia torosa (648, 650)*. The (3S)-stereochemistry of torosachrysone has been assigned using the exciton coupling method on the benzoate derivative of torosachrysone-8-*O*-methyl ether (**324**) (*214*). Asperflavin (**329**), an isomer of torosachrysone, has been isolated from cultures of *Aspergillus flavus (331)*.

(**328**)   (**329**)

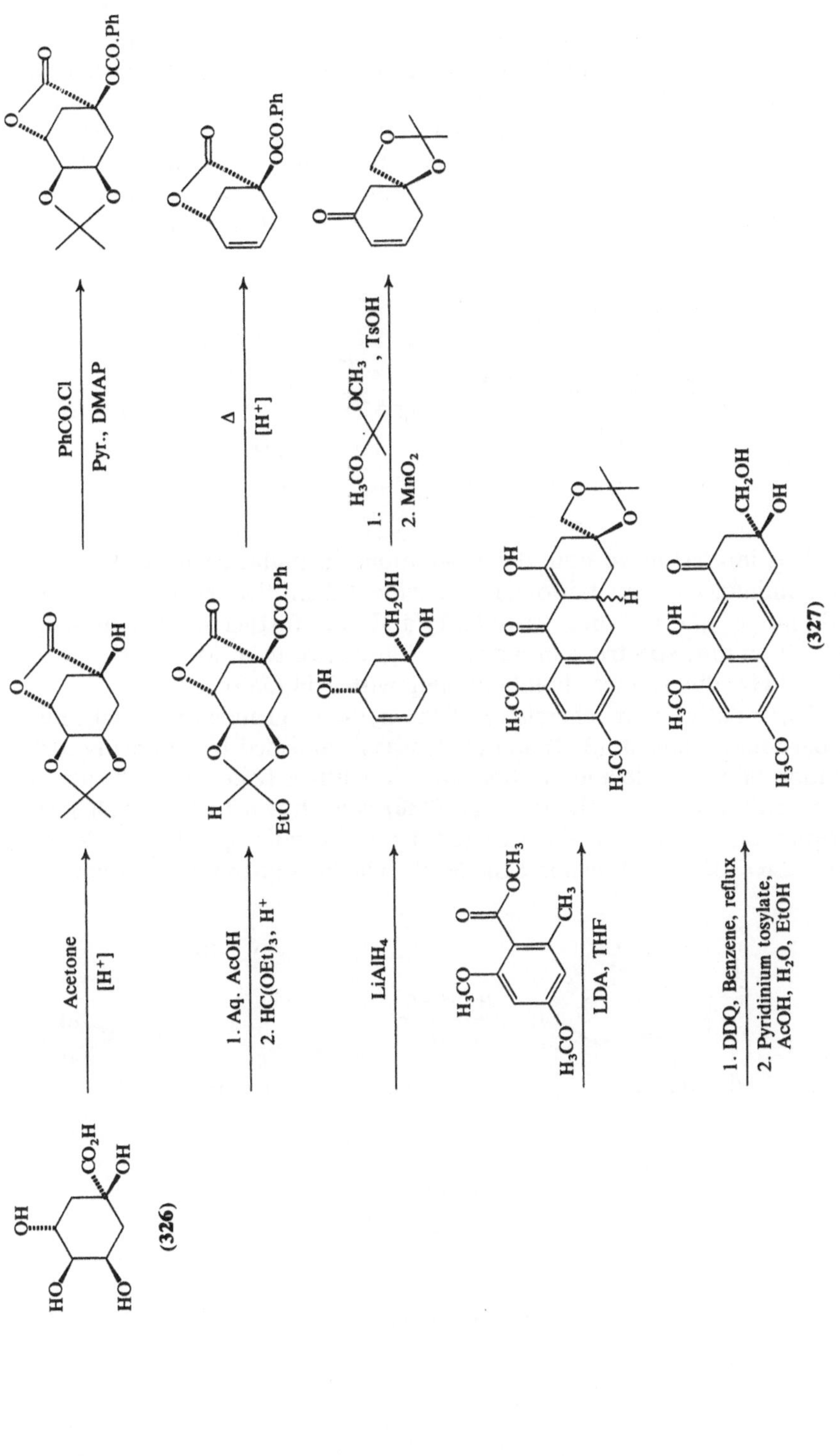

Scheme 57. Synthesis of chiral reference compound (**327**)

The bright red and yellow colours of the cap and stipe base, respectively, of an Australian dermocybe are due to the presence of the tetrahydroanthraquinones (330) and (331) (285). These novel quinones, which must be biogenetically closely related to torosachrysone, exhibit strong antibacterial and antifungal activity (24). The structures followed principally from the electronic and ^{1}H-n.m.r. spectra and from chemical correlation with the acetyl derivatives of austrocortinin (301) [from (330)] and 6-methylxanthopurpurin-3-O-methyl ether (300) [from (331)].

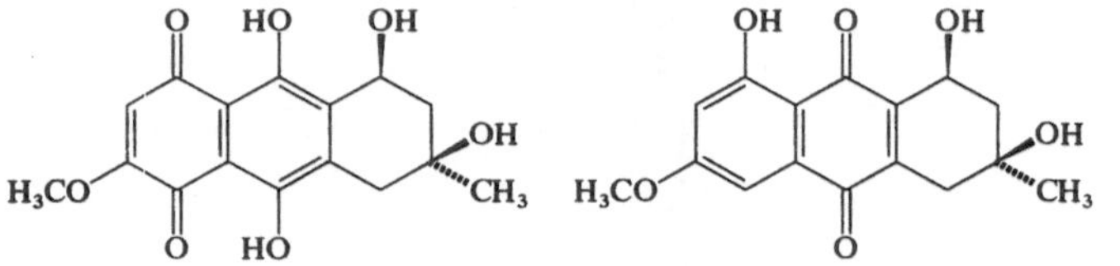

(330)   (331)

It is interesting to note the 5,8-quinone formulation for (330). The predominance of this tautomer was evident from the ^{1}H-n.m.r. chemical shift of the C-7 proton (δ 6.21) and was further consistent with the ^{13}C-n.m.r. spectrum in which C-5 gave rise to a doublet (δ 177.9, $J = 7.3$ Hz) due to three bond coupling with 7-H (585).

The relative stereochemistry of the hydroxy groups in the tetrahydroaromatic rings in (330) and (331) was established by the ready formation of acetonide and arylboronate derivatives (285, 585). The absolute configuration of the quinone (330) was shown to be (1S,3S) by degradation to the (R)-lactone (333) via the corresponding 1-deoxy derivative (332) as shown in Scheme 58. The lactone (333) was prepared

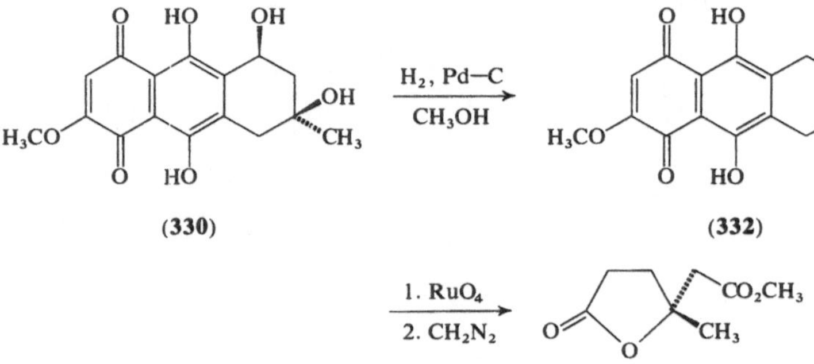

(330)   (332)

(333)

Scheme 58. Degradation of austrocortirubin

for comparison from geraniol, chirality being introduced unequivocally using the Sharpless asymmetric epoxidation process (Scheme 59) (585).

The tetrahydroanthraquinone (332) and the corresponding derivative of austrocortilutein have also been isolated and characterised from the red Australian dermocybe (283 a) and have subsequently been detected chromatographically in D. splendida and D. umbonata (402).

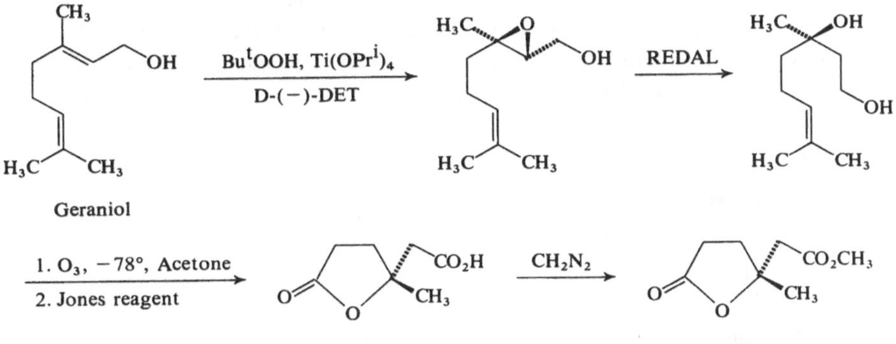

Scheme 59. Synthesis of (R)-lactone (333)

The quinones (330) and (331) have been detected in only a few Dermocybe species (Table 29), their known distribution to date being limited to the Southern Hemisphere.

Table 29. Occurrence of Austrocortirubin and Austrocortilutein

Austrocortirubin (330)	Dermocybe sp. [0.52%, dried fungus] (585), (285); TLC: D. splendida, D. umbonata (402)
Austrocortilutein (331)	Dermocybe sp. [0.21%, dried fungus] (585), (285); TLC: D. splendida, D. umbonata (402)

Plausible biosynthetic relationships between the various pre-anthraquinones discussed above and some of the anthraquinones of Section 3.5.1 are depicted in Scheme 60. Atrochrysone (321) may be considered as the biogenetic precursor of all of the anthraquinones of the emodin family (598). Thus, if water is lost from atrochrysone, emodin anthrone (335) is produced which could then lead by oxidation to emodin (288). The hypothetical β-keto carboxylic acid precursor (334) to atrochrysone could provide the link between the acidic and the neutral fungal anthraquinones since dehydration followed by oxidation would lead from (334) to endocrocin (303). Torosachrysone (328) appears to be the logical precursor to austrocortilutein (331) and thence to austrocortirubin (330). Interestingly, the possible formation of 6-methylxanthopurpurin-3-O-methyl ether (300) and austrocortinin (301) from torosachrysone *via* the tetrahydroanthraquinones shown would involve oxidation as a prelude to dehydration, a sequence in contrast with that leading from atrochrysone to emodin.

In addition to those transformations shown in Scheme 60, both atrochrysone (321) and its methyl ether, torosachrysone (328), are the precursors of a wide variety of oxidatively coupled products which are discussed in the following section.

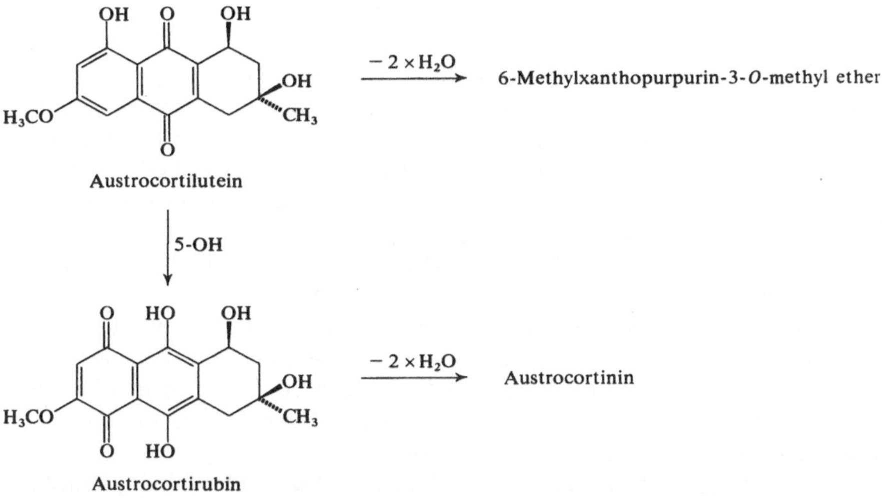

Scheme 60. Possible relationships between monomeric pre-anthraquinones and anthraquinones

The chromatographic properties of the monomeric pre-anthraquinones are collected in Table 30.

Table 30. *Chromatographic Properties of Monomeric Pre-anthraquinones*

Pigment	TLC R_f[a]	Colour of Pigment on TLC			References
		Daylight	UV (366 nm)	+ Reagent	
Torosachrysone-8-*O*-methyl ether (324)	0.37	Pale yellow	Greenish-white	–	(515, 617)
Austrocortilutein (331)	0.36	Orange	Brown	Orange-red[b]	(585)
Atrochrysone (321)	0.35	Greenish-yellow	Bright ochre	Olive brown[c]	(515, 617)
Austrocortirubin (330)	0.32	Red	Orange	Purple[b]	(585)
4-Hydroxytorosachrysone-8-*O*-methyl ether (325)	0.28	Pale yellow	Greenish-white	–	(515, 617)
cis-4-Hydroxyatrochrysone (322)	0.26	Greenish-yellow	Ochre	Yellow-orange[c]	(515, 617)
trans-4-Hydroxyatrochrysone (323)	0.26	Greenish-yellow	Ochre	Yellow-orange[c]	(515, 617)

[a] Silica gel 60, Merck precoated plates (0.25 mm). Solvent – benzene:ethyl formate:formic acid = 10:5:3.
[b] Plate exposed to ammonia vapour.
[c] Plate sprayed with concentrated sulphuric acid.

3.5.3. Dimeric Pre-anthraquinones

Many of the metabolites in this class have come to light as the result of intensive chemical studies on the pigments of toadstools belonging to *Cortinarius*, subgenus *Phlegmacium* (*250, 252, 384, 387, 515, 617, 657*). The majority have structures composed of linked 3,4-dihydroanthracen-1(2*H*)-one subunits. Although most structural variation is encountered among members of *Phlegmacium* itself, several pigments, e.g. flavomannin-6,6'-di-*O*-methyl ether (**348**) and anhydroflavomannin-9,10-quinone-6,6'-di-*O*-methyl ether (**353**), are also encountered widely in *Dermocybe* and in *Tricholoma* (Table 33). The bianthraquinone skyrin (**339**), a common metabolite of moulds and lichens (*667, 668*), is also produced by *Hypomyces* species, ascomycetes which invade and parasitise certain mushrooms reducing their fruit bodies to unrecognisable masses of orange mycelium.

Dimeric pre-anthraquinones probably are formed by initial phenolic coupling of two dihydroanthracenone units. The sites in the respective aromatic nuclei at which this coupling occurs form the basis for the classification of pigments followed below.

Oxidative dimerisation at the dihydroanthracenone level may be followed by enzymatic or chemical modifications which transform portions of the molecule into anthrone or anthraquinone moieties. The majority of the dimers is present in the form of their di-, tri- or tetra-*O*-methyl ethers, in which case the pigments become more stable as the methylation increases. Ultimately, bianthraquinones may result and these compounds too are conveniently included in this section.

During the enzymatic oxidative dimerisation process there is always formed a linkage whose rotation is sterically hindered. This leads to the occurrence of atropisomers which, coupled to the possibility that each alicyclic section of the molecule is capable of harbouring several chiral centres, creates formidable stereochemical problems which are not yet entirely solved. In the meantime a proposal to classify the various stereoisomers on the basis of their chiroptical properties has been formulated (*617, 627*). Accordingly, those compounds which exhibit a negative Cotton effect to longer wavelength and a positive Cotton effect to shorter wavelength in the vicinity of 275 nm in the CD spectrum are designated type A, while in the CD spectra of type B molecules the signs of the Cotton effects are inverted. Diastereoisomerism created by additional chiral centres is accommodated where necessary by the use of subscript numerals, A_1, A_2, and so forth. According to this system, diastereoisomeric molecules described by the same letter will possess the same axial chirality, since it is this structural feature which dominates the CD spectrum.

The stereochemical complexity and chemical sensitivity of many dimeric pre-anthraquinones makes them less amenable to chromatographic identification than the anthraquinones themselves. Consequently, they have proved somewhat more difficult to use for the taxonomy of *Phlegmacium* (*118, 265, 371, 509, 515, 617*) than have the anthraquinones to the systematics of *Dermocybe*. Chemical instability of many pigments causes marked colour changes when sporophores of certain *Phlegmacium* species are stored in the herbarium (*617*).

A. Atrovirin Group (Initial 5,5'-coupling)

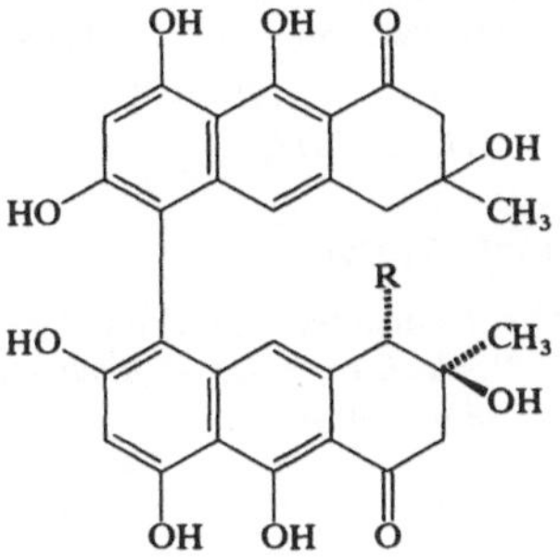

(336) R = H
(337)* R = OH

* Relative configuration only

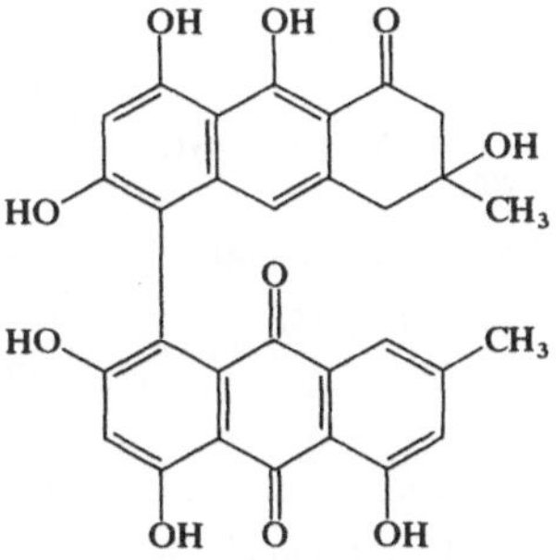

(338)

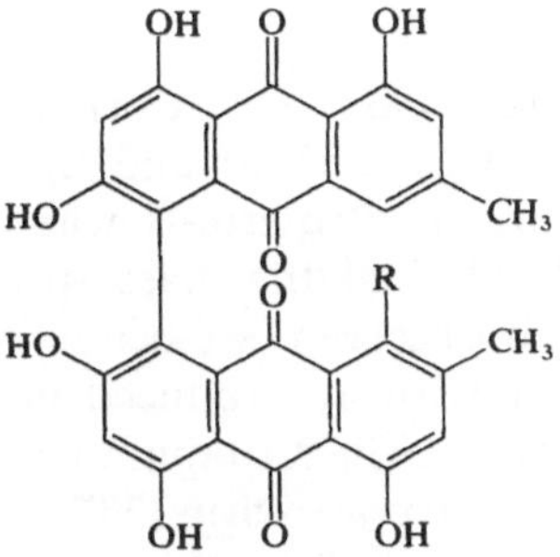

(339) R = H
(340) R = OH

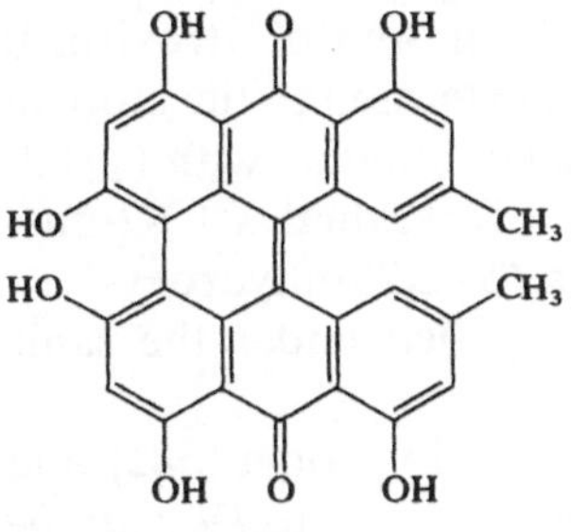

(341)

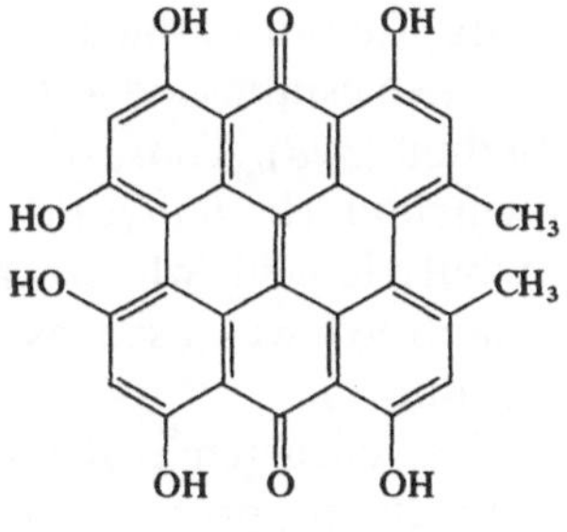

(342)

Table 31. *Occurrence of Pigments of the Atrovirin Group*[a]

(+)-Atrovirin-B (**336**)	*Cortinarius atrovirens* [8×10^{-3}%] (*515*), (*384, 617*)
trans-4-Hydroxy-atrovirin-B (**337**)	*Cortinarius atrovirens* [2.6×10^{-3}%] (*515*), (*617*)
Anhydroatrovirin-9,10-quinone (**338**)	*Cortinarius atrovirens* (*384, 617*), *C. odoratus* (*515, 617*)
Skyrin (**339**)	(+)-(*S*)-Enantiomer: *Cortinarius atrovirens* [1×10^{-3}%] (*515*), (*384, 617*), *C. ionochlorus*, *C. odoratus* (*515, 617*), *Dermocybe austroveneta* [7×10^{-4}%] (*279*), *Hypomyces aurantius* (*87*), *H. lactifluorum* (*352, 377*), *H. trichothecoides* (*154*); Undetermined, TLC: *Dermocybe alcalisensibilis*, *D. alienata*, *D. aurea*, *D. luteostriatula*, *D. obscuroolivea* var. *brunnea*, *D. olivipes* (*402*)
(+)-(*S*)-Aurantioskyrin (**340**)	*Cortinarius atrovirens* [4×10^{-4}%] (*515*), (*617*)
Protohypericin (**341**)	*Dermocybe austroveneta* [5×10^{-5}%] (*279*)
Hypericin (**342**)	*Dermocybe austroveneta* [artefact formed from (**341**) on exposure to light] (*279*); TLC: *D. alcalisensibilis*, *D. alienata*, *D. aurea*, *D. luteostriatula*, *D. obscuroolivea* var. *brunnea*, *D. olivipes* (*402*)

[a] Yields refer to percentage of fresh weight of fungus.

In addition to atrochrysone (**321**) *Cortinarius atrovirens* contains the 5,5′-linked dimer atrovirin-B (**336**), and pigments derived therefrom (Table 31). The structure of atrovirin-B was deduced from the [1]H-n.m.r. spectrum (Table 32) and the mass spectrum in which only a weak dibenzofuran ion is observed (see below) (*384, 515*).

(+)-Skyrin (**339**) is obviously produced in *Cortinarius* from atrovirin (**336**) *via* anhydroatrovirin-9,10-quinone (**338**), as is (+)-aurantioskyrin (**340**) from 4-hydroxyatrovirin (**337**).

The absolute configuration of the chiral axis in the naturally occurring skyrin molecule has remained an outstanding question for many years. Recently, the stereochemistry of (+)-skyrin and of atrovirin-B has been established as (*S*) by making use of an extension to the Horeau method (*398*). Thus, reaction of each of these compounds with (±)-2-phenylbutyric anhydride led to an excess of residual (+)-2-phenylbutyric acid, which was also the case when (*S*)-2,2′-dihydroxy-1,1′-binaphthyl was used as the substrate and esterified under the same conditions.

The occurrence of the photodynamic pigment hypericin (**342**) and its precursor, protohypericin (**341**), seems to be restricted to *Dermocybe* species of the Southern Hemisphere.

Table 32. *^{1}H-N.m.r. Data for Atrochrysone* (**321**), *Atrovirin-B* (**336**), *FDM* (**348**), *Pseudo-phlegmacin-A* (**367**), *Phlegmacin-B$_1$* (**371**), *Tricolorin-A* (**377**), *Cortinarin* (**381**), *and Rufooli-vacin* (**382**) *(δ values, with TMS as internal standard)*

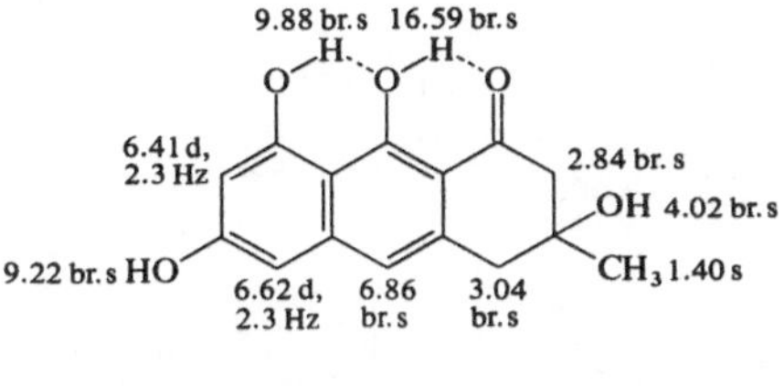

(**321**) [D$_6$] acetone

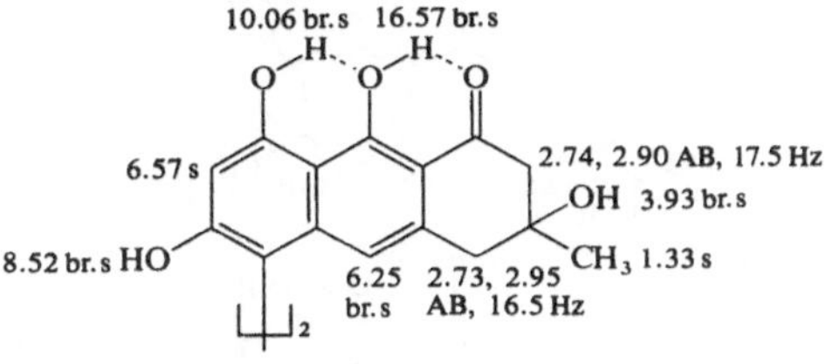

(**336**) [D$_6$] acetone

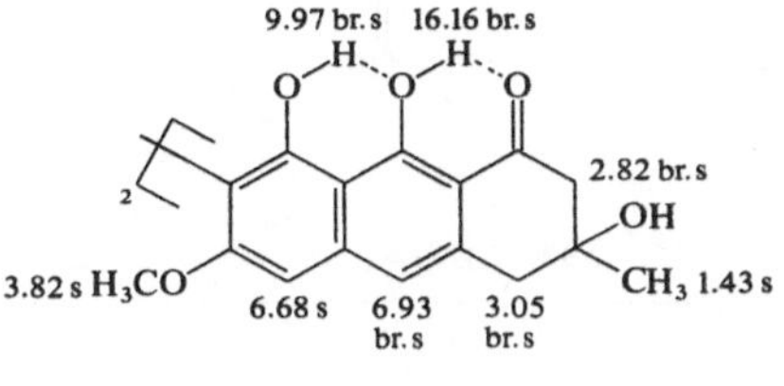

(**348**) CDCl$_3$

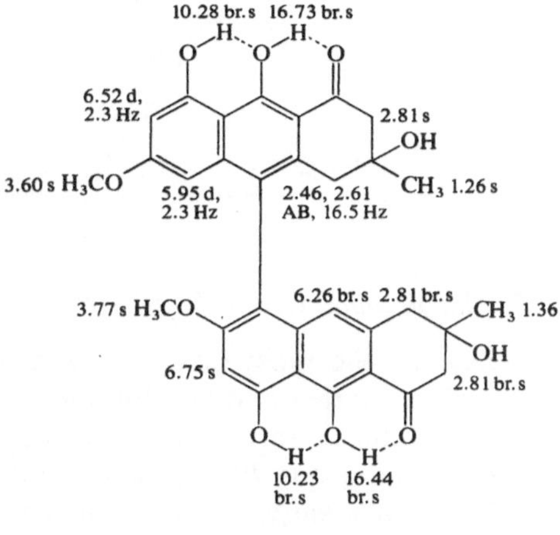

(**367**) CDCl$_3$

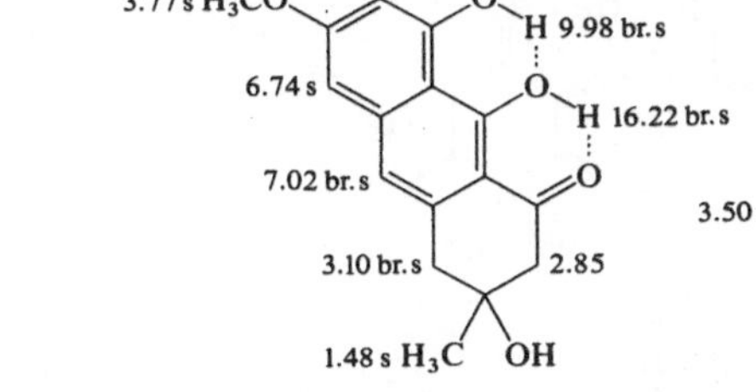

(**371**) CDCl$_3$

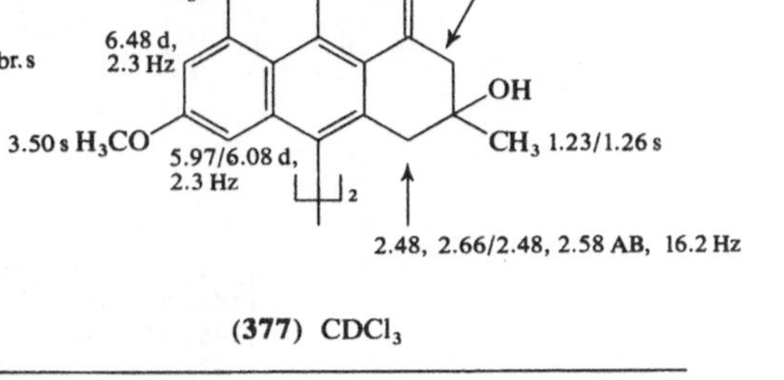

(**377**) CDCl$_3$

Table 32 (*continued*)

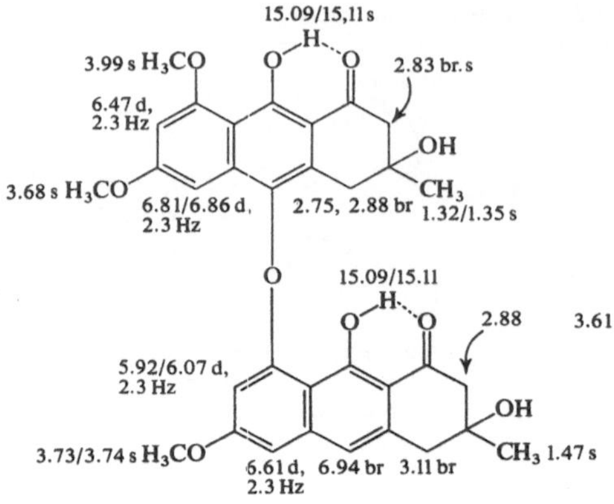

15.09/15,11 s

3.99 s H₃CO   2.83 br. s

6.47 d,
2.3 Hz

3.68 s H₃CO   6.81/6.86 d,   2.75, 2.88 br   1.32/1.35 s
2.3 Hz

CH₃

OH

15.09/15.11

2.88

5.92/6.07 d,
2.3 Hz

3.73/3.74 s H₃CO   6.61 d, 6.94 br 3.11 br   CH₃ 1.47 s
2.3 Hz

OH

(381) CDCl₃

9.81 s

4.07 s H₃CO   CH₃ 2.69 s

6.52 d,
2.3 Hz

H₃CO   6.06 d,   CH₃ 1.97 s
3.55 s   2.3 Hz

6.09 d,
2.3 Hz

3.61 s H₃CO   CH₃ 1.89 d, 1.5 Hz

6.58 d,
2.3 Hz   6.69 q, 1.5 Hz

4.11 s H₃CO

16.58 s

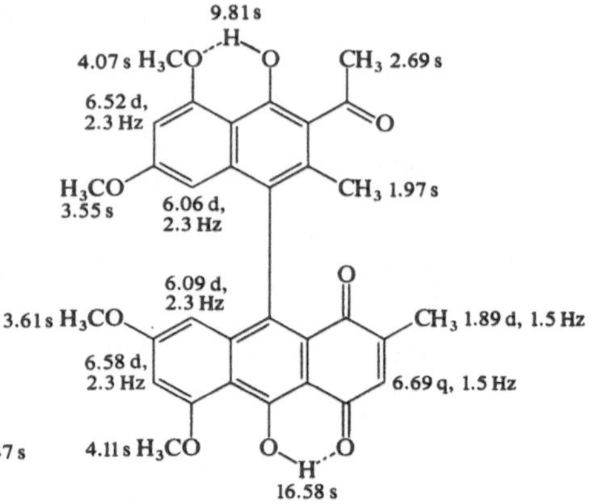

(382) CDCl₃

B. Flavomannin Group (Initial 7,7'-coupling)

By far the most diverse group of dimeric pre-anthraquinones in regard to number and distribution are the flavomannin derivatives (Table 33). Interestingly, compounds of this type are not restricted to *Cortinarius*, subgenus *Phlegmacium*, but also occur in *Dermocybe* and in yellow *Tricholoma* species. Whereas *Cortinarius* and *Dermocybe* are closely related, the systematic relationship to *Tricholoma* is less obvious.

The parent pigment of this group, flavomannin (**343**), the A type atropisomer of which has been isolated from *Cortinarius odoratus*, was first described as a metabolite of the mould *Penicillium wortmanni* (*53*).

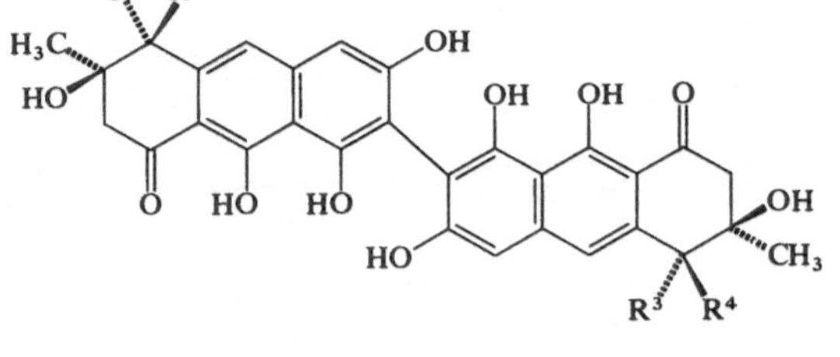

	R¹	R²	R³	R⁴
(343)	H	H	H	H
(344)	H	H	H	OH
(345)	H	OH	H	OH
(346)	—O—		—O—	

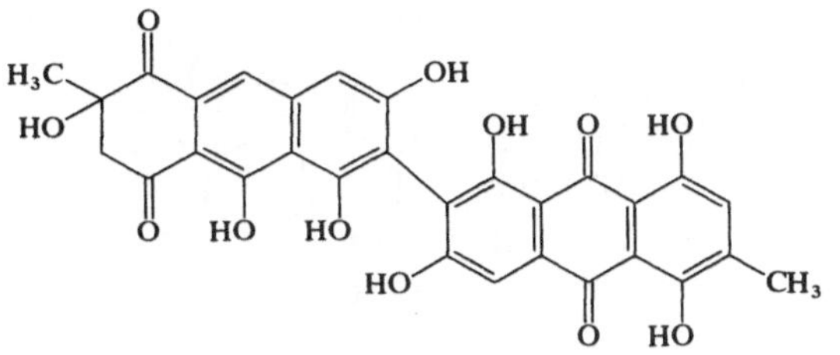

(**347**)

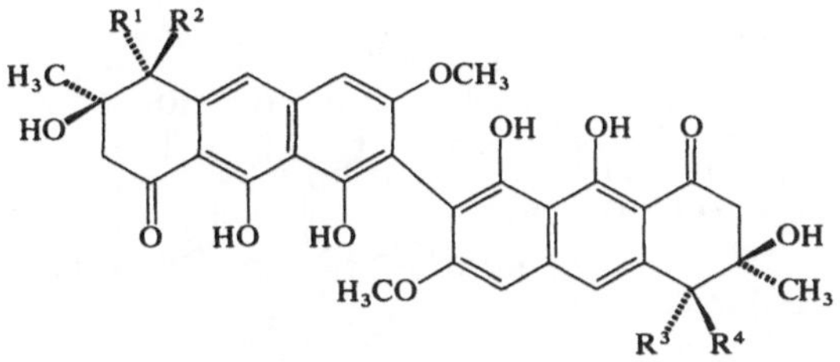

	R¹	R²	R³	R⁴
(**348**)	H	H	H	H
(**349**)	H	H	OH	H
(**350**)	H	H	—O—	
(**351**)	OH	H	H	OH
(**352**)	H	OH	—O—	

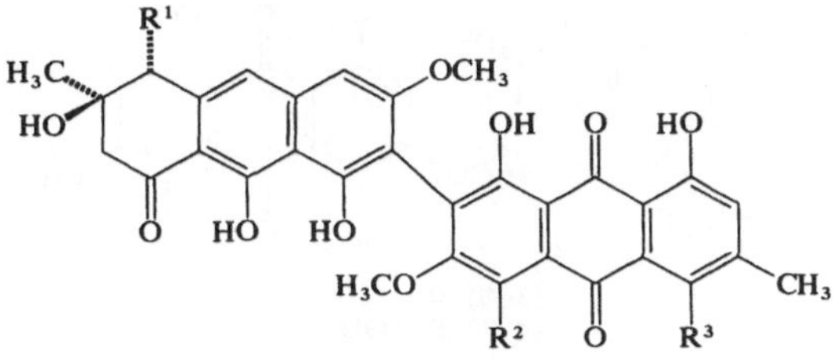

	R¹	R²	R³
(**353**)	H	H	H
(**354**)	H	H	OH
(**355**)	H	OH	H
(**356**)	OH	H	H

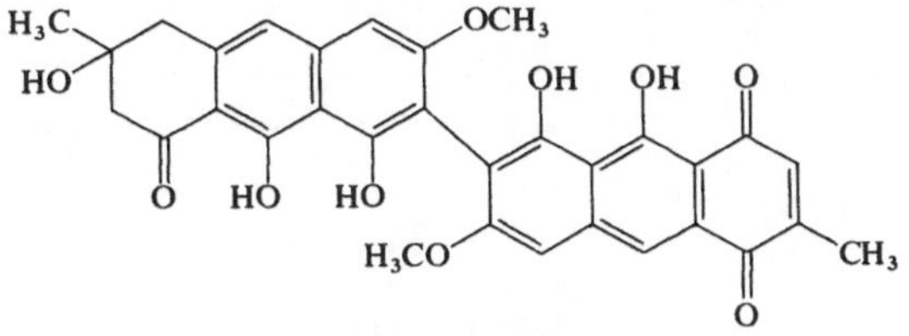

(**357**)

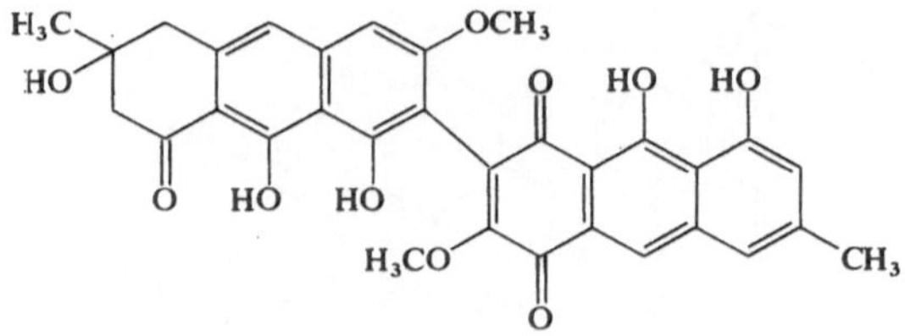

(358)

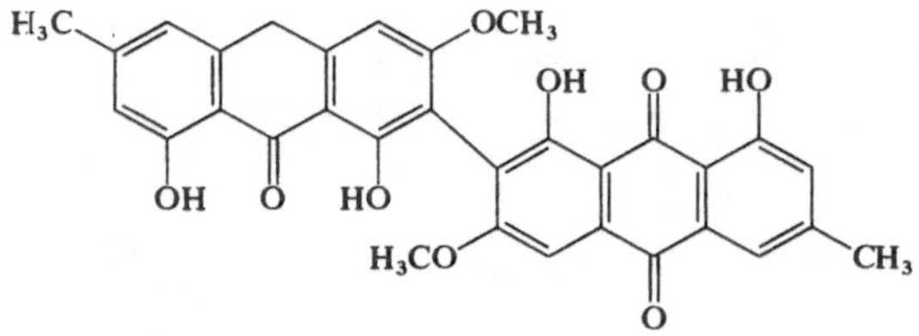

(359)

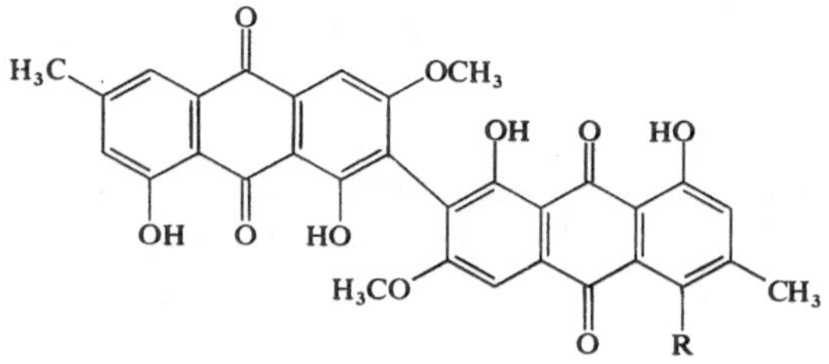

(360) R = H
(361) R = OH

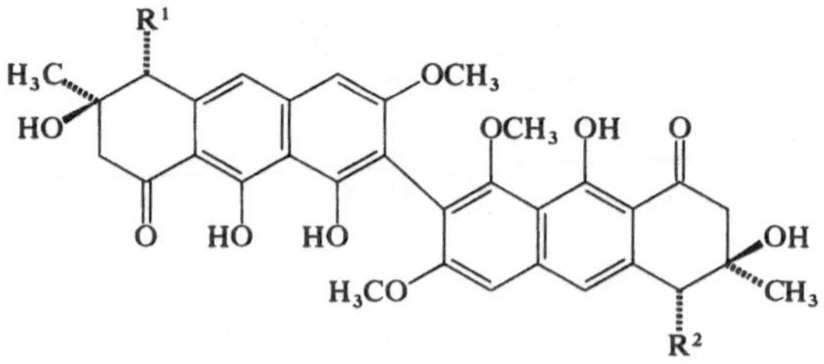

R¹   R²

(362) H   H
(363) H   OH
(364) OH   H

(365)

Table 33. *Occurrence of Pigments of the Flavomannin Group*[a, b]

Flavomannin-A (343)	*Cortinarius odoratus* $[1.5 \times 10^{-2}\%]$ (515), (617)
cis-4-Hydroxy-flavomannin-A (344)	*Cortinarius odoratus* (515, 617)
cis,cis-4,4′-Dihydroxy-flavomannin-A (345)	*Cortinarius odoratus* $[1.5 \times 10^{-3}\%]$ (515), (617)
4,4′-Dioxo-flavomannin-A (346)	*Cortinarius atrovirens, C. ionochlorus* (516, 617)
4-Hydroxy-4′-oxoanhydro-flavomannin-9,10-quinone (347)	*Cortinarius atrovirens* (515, 617)
Flavomannin-6,6′-di-*O*-methyl ether (FDM) (348)	Type A: *Cortinarius citrinus* $[5.2 \times 10^{-2}\%]$ (515), (617), *Dermocybe crocea* (= D. cinnamomeolutea = D. saligna) $[5 \times 10^{-2}\%]$ (657), (400, 401, 629); A (+B): *Cortinarius fulmineus* $[1.6 \times 10^{-2}\%]$ (515), (617), *C. subfulgens* (515, 617); A+B: *Tricholoma auratum* (= T. equestre) $[7 \times 10^{-3}\%]$ (515), (629); B: *Cortinarius pseudosulphureus* [0.13%] (515), (617), *Tricholoma sulphureum* $[1 \times 10^{-2}\%]$ (515), (293, 294); Undetermined: *Dermocybe palustris* var. *sphagneti, D. uliginosa* (629); TLC: *Cortinarius citrinolilacinus* (371), *C. incognitus, C. subcroceofolius* (400), *C. sulfurinus, C. xanthochlorus* (371), *Dermocybe alnophila* (401), *D. aspenensis* (403), *D. cinnamomea, D. holoxantha* (401), *D. leptospermarum* (402), *D. malicoria, D. parientalis* (402), *D. phoenicea* (401), *D. phoenicea* var. *occidentalis* (403), *D. schaefferi* (= D. carpineti) (401), *D. semisanguinea* (401, 403), *D. sommerfeltii* (= D. cinnamomeobadia), *D. sphagneti* (401), *D. sphagnogena* (400), *Tricholoma flavobrunneum, T. malluvium* (86)
trans-4-Hydroxy-flavomannin-6,6′-di-*O*-methyl ether (349)	Type A: *Cortinarius splendens* [0.3%] (515), *C. vitellinus* [0.12%] (515), $[5 \times 10^{-2}\%]$ (627); B: *Tricholoma sulphureum* $[1.7 \times 10^{-2}\%]$ (515), (293, 294)
4-Oxoflavomannin-6,6′-di-*O*-methyl ether (350)	Type A: *Cortinarius vitellinus* (294, 627); B: *Tricholoma sulphureum* $[1.6 \times 10^{-2}\%]$ (515), (293, 294)
cis,trans-4,4′-Dihydroxy-flavomannin-6,6′-di-*O*-methyl ether (351)	Type B: *Tricholoma sulphureum* $[1.1 \times 10^{-2}\%]$ (515), (293, 294) [accompanied by 15–20% of the *trans-trans* epimer] (515)

Table 33 (continued)

4-Oxo-4'-*cis*-hydroxy-flavomannin-6,6'-di-*O*-methyl ether (352)	Type B: *Tricholoma sulphureum* [2 × 10⁻³%] (515)

4-Oxo-4'-*cis*-hydroxy-
flavomannin-6,6'-di-*O*-methyl
ether (**352**)

Type B: *Tricholoma sulphureum* [2 × 10⁻³%] (*515*)

Anhydroflavomannin-9,10-
quinone-6,6'-di-*O*-methyl
ether (AFDM) (**353**)

Type A: *Cortinarius citrinus* (*515*), *Dermocybe crocea*
[8.6 × 10⁻³%] (*657*), (*400, 401, 515, 629*); A (+B): *Cortinarius subfulgens, Tricholoma auratum* (*515*); A + B: *Cortinarius elegantior* (*515*); Undetermined: *C. fulmineus, C. pseudosulphureus, Tricholoma sulphureum* (*515*); TLC: *Cortinarius citrinolilacinus* (*371*), *C. incognitus, C. subcroceofolius* (*400*), *C. sulfurinus* (*371*), *C. tubarius* (*400*), *C. xanthochlorus* (*371*), *Dermocybe alnophila* (*401*), *D. aspenensis* (*403*), *D. cinnamomea, D. holoxantha* (*401*), *D. leptospermarum* (*402*), *D. malicoria* (*401*), *D. parientalis* (*402*), *D. phoenicea* (*401*), *D. phoenicea* var. *occidentalis* (*403*), *D. schaefferi* (*401*), *D. semisanguinea* (*401, 403*), *D. sommerfeltii, D. sphagneti* (*401*), *D. sphagnogena* (*400*), *D. uliginosa* (*401*), *Tricholoma sulphureum* (*515*)

4-Hydroxyanhydroflavo-
mannin-
9,10-quinone-6,6'-
di-*O*-methyl ether (**354**)

Cortinarius splendens (*515*), *C. vitellinus* (*515, 627*)

5-Hydroxyanhydroflavo-
mannin-
9,10-quinone-6,6'-
di-*O*-methyl ether (**355**)

Cortinarius citrinus, C. fulmineus (*515*)

trans-4'-Hydroxyanhydro-
flavomannin-9,10-quinone-
6,6'-di-*O*-methyl ether (**356**)

Cortinarius splendens (*515*)

Anhydroflavomannin-1,4-
quinone-6,6'-di-*O*-methyl
ether (**357**)

Cortinarius splendens (*515, 617*), *C. vitellinus* (*515, 627*)

Anhydroflavomannin-5,8-
quinone-6,6'-di-*O*-methyl
ether (**358**)

Cortinarius citrinus (*515, 617*)

Dianhydroflavomannin-9,10-
quinone-6,6'-di-*O*-methyl
ether (**359**)

Dermocybe crocea [trace, MS] (*657*)

7,7'-Biphyscion (**360**)

Racemic form: *Tricholoma auratum* (*291*), [artefact] (*629*);
Undetermined, TLC: *Cortinarius citrinolilacinus, C. subfulgens, C. sulfurinus, C. xanthophyllus* (*371*), *Dermocybe alnophila, D. cinnamomea, D. crocea, D. holoxantha, D. malicoria, D. schaefferi, D. sphagnogena, D. uliginosa* (*400*)

4-Hydroxy-7,7'-
biphyscion (**361**)

Cortinarius vitellinus (*617, 627*)

Table 33 (*continued*)

Flavomannin-6,6',8-tri-*O*-methyl ether (FTM) (**362**)	Type A: *Dermocybe semisanguinea* [$7.6 \times 10^{-4}\%$] (*515*), (*94*); B: *Cortinarius elegantior* [$7.6 \times 10^{-2}\%$] (*384*), (*617, 657*), *C.* spec. 2 (*515, 617*), *Tricholoma auratum* [$1.5 \times 10^{-3}\%$], *T. sulphureum* [$2 \times 10^{-4}\%$] (*515*); Undetermined: *Cortinarius fulmineus, C. subfulgens* (*515*)
trans-4-Hydroxyflavomannin-6,6',8-tri-*O*-methyl ether (**363**)	Type A: *Cortinarius splendens* (*515, 617*)
trans-4-Hydroxyflavomannin-6,6',8'-tri-*O*-methyl ether (**364**)	Type B: *Tricholoma sulphureum* (*515*)
Anhydroflavomannin-9,10-quinone-6,6',8'-tri-*O*-methyl ether (**365**)	*Cortinarius elegantior* (*384, 515, 617*), *C.* spec. 2 (*515, 617*)

[a] Yields refer to percentage of fresh weight of fungus.
[b] Within the formulae (**343**)-(**365**) stereochemical detail relates to the relative configuration of centres in the same half of a molecule; no absolute configurations or relative stereochemistries of centres in different halves of any molecule are implied.

However, only very few toadstools produce pigments such as (**343**) in which both the 6- and 6'-hydroxy groups remain free. Notable among these is *Cortinarius atrovirens* which owes much of its dark green colour to the presence of 4,4'-dioxoflavomannin-A (**346**).

The most common flavomannin derivative is the 6,6'-di-*O*-methyl ether, FDM (**348**), which occurs in both of its atropisomeric forms in *Cortinarius*. In *Tricholoma auratum* (= *T. equestre*) a mixture of FDM diastereoisomers, racemic at the biaryl linkage, is present.

FDM (**348**) has often been found accompanied by its oxidation product anhydroflavomannin-9,10-quinone-6,6'-di-*O*-methyl ether, AFDM (**353**). In some cases AFDM and some of the other related oxidation products of FDM (**348**) listed in Table 33 may have been formed on ageing of the toadstools or during the isolation of the pigments themselves. The remarkable colour changes to green, red or violet produced on addition of aqueous base to different parts of the fruit bodies is of taxonomic value in identifying certain *Phlegmacium* species (*516, 617, 627*). These changes too may be attributed to oxidative processes effecting the labile pre-anthraquinone constituents.

The isolation of 7,7'-biphyscion (**360**) in racemic form from powdered, dried sporophores of *Tricholoma auratum* suggested that this compound was an artefact formed during the isolation procedure from flavomannin type precursors; a similar origin seems likely for the 7,7'-

biphyscion which has been detected chromatographically in many *Cortinarius (371)* and *Dermocybe (400)* species.

Although a detailed analysis of the spectroscopic data which allowed structural assignment to the pigments (**343**) to (**365**) is beyond the scope of this review some useful comments are nevertheless in order.

Mass spectrometry has served as a valuable diagnostic tool during studies of the complex pre-anthraquinone pigments (*515, 627*). Flavomannin (**343**) and its derivatives undergo characteristic fragmentations on electron impact which lead to intense dibenzofuran ions. Thus, FDM (**348**) and related methyl ethers are degraded on electron impact according to Scheme 61, whereas pigments bearing free hydroxy groups at C-6 suffer consecutive loss of three molecules of water from the parent ion leading to dibenzofuran ions of type (**366**). In contrast, pigments of the atrovirin group produce only very weak dibenzofuran ions and in the spectra of pseudophlegmacin, phlegmacin, and tricolorin type pigments such ions are not visible at all. In these latter cases the most prominent ions are formed by cleavage of the biaryl bond.

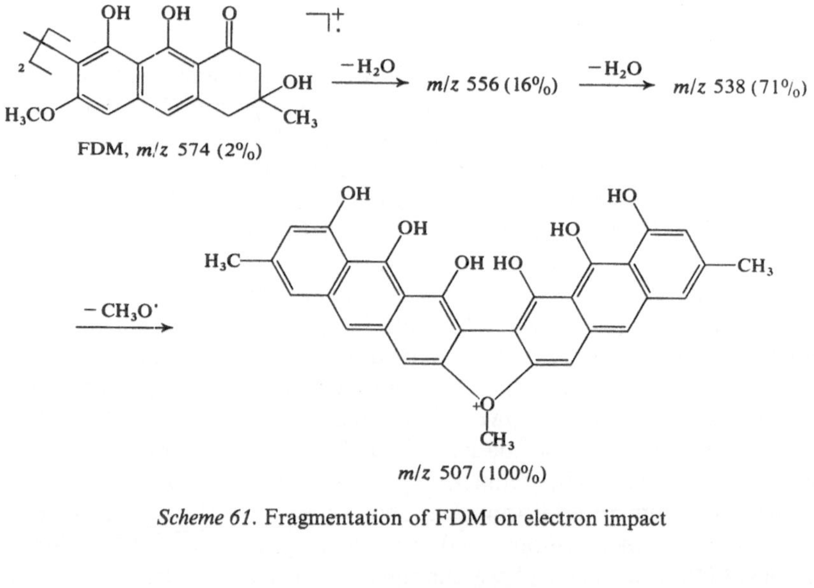

Scheme 61. Fragmentation of FDM on electron impact

Of paramount importance to structure elucidation of the flavomannin derivatives and of the other groups of dimeric pre-anthraquinones discussed below was the use of n.m.r. spectroscopy. Of special significance in this regard have been the shifts arising as a consequence of the anisotropic influence of the biaryl linkages and the position of

(366)

signals due to chelated hydroxy groups. The chemical shifts and multi-plicities of resonances in the ^{1}H-n.m.r. spectra of the parent pigment in each of the groups discussed here are presented in Table 32.

Although a considerable amount of structure information has been obtained using these and other spectroscopic techniques, the elucida-tion of the absolute configuration at the various stereocentres in these molecules has created considerable problems and awaits the results of further investigation. A detailed discussion of the stereochemical aspects is therefore best postponed pending the clarification which these studies should eventually produce.

Further pigments of the flavomannin group contain additional hy-droxy groups in the 4- and 4'-positions (Table 33). The relative stereo-chemistry in such cases has been established by ^{1}H-n.m.r. studies in-volving acetonide derivatives formed by treatment of the pigments with 2,2-dimethoxypropane in the presence of an acid catalyst (294, 296). In the case of *trans* disposed 3,4-dihydroxy derivatives the formation of the cyclic acetal forces the cyclohexene ring into a pseudo-boat conformation which poses the 4-methyl group in the shielding zone above the plane of the naphthalene nucleus. In the acetonide derivative of a *cis*-glycol, however, the 4-methyl group occupies a near-equatorial orientation remote from the influence of the naphthalene system. These effects may be most vividly demonstrated by using the case of *cis,trans*-4,4'-dihydroxyflavomannin-6,6'-di-*O*-methyl ether (351). This pigment formed a bis-acetonide derivative which in the ^{1}H-n.m.r. spectrum ex-hibited signals at δ 1.06 and 1.31 due to the methyl groups in the *trans*- and *cis*-hydroxylated rings, respectively.

Subtle conformational changes brought about by alteration in the relative stereochemistry of hydroxy substituents at C-3 and C-4 are also translated into diagnostic differences in the ^{1}H-n.m.r. spectra of the 3,4-dihydroxylated pigments themselves (515). Thus, a *trans* ar-rangement of hydroxy groups as in (349) leads to a well separated AB-quartet for the methylene protons at C-2 and a small benzylic coupling ($J = 0.6$ Hz) between 4-H and 10-H. In contrast, the protons of the methylene group in the *cis*-glycol ring in the pigment (351) appear as a broad singlet and the benzylic coupling experienced by 4-H is considerably larger ($J = 0.9$ Hz).

The occurrence in *Phlegmacium* of the 1,4-anthraquinone derivatives (357) and (358) is notable. These compounds may be easily recognised in thin layer chromatograms by their violet colour. The 1,4-anthraquinone sub-unit found in the pigment (357) has been isolated recently from *Aspergillus cristatus* (447).

From some toadstools, e.g. *Cortinarius splendens*, flavomannin derivatives have been isolated in very high yields. A single fruit body of this fungus for example has afforded *ca.* 200 mg of *trans*-4-hydroxy-flavomannin-6,6'-di-*O*-methyl ether (349) (515). The pigment (349) exhibits strong antibiotic activity against several bacteria (24) and causes pulmonary congestion and hepatic necrosis in mice (698). When administered interperitonally, the LD_{50} is as low as 10 mg/kg/day (434). In view of the high concentrations of pigments in the fruit bodies this toxicity may at least in part be responsible for several severe poisonings which have been attributed to the ingestion of *C. splendens* (129).

C. Pseudophlegmacin Group (Initial 5,10'-coupling)

The pseudophlegmacin group represents a rare coupling type so far observed in only a limited number of *Cortinarius* species (Table 34). Interestingly, both atropisomers of pseudophlegmacin (367) co-occur in *Cortinarius prasinus* and have been separated by chromatography. A comparison of CD spectra with those cited by MASON *et al.* (475) for 1,1'-binaphthyls suggests for pseudophlegmacin-A the (*S*)-configuration at the chiral axis.

The structures of the pseudophlegmacin diastereoisomers and of the corresponding methyl ethers (368) and (369) were deduced from

	R¹	R²
(367)	H	H
(368)	CH₃	H
(369)	H	CH₃

(370)

Table 34. *Occurrence of Pigments of the Pseudophlegmacin Group*[a]

Pseudophlegmacin (**367**)	Type A: *Cortinarius prasinus* [2.9×10^{-3}%] (*515*), (*617*); B: *Cortinarius prasinus* [2.9×10^{-3}%] (*515*), (*617*)
Pseudophlegmacin-8'-O-methyl ether (**368**)	Type B (+A): *Cortinarius russeoides* (*515*)
Pseudophlegmacin-8-O-methyl ether (**369**)	Type B (+A): *Cortinarius russeoides* (*515*)
Prasinone (**370**)	*Cortinarius nanceiensis, C. prasinus* [1×10^{-3}%], *C. russeoides* (*515*)

[a] Yields refer to percentage of fresh weight of fungus.

the [1]H-n.m.r. spectra, with special reference being made to the shielding effects caused by the biaryl linkage. Details of the [1]H-n.m.r. spectrum of pseudophlegmacin-A are presented in Table 32.

An interesting blue-green pigment, prasinone (**370**), has been isolated from the green fruit bodies of *C. prasinus* and other species. It exhibited an absorption maximum at 614 nm in the electronic spectrum and showed purple fluorescence under UV light. From the spectroscopic data, and its co-occurrence with pseudophlegmacin (**367**), the dibenzo[*a,j*]perylenequinone structure (**370**) has been tentatively proposed (*515*).

D. Phlegmacin Group (Initial 7,10'-coupling)

Phlegmacin (**371**) and its methyl ether (**372**) have proved of value in the taxonomy of *Phlegmacium* where they are each representative of two different groups of closely related fungi (*515, 617*). As in pre-

(**371**) R = H
(**372**) R = CH₃

(**373**)

(374) R = H
(375) R = CH₃

vious cases the parent pigment, phlegmacin (371), has often been found accompanied by its oxidation products (Table 35).

It is interesting to note from Table 35 that phlegmacin (371) is produced by the various species of *Cortinarius* with differing degrees

Table 35. *Occurrence of Pigments of the Phlegmacin Group*[a]

Phlegmacin (371)	Type B_1 : *Cortinarius aureofulvus* (617, 657), *C. auroturbinatus* [3.3×10^{-2}%] (515), (384, 617); B_1 (+A_1): *Cortinarius odorifer* [5×10^{-2}%] (626), (617, 628), *C. prasinus* [3.3×10^{-2}%] (515), (617); Undetermined, TLC: *Cortinarius cedretorum* (371), *C. orichalceus* (118, 371), *C. prasinocyaneus* (371), *C. rufoolivaceus* (120), *C. xanthophyllus* (371)
Phlegmacin-8′-O-methyl ether (372)	Type A_1 (+B_1): *Cortinarius nanceiensis* (515, 617, 657), *C. percomis* [1.7×10^{-2}%] (628), (617), *C. russeoides* (371, 515); B_2: *Cortinarius guttatus* (515, 617, 657); B_2 (+A_2): *Cortinarius olivellus* [5.6×10^{-3}%], *C. personatus* (= *C.* spec. 1) (384, 515, 617, 657); Undetermined, TLC: *Cortinarius claroflavus*, *C. eufulmineus*, *C. russeoides*, *C. superbus* (371)
Anhydrophlegmacin (373)	*Cortinarius odorifer* [1.5×10^{-2}%] (626)
Anhydrophlegmacin-9,10-quinone (374)	*Cortinarius aureofulvus* (617, 657), *C. auroturbinatus* [3.8×10^{-3}%] (515), (384, 617), *C. odorifer* [1.2×10^{-2}%] (515), (617), *C. prasinus* (515, 617); TLC: *C. cedretorum*, *C. prasinocyaneus* (371)
Anhydrophlegmacin-9,10-quinone-8′-O-methyl ether (375)	Type A_1 (+B_1): *Cortinarius nanceiensis* (515, 617, 657), *C. percomis* [1.2×10^{-2}%] (628), (515, 617), *C. russeoides* (371, 515); B_2 (+A_2): *Cortinarius guttatus* (657), *C. olivellus* (515, 617), *C. personatus* (384, 515, 617, 657); TLC: *Cortinarius claroflavus*, *C. eufulmineus*, *C. superbus* (371)

[a] Yields refer to percentage of fresh weight of fungus.

References, pp. 253–286

of axial stereoselectivity. Thus, *C. aureofulvus* and *C. auroturbinatus* contain pure phlegmacin-B_1 while in species such as *C. odorifer*, phlegmacin-B_1 is produced in admixture with minor amounts of the A_1 atropisomer. A mixture of the phlegmacins-A_2 and -B_2 has been isolated from *Cassia torosa* (Leguminosae) by SHIBATA and coworkers (*647*). The Japanese group was able to separate both stereoisomers of this plant metabolite as well as to purify by column chromatography each of the phlegmacin atropisomers isolated by STEGLICH *et al.* from *Cortinarius odorifer*. It has been deduced from their respective chiroptical properties that the pigments from the higher plant are enantiomeric with the *Cortinarius* metabolites (*647*).

From fruit bodies of *Cortinarius auroturbinatus* an interesting 'tetramer' has been isolated for which structure (**376**) has been proposed (*515*). Due to restricted rotation about the central C-C bond the pigment was isolated in the form of two rotational isomers.

(**376**)

The phlegmacin pigments show a close structural similarity to the pigments isolated from the poisonous Mexican shrub *Karwinskia humboldtiana* (Rhamnaceae) (*196*) which cause segmented demyelination of peripheral nerves. According to preliminary pharmacological assays, phlegmacin-B_1 (**371**) is of comparable toxicity and causes similar effects to the *Karwinskia* toxins (*698*).

E. Tricolorin Group (Initial 10,10'-coupling)

BESL and BRESINSKY (*86*) discovered in *Leucopaxillus tricolor* (Tricholomataceae), besides endocrocin (**303**), two yellow pigments which bore superficial resemblances to asperflavin (**329**) and to phlegmacin (**371**). Closer chemical investigations, however, have revealed that these substances represent a new type of dimeric pre-anthraquinone (*94, 515*).

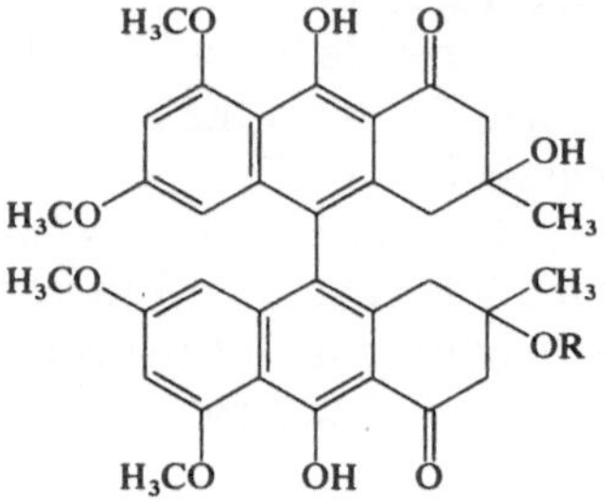

(**377**) R = H
(**378**) R = Ac

The structures of tricolorin-A (**377**) and its 3-*O*-acetyl derivative (**378**) followed from the spectroscopic data. Once again, the effect on the ¹H-n.m.r. spectra of shielding caused by the biaryl linkage was of particular importance (*515*). The chemical shift data for tricolorin-A (**377**) are presented in Table 32 and appear to be in close agreement with those of singueanol-I (**379**), a constituent of *Cassia singueana* (*214*). As the data in Table 32 reveal, the ¹H-n.m.r. spectrum of tricolorin-A showed a doubling of most signals, a phenomenon which can best be explained by the presence in (**377**) of two chiral centres of opposite absolute configuration (*515*). As the formulation (**380**) for tricolorin-A demonstrates, such a stereochemical situation imparts a magnetic nonequivalence to the individual halves of the molecule which

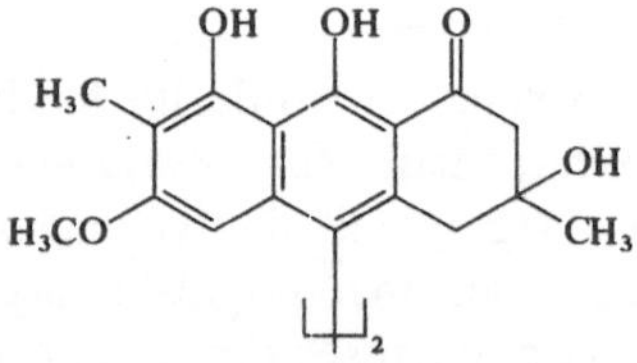

(**379**)

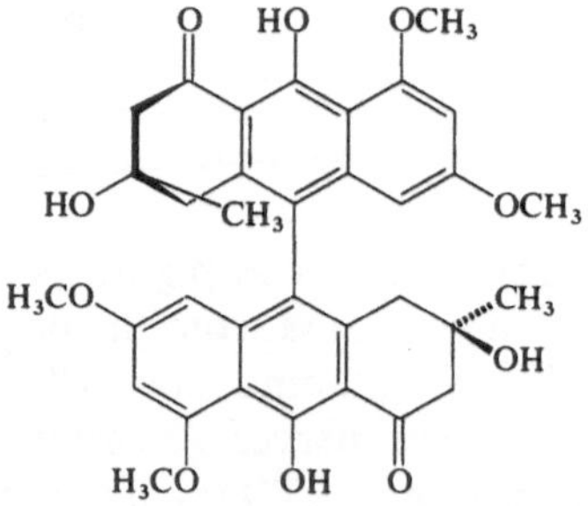

(**380**)

would not otherwise be the case. The possibility that tricolorin-A (377) exists instead as an equimolar mixture of two diastereoisomers is less likely since the ^{1}H-n.m.r. spectrum of the acetate (378) showed only a single set of signals in accord with the proposed structure.

Both tricolorin pigments exhibit A type CD spectra (i.e. negative Cotton effect at longer wavelength) which by comparison with the CD spectrum of singueanol-I (379) points to the (R)-absolute configuration for the chiral axis in the pigments (377) and (378), as depicted in stereostructure (380) (515).

F. Cortinarin Group[1] (Initial 8-O,10′-coupling)

Cortinarius elegantior contains among its neutral pigments a pale yellow compound, cortinarin, to which the unique dimeric structure (381) has been assigned from spectroscopic data (Table 32) and from its cleavage with dilute aqueous sodium hydroxide into physcion (290) and emodin-6,8-di-O-methyl ether (292) (387, 515, 617).

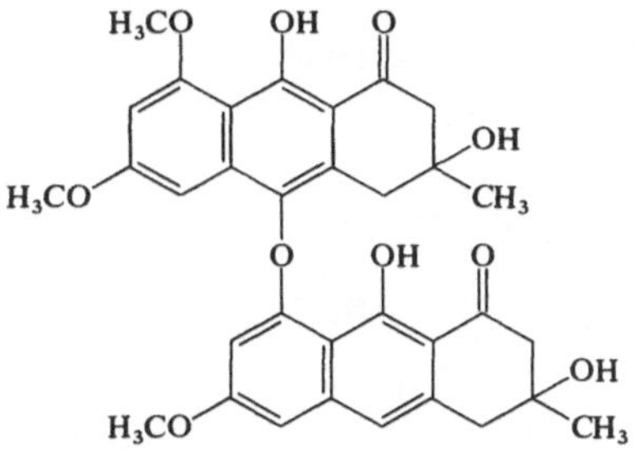

(381)

On chromatography plates cortinarin appeared as two discrete but closely spaced spots both of which, incidentally, exhibit strong greenish-white fluorescence under UV light. The chromatographic separation has been explained by the existence of two diastereoisomers of (381) differentiated by restricted rotation about the biaryl ether bridge, and is clearly demonstrable by ^{1}H- and ^{13}C-n.m.r. experiments (387). Thus, on heating a solution of (381) in the probe of the n.m.r. spectrometer, signals which appear twinned at room temperature (Table 32) gradually coalesce. A wider distribution of cortinarin (381) is apparent from its recent isolation from the North American toadstool Cortinarius mallochii (405).

---

[1] Unfortunately, the name cortinarin, coined originally in ref. 384, has subsequently been assigned by TEBBETT and coworkers (651) to a supposed cyclopeptide toxin from Cortinarius orellanus.

G. Rufoolivacin Group

Young sporophores of *Cortinarius rufoolivaceus* display inconspicuous grey-green colours which turn to red-violet on ageing. Normal extraction of the fruit bodies of *C. rufoolivaceus* gave an intensely violet solution from which rufoolivacin (**382**) and dirufoolivacin (**383**) have been separated by chromatography (*250, 252, 515*). Although these

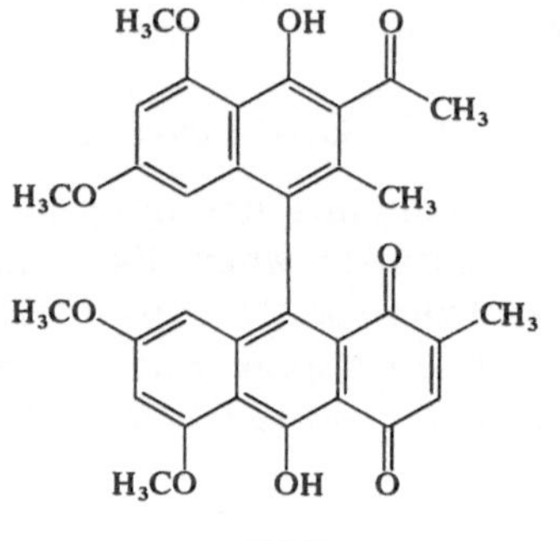

(**382**)

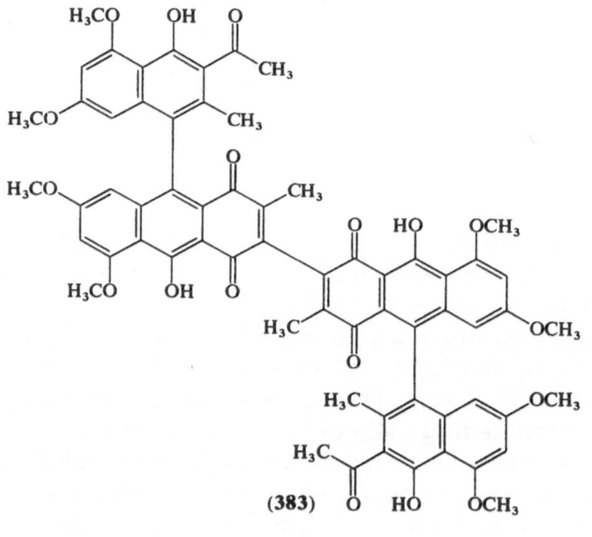

(**383**)

Table 36. *Occurrence of Pigments of the Rufoolivacin Group*

Rufoolivacin (**382**)	*Cortinarius odorifer* (*515, 617*), *C. rufoolivaceus* (*250, 252, 515, 617*)
Dirufoolivacin (**383**)	*Cortinarius orichalceus* [c] (*388, 617*), *C. rufoolivaceus* (*250, 252, 515*)

pigments are not strictly pre-anthraquinones, their biogenesis involving an anthracenone sub-unit is obvious and they are best included here.

The 1,4-anthraquinone structures of (382) and (383) have been firmly established by extensive n.m.r. investigations, including n.O.e. experiments, and the analysis of proton coupled ^{13}C-n.m.r. spectra. ^{1}H-N.m.r. data for rufoolivacin (382) form part of Table 32. It was conveniently analysed by reference to the spectra of known 1,4-anthraquinones (447) and of torachrysone-8-O-methyl ether (284) (577). From a biogenetic viewpoint it is interesting to note that torachrysone-8-O-methyl ether (284) itself, and several closely related naphthoquinones are also found in extracts of C. rufoolivaceus (see Section 3.4).

Dirufoolivacin (383) has been separated using high performance liquid chromatography into three stereoisomers (ratio 1:2:1) which exhibited ^{1}H-n.m.r. signals in accord with their identity as the three diastereoisomers (trans-trans, trans-cis = cis-trans, cis-cis) derivable from structure (383) by assuming restricted rotation about each of the biaryl linkages (250, 252, 515). Dirufoolivacin is thus a rare example of a natural product which occurs in several diastereoisomeric modifications due to the presence of three centres of axial chirality while lacking even a single asymmetric carbon atom.

(384) R = H
(385) R = OH

Extraction of young C. rufoolivaceus fruit bodies under an inert atmosphere led to the isolation of several strongly fluorescent leuco compounds such as (384) and (385) which are the immediate precursors of the violet pigments (250, 384). On exposure to air, and during routine chromatography, compounds such as (384) and (385) were converted into rufoolivacin (382) and dirufoolivacin (383).

The chromatographic properties of the dimeric pre-anthraquinones are given in Table 37 (515, 617).

Table 37. *Chromatographic Properties of Dimeric Pre-anthraquinones*

Pigment (structure number)	TLC $R_f{}^a$	Colour of Pigment on TLC		
		Daylight	UV (366 nm)	$+NH_3$
(361)	0.72	Red	–	–
(340)	0.55	Red	Brown	Purple
(339)	0.54	Orange-yellow	Brown	Lilac-red
(374)	0.51	Orange-yellow	Black	Brown
(353)	0.48	Orange-yellow	Black	Red-brown
(354)	0.48	Red	Black	Violet
(355)	0.48	Red	Black	Lilac
(383)	0.46^b	Red-violet	Red	Blue-violet
(365)	0.45	Orange-yellow	Black	Brown
(357)	0.45	Violet	Black	Blue-violet
(358)	0.43	Violet	Black	Blue-violet
(375)	0.43	Orange-yellow	Black	Brown-red
(338)	0.39	Yellow	–	Purple-brown
(356)	0.37	Lemon yellow	–	–
(367)	0.37	Orange-yellow	–	–
(350)	0.36	Lemon yellow	–	–
(382)	0.35	Red-violet	Red	Blue-violet
(348)	0.34	Lemon yellow	Brown	–
(371)	0.34	Lemon yellow	–	–
(362)	0.31	Lemon yellow	Pale ochre^c	–
(372)	0.31	Lemon yellow	Ochre^c	–
(336)	0.30	Green-yellow	Yellow-brown	–
(381)	0.30^b	Pale yellow	Greenish-white^c	–
(349)	0.27	Lemon yellow	–	–
(346)	0.24	Green-yellow	–	–
(343)	0.24	Green-yellow	Orange	–
(364)	0.23	Lemon yellow	Ochre^c	–
(337)	0.22	Green-yellow	Yellow-brown	–
(344)	0.16	Green-yellow	Orange	–
(345)	0.08	Green-yellow	Orange	–

^{a} Kieselgel 60 F_{254}, Merck precoated plates. Solvent – benzene:ethyl formate:formic acid = 10:5:3.
^{b} Two overlapping spots due to the presence of diastereoisomers.
^{c} Fluorescent under UV light.

3.5.4. Further Octaketides

Fruit bodies of the slime mould *Metatrichia vesparium* contain homotrichione (386) as the major pigment along with the arcyriaflavins-B and -C (see Section 5.3), and minor amounts of vesparione (387) and trichione (280) (*436, 438*). The structure of homotrichione (386) followed from the spectroscopic data and by comparison with trichione

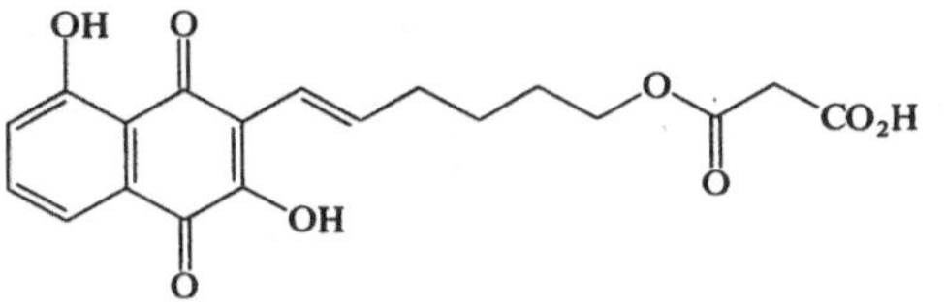

(386) 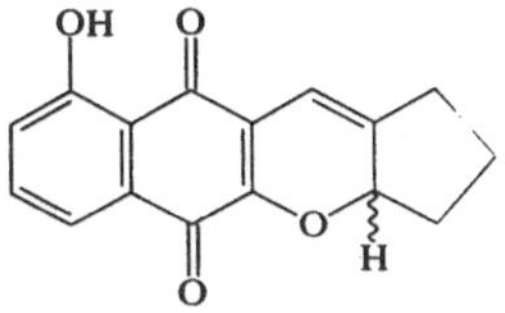   (387)

(280), already mentioned (Section 3.4) as the principal pigment of another myxomycete, *Trichia floriformis* (438).

The interesting red pigment vesparione (387) [UV/vis. (methanol): λ_{max} (log ε) = 232 (3.90, br.), 260.5 (4.08), 296 (sh., 3.89), 415 nm (3.56)], which was isolated in optically inactive form, seems to be related biosynthetically to homotrichione (386) according to Scheme 62.

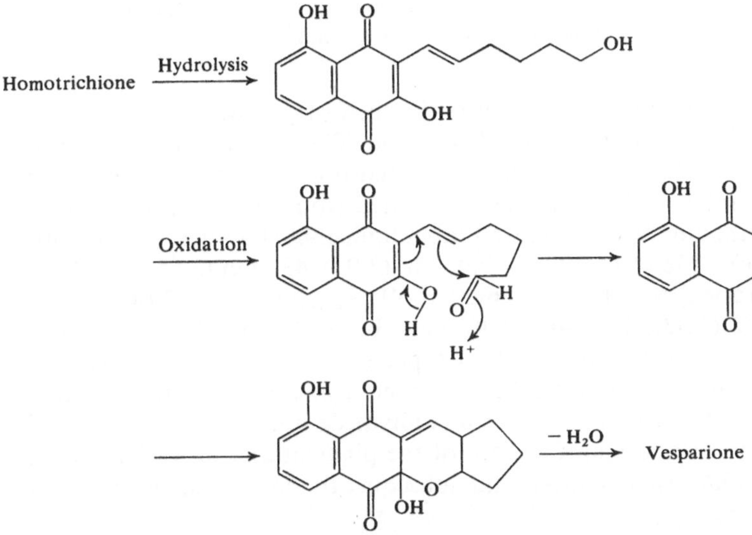

Scheme 62. Possible biogenesis of vesparione (387) from homotrichione (386)

The structure (387) for vesparione was deduced from spectroscopic data and biogenetic considerations and was confirmed by synthesis (Scheme 63) (436). In accord with the structure of the synthetic material, the carbonyl signals at δ 178.7 (C-5) and δ 188.0 (C-10) in the ^{13}C-n.m.r. spectrum are split by ^{3}J-couplings into a doublet ($J = 4$ Hz) and a broadened doublet, respectively (cf. ref. 370).

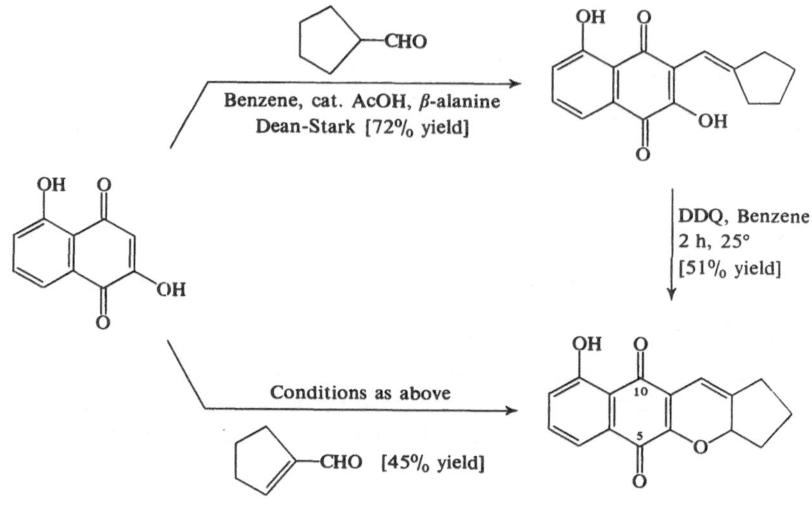

Scheme 63. Synthesis of vesparione

Further interesting variations in the chemistry of Myxomycetes have been encountered in the genus *Lindbladia*. The almost black fructifications of *Lindbladia tubulina* assume a red-brown colour on treatment with mineral acid. From extracts obtained from this slime mould with acidic methanol, two pigments, lindbladione (**388**) [UV/vis. (methanol): λ_{max} (log ε) = 252 (4.14), 277 (4.25), 348 (3.90), 435 nm (3.65)] and lindbladiapyrone (**389**) [UV/vis. (methanol): λ_{max} (log ε) = 215 (4.22), 237 (4.14), 272 (4.02), 445 nm (3.60)], have been obtained in yields corresponding to 0.12% and 0.03%, respectively, of the fresh weight of the organism (*456*). Lindbladione formed a mixture of tetra-, tri-, and diacetates when treated with acetic anhydride and sodium acetate and from the mass and n.m.r. spectra of the pigment and of its derivatives structure (**388**) was deduced and subsequently confirmed by synthesis (Scheme 64).

The structure (**389**) of lindbladiapyrone followed from analysis of the ¹H-n.m.r. spectrum which clearly revealed the presence of an n-

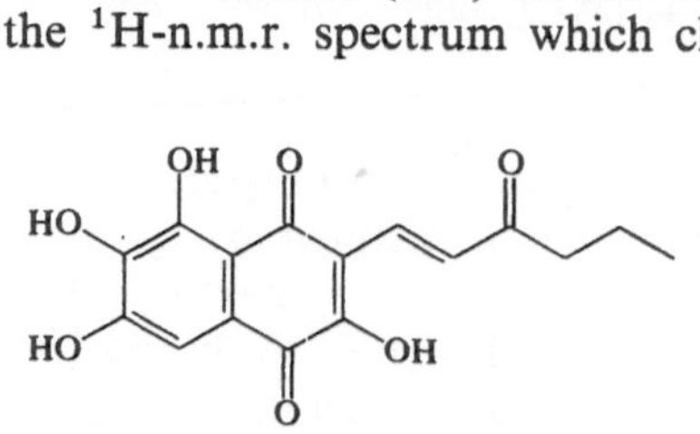

(388)

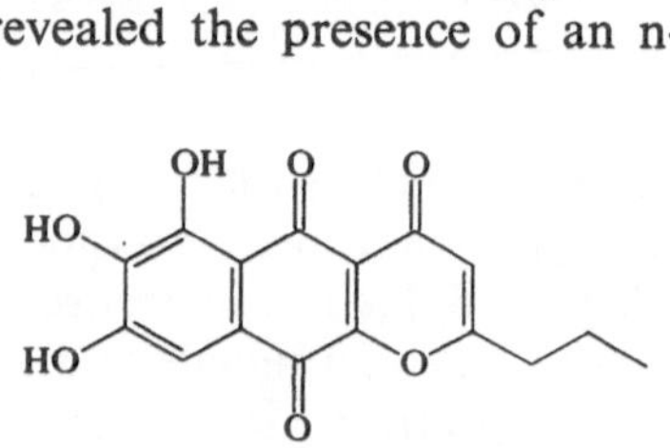

(389)

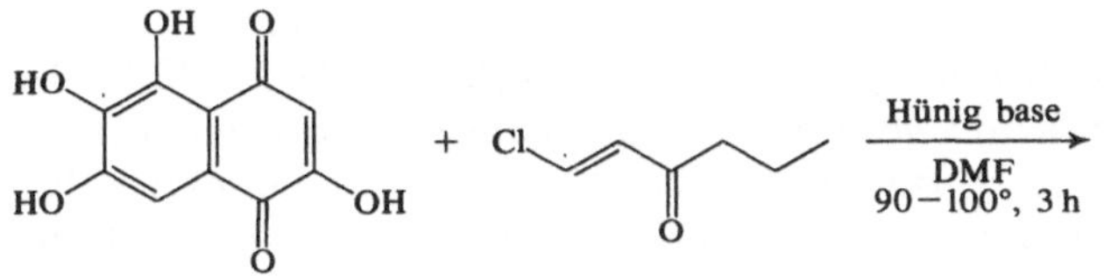

Scheme 64. Synthesis of lindbladione

propyl side chain, two aromatic protons, and three phenolic hydroxy groups and from the molecular formula which contains two hydrogen atoms less than that of lindbladione (456). It is interesting that the proton in the pyrone ring in (389) exchanges rapidly with D_2O.

Fermentations of the ascomycete *Sepedonium chrysospermum* produce large quantities of (−)-chrysodin [UV/vis. (chloroform): λ_{max} (log ε) = 243 (3.99), 308.5 (4.12), 380 nm (4.61)] for which the azaphilone structure (390) has been proposed on the basis of chemical transformations and physical measurements (170). The absolute configuration shown for chrysodin (390) may be assigned following more recent work on the stereochemistry of azaphilone pigments (632, 714).

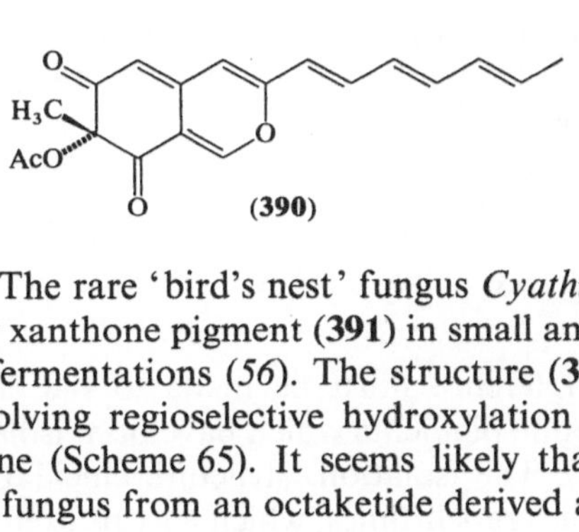

(390) 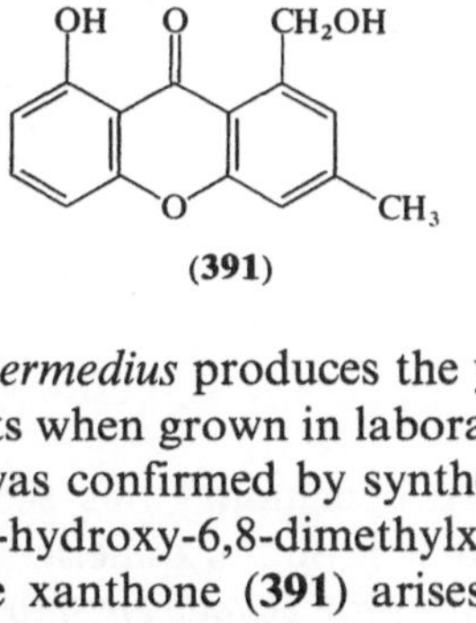   (391)

The rare 'bird's nest' fungus *Cyathus intermedius* produces the yellow xanthone pigment (391) in small amounts when grown in laboratory fermentations (56). The structure (391) was confirmed by synthesis involving regioselective hydroxylation of 1-hydroxy-6,8-dimethylxanthone (Scheme 65). It seems likely that the xanthone (391) arises in the fungus from an octaketide derived anthraquinone precursor.

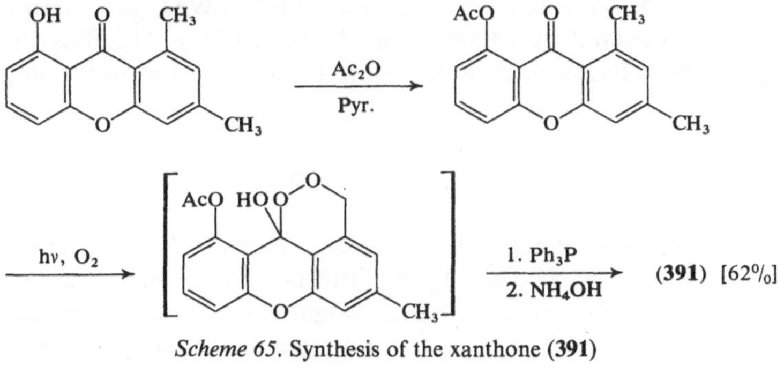

Scheme 65. Synthesis of the xanthone (391)

A completely different octaketide, obviously formed by oxidative dimerisation of a unique precursor, is xylindein (**392**) (*103, 104, 207*). It is produced by *Chlorosplenium aeruginascens* (= *Chlorociboria aeruginascens*) both in fructifications and in culture and imparts a beautiful blue-green colour to the fruit bodies and to the dead wood on which the fungus lives. KÖGL's early work on the isolation and structure of xylindein and some interesting historical facts about the pigment have been brought together by THOMSON (*655*).

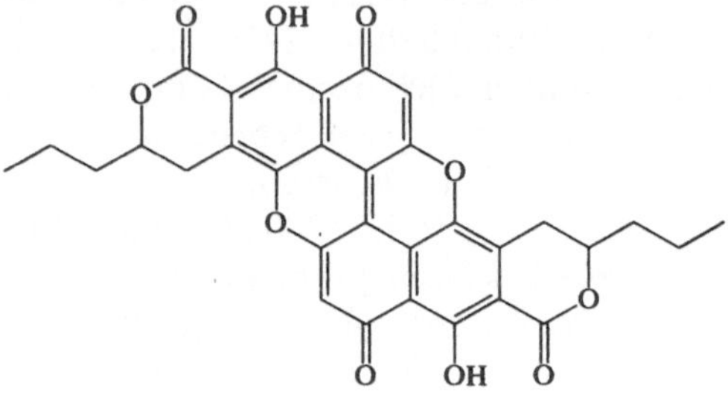

(**392**)

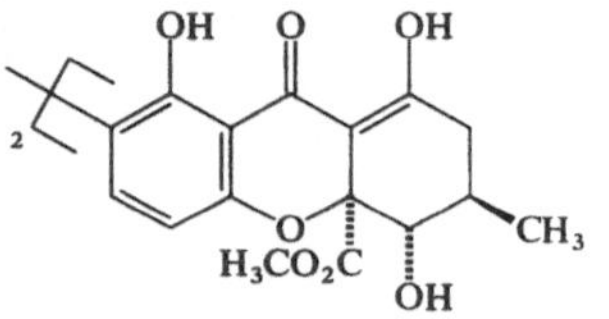

(**393**)

In concluding this section reference must be made to the ergochrome group of dimeric octaketide pigments which have been isolated from ergot (*Claviceps purpurea*). The isolation, structure elucidation, biosynthesis and physiology of the ergochromes, which may be exemplified here by ergochrome-AA (secalonic acid-A) (**393**), have been comprehensively covered by FRANCK and FLASCH (*245*). The biosynthesis of ergochromes has been reviewed more recently by FRANCK (*244*).

3.6. Nonaketides

Some fungi belonging to *Cortinarius*, subgenus *Leprocybe*, exhibit yellow-green fluorescence when their fruit bodies are viewed under UV light. This phenomenon was first investigated chromatographically by GABRIEL (*264, 265*) who detected and designated the major fluorescent

principles as 'pigment-A' and 'pigment-B'. It was further concluded
from preliminary chemical and spectroscopic studies that 'pigment-B'
was a glycoside derivative of 'pigment-A', and that both compounds
contained a xanthone chromophore.

These fluorescent compounds have now been isolated from *Cortinarius cotoneus*, *C. melanotus*, and *C. venetus* and characterised by spectroscopic and chemical methods as the nonaketide leprocybin (**394**)
(GABRIEL's 'pigment-B'), and its aglycone, leprocyboside (**395**) ('pigment-A') (*437*).

Leprocybin (**394**) was cleaved to leprocyboside (**395**) and D-glucose

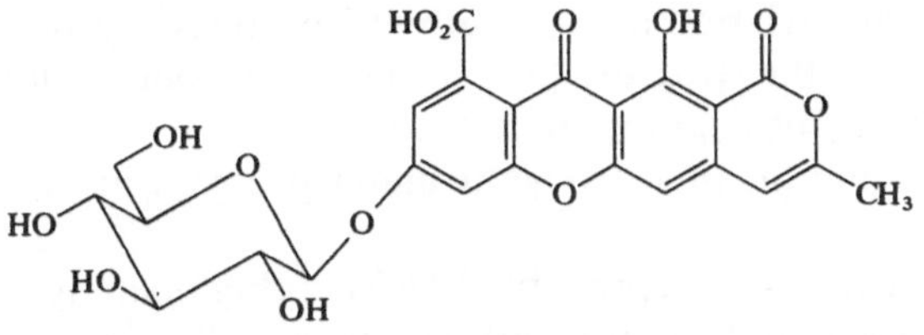

(**394**)

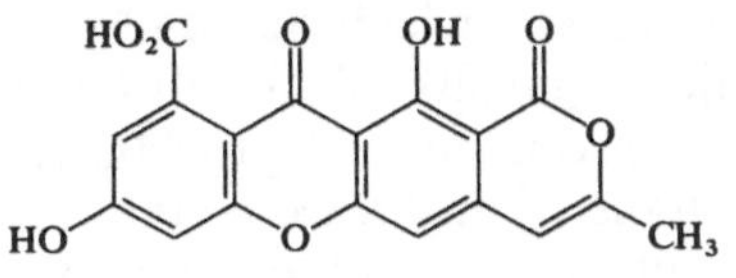

(**395**)

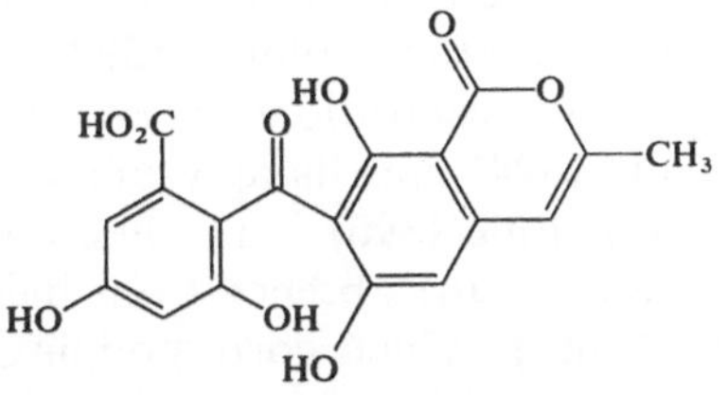

(**396**)

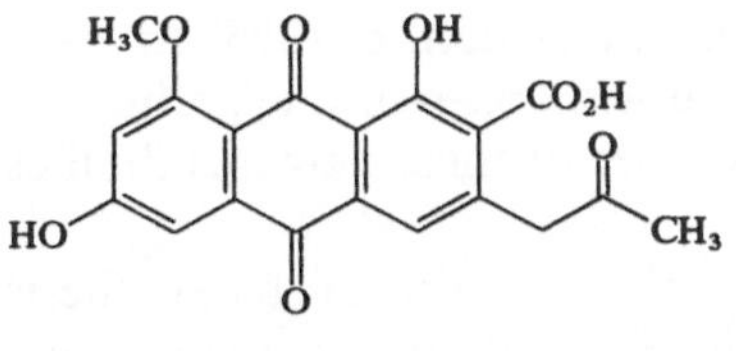

(**397**)

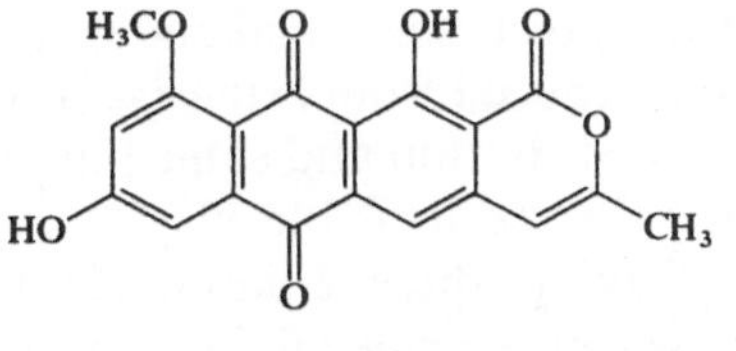

(**398**)

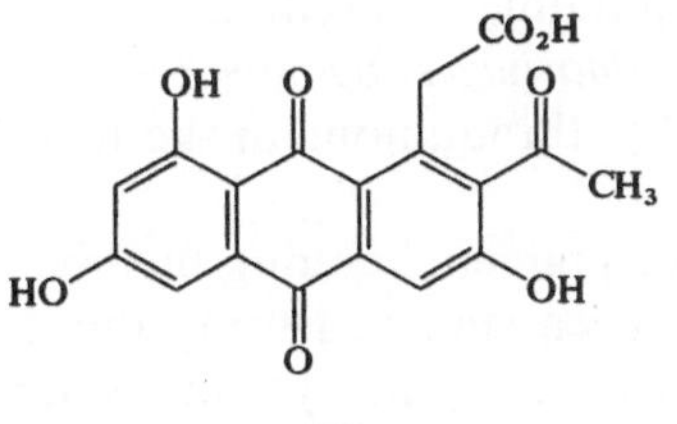

(**399**)

Table 38. *Occurrence of Leprocybin* (**394**) *and Leprocyboside* (**395**)

Cortinarius ahsii, C. annulatus, C. betuletorum, C. clandestinus, C. corrugatus, C. cotoneus [1.2%][a]*, C. croceocolor, C. flavifolius, C. isabellinus, C. melanotus* [1.0%][a]*, C. mellinus, C. nothoraphanoides, C. psittacinus, C. raphanoides, C. renidens, C. subannulatus, C. valgus, C. variipes, C. venetus* [1.8%][a]*, C. venetus* var. *montanus* [0.8%][a]*, C. zinziberatus*

---

[a] Refers to proportion of leprocybin isolated as a percentage of the dry weight of the fungus (*437*). Occurrence in other species was determined by TLC (*406*).

using β-glucosidase and the constitution of the aglycone was elucidated by several derivatisations and degradations, including decarboxylation followed by methylation to 8-*O*-methyl(decarboxy)leprocyboside. Final confirmation of the structure (**395**) of leprocyboside itself was obtained by synthesis of this degradation product.

The distribution of leprocybin and leprocyboside is given in Table 38.

The biosynthesis of leprocybin (**394**) *via* ring opening of a precursor anthraquinone may be confidently assumed following isolation of the diaryl ketone leprophenone (**396**). This ketone, together with the anthraquinones leprolutein (**397**) and anhydroleprolutein (**398**), is present in low concentrations in several *Leprocybe* species which contain leprocybin (**394**) and its aglycone as the major pigments (Table 39) (*590*). The ketone (**396**) is the likely immediate precursor of leprocyboside (**395**), but the absence from these toadstools of any trace of the ring cleavage products corresponding to leprolutein (**397**) and to anhydroleprolutein (**398**) suggests that these metabolites lie on a side track of leprocybin biosynthesis. A third minor anthraquinone from *Leprocybe*, leprovenetin, possesses the structure (**399**) formed by a unique folding of the precursor nonaketide chain (*590*). The structures (**396**) to (**399**) were deduced from extensive proton decoupled ^{13}C-n.m.r. experiments. The distribution and some physical properties of these new metabolites are collected in Table 39.

Taking these compounds together, the pattern of biosynthesis shown in Scheme 66 for the nonaketide pigments of *Leprocybe* may be advanced.

It is of interest that alongside predominant nonaketides several species belonging to *Cortinarius*, subgenus *Leprocybe*, also manufacture an octaketide derived anthraquinone in the form of dermolutein (**305**) (Table 26).

The distribution of pigments among members of *Cortinarius*, subgenus *Leprocybe*, has been surveyed using thin layer chromatography and is of some importance to the systematic taxonomy of the group (*264, 332, 333, 372, 406*).

Table 39. *Occurrence and Properties of Leprocybe Nonaketides*

Pigment	Occurrence[a]	Colour M.p.	TLC R_f	Colour $+NH_3$	Ultraviolet/Vis. [λ_{max} (log ε)]	Infrared [ν_{col}]
Leprocybin (394)	see Table 38	Yellow-green >240° (dec.)	0.46[b]	–	228(4.12), 274(4.53), 310(4.32), 362 nm(4.10) [water]	1730, 1715 (sh.) cm^{-1}
Leprocyboside (395)	see Table 38	Olive green >350°	0.31[c]	–	298(sh., 4.35), 307(4.42), 339(4.27), 360 nm (4.36)[DMF]	1730, 1700 cm^{-1}
Leprophenone (396)	*Cortinarius cotoneus* [8×10^{-4}%], *C. psittacinus* (590)	Colourless 198°	0.52[c]	Yellow	215(3.81), 242(4.04), 269(3.83), 313 nm (3.69) [methanol]	1695, 1680, 1658, 1693–1594 cm^{-1}
Leprolutein (397)	*Cortinarius cotoneus* [4.8×10^{-3}%], *C. psittacinus, C. subannulatus* (590)	Red >350°	0.20[c]	Red	229(2.94), 261(2.82), 311(sh., 2.68), 399(2.33), 422 nm(2.27) [methanol]	1730, 1681, 1629 cm^{-1}
Anhydro-leprolutein (398)	*Cortinarius cotoneus* [5×10^{-4}%], *C. psittacinus, C. subannulatus* (590)	Orange >320°	0.28[c]	Violet	–	–
Leprovenetin (399)	*Cortinarius cotoneus* [8×10^{-4}%], *C. psittacinus, C. subannulatus, C. venetus* [4.8×10^{-3}%], *C. venetus* var. *montanus* [4.8×10^{-3}%] (590)	Yellow 270° (dec.)	0.52[c]	Red-violet[d]	218(4.54), 270(4.57), 285(4.65), 311(sh., 4.29), 341(sh., 3.89), 429 nm(3.98) [methanol]	1725, 1720(sh.), 1700, 1670, 1635, 1600 cm^{-1}

[a] Isolated yields refer to the weight of dried fungus.
[b] Cellulose F$_{254}$ Merck precoated plates (methanol:water=4:1) (437).
[c] Kieselgel 60 F$_{254}$ Merck precoated plates (toluene:ethyl formate:formic acid=10:10:3) (437, 590).
[d] With concentrated sulphuric acid leprovenetin produces a vivid green colour reminiscent of that shown by skyrin.

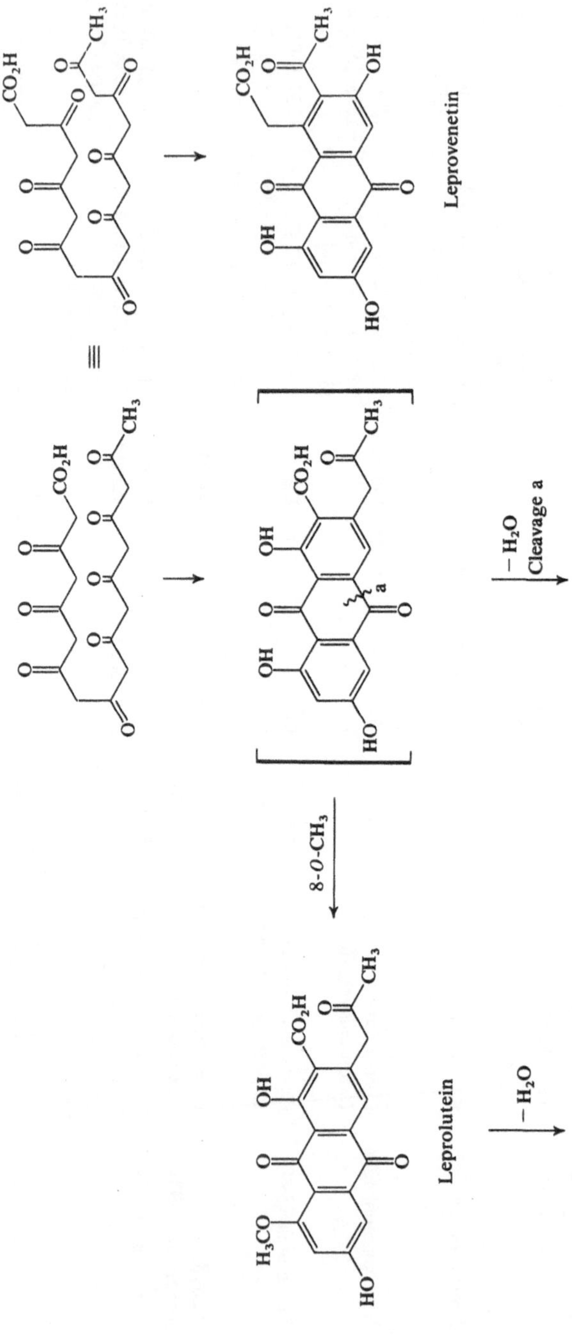
Leprovenetin
Cleavage a
$-H_2O$
8-O-CH$_3$
Leprolutein
$-H_2O$

References, pp. 253–286

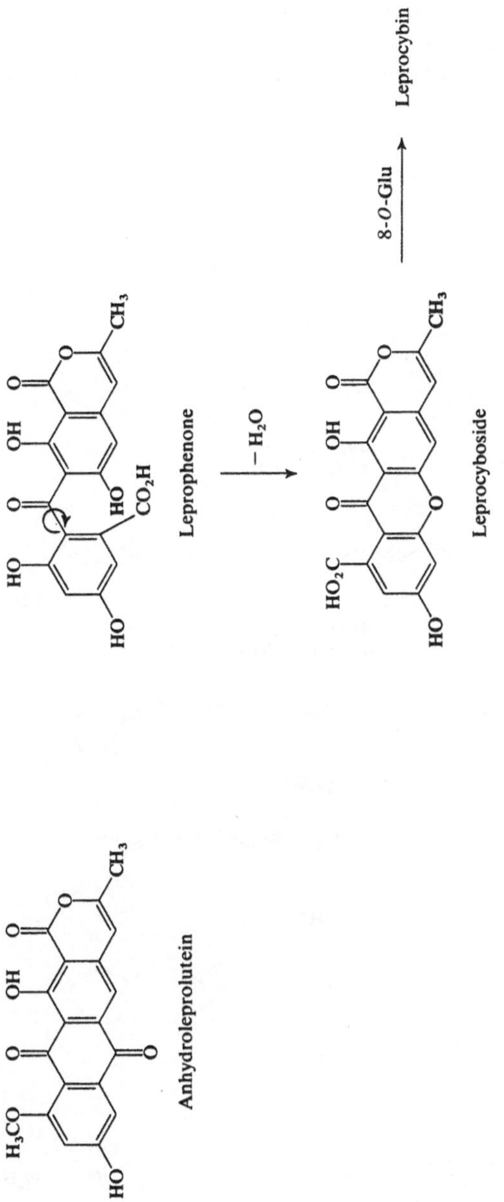

Scheme 66. Possible biogenetic inter-relationships of *Leprocybe* pigments

3.7. Compounds of Fatty Acid or Higher Polyketide Origin

3.7.1. Nonisoprenoid Polyene Pigments

Although carotenoid pigments (Section 4.2) and (generally) colour-less poly-yne metabolites occur in many fungi (*110, 667, 668*), non-isoprenoid polyene pigments are to be found in only a few macromycetes (Table 40).

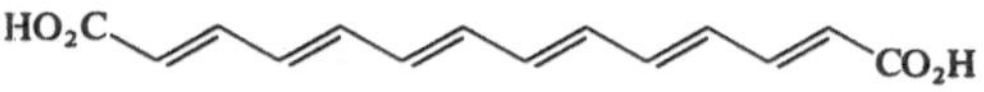

(400)

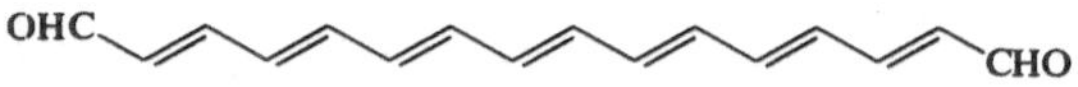

(401)

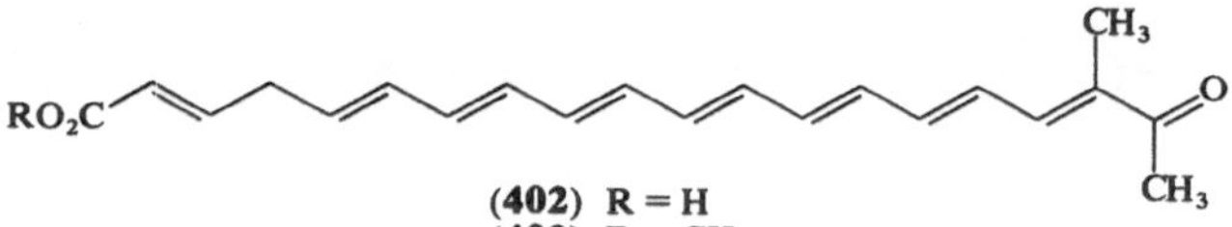

(402) R = H
(403) R = CH₃

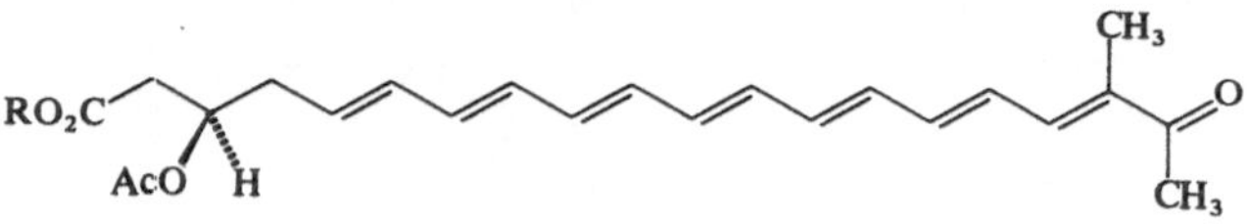

(404) R = H
(405) R = CH₃

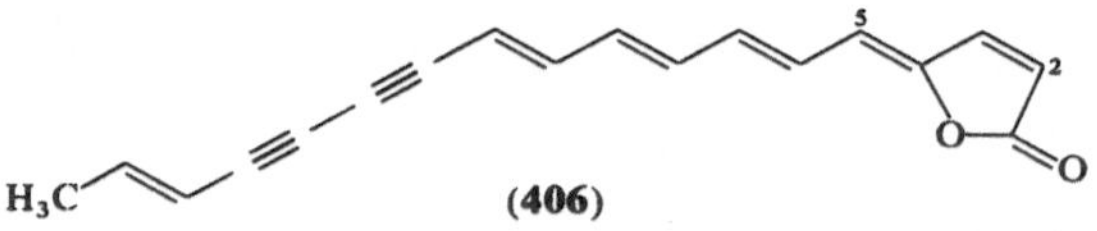

(406)

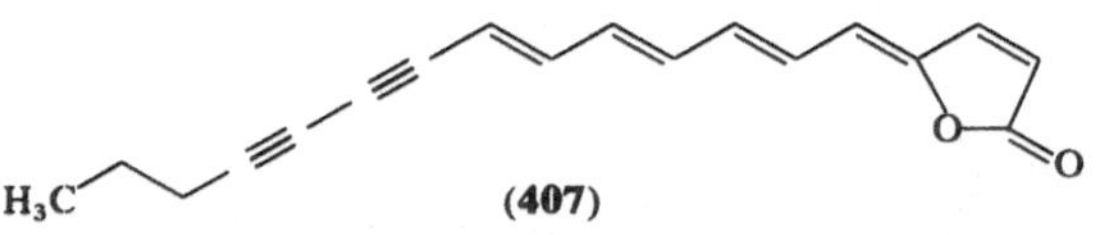

(407)

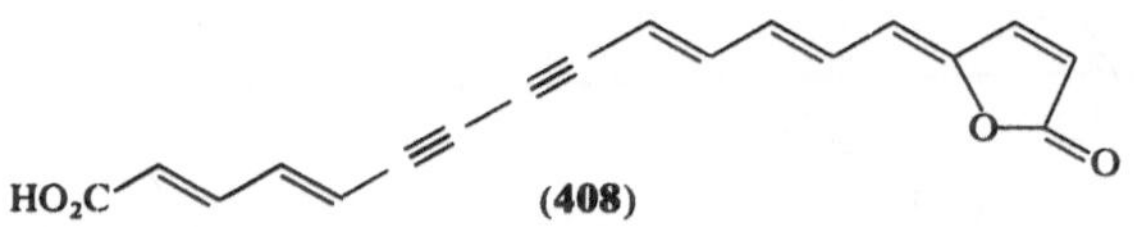

(408)

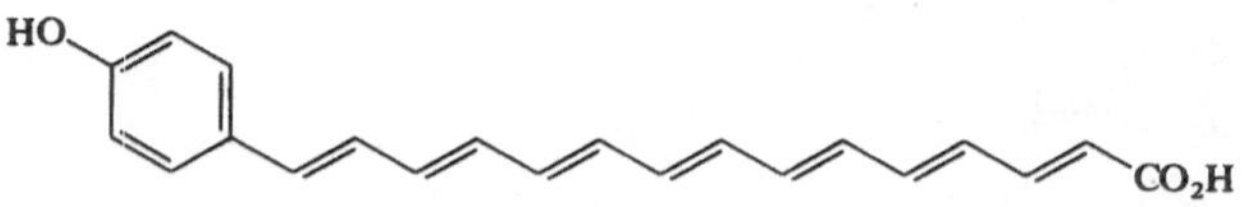

(220)

Table 40. Occurrence and Properties of Polyene Pigments

Pigment	Occurrence	Colour M.p.	TLC R_f	Ultraviolet/Vis. [λ_{max} (log ε)]
Corticrocin (400)	Piloderma croceum (= Corticium croceum) [mycelium] (219)	Yellow 270° (subl.) 317° (dec.)	–	374 (4.79), 393 (4.95), 416 nm (4.92) [ethanol][a]
Cortisalin (220)	Cytidia salicina (= Corticium salicinum) (306)	Red-violet >290° (dec.)	–	450 nm (4.90) [pyridine]
Verpacrocin (401)	Verpa digitaliformis [cultures] (93, 490)	Orange 208° (dec.)	0.54[b]	404 (sh., 5.04), 423 (5.22), 448 nm (5.21) [acetone]
Piptoporic acid (402)	Piptoporus australiensis (277)	Orange-red oil	0.53[c]	249 (3.94), 305 (3.90), 404 (4.68), 415 nm (sh., 4.67) [methanol]
Methyl piptoporate (403)	Piptoporus australiensis (277)	Orange-red oil	0.95[c]	245 (4.02), 290 (3.87), 302 (3.99), 378.5 (4.74), 398.5 (4.89), 422.5 nm (4.82) [hexane]
(3R)-3-Acetoxy-2,3-dihydro-piptoporic acid (404)	Piptoporus australiensis (277)	Orange-yellow oil	0.35[c]	248 (3.16), 305 (4.23), 403 (4.95), 415 nm (sh., 4.94) [methanol][a]
Xerulin (406) + Dihydroxerulin (407)	Xerula melanotricha (= Oudemansiella badia) [c] (25)	Yellow	–	400, 412 nm [methanol]
Xerulinic acid (408)	Xerula melanotricha [c] (490)	Orange-yellow >320° (dec.)	0.29[d]	225 (3.82), 280 (3.74), 416 (4.88), 432 nm (4.46) [methanol]

a UV data applies to the methyl ester derivative.
b Merck Kieselgel 60 (benzene:ethyl formate:formic acid = 10:5:3).
c Merck Kieselgel 60 (ethyl acetate:light petroleum:methanol = 40:10:1).
d Merck Kieselgel 60 (cyclohexane:ethyl acetate:formic acid = 120:40:5).

Corticrocin (**400**), the first pigment of its type to be found in Nature, was isolated from the mycelium of the mycorrhizal fungus *Piloderma croceum* (= *Corticium croceum*) and characterised by ERDTMAN (*219*). Extreme purification problems were encountered due to the acute insolubility of the substance in organic solvents. It was ultimately identified from the electronic spectrum of its less resiliant methyl ester derivative and by catalytic hydrogenation to the dimethyl ester of dodecanedioic acid.

Similar solubility properties are exhibited by cortisalin (**220**), the pigment from the red fruit bodies of *Cytidia salicina* (= *Corticium salicinum*) (*306*), and by verpacrocin (**401**), an orange-red dialdehyde from cultures of the ascomycete *Verpa digitaliformis* (*93*).

Syntheses of both corticrocin (**400**) (*576, 709*) and cortisalin (**220**) (*472*) have been reported.

The distinctive bright orange brackets produced by *Piptoporus australiensis* contain a series of pigments each possessing a heptaenone chromophore (*277*). Piptoporic acid (**402**) is accompanied by its methyl ester (**403**) which on catalytic hydrogenation afforded methyl 18-methyl-19-oxoicosanoate. The structures (**402**) and (**403**) followed from the spectra of the pigments and from those of their perhydro derivatives. The absolute configuration of the pigment (**404**) was determined by chemical correlation of the methyl ester derivative (**405**) with methyl (3 *R*)-3-acetoxy-4-carboxybutanoate.

The presence of these polyene pigments in *P. australiensis* may protect the fruit bodies from attack by insects; sporophores of the closely related *Piptoporus portentosus* which are devoid of pigmentation in the flesh are rapidly invaded and voraceously devoured by larvae. In this context, however, it must be noted that *P. australiensis* also produces considerable quantities of the interesting citraconic anhydride derivative (**409**), which is also lacking in *P. portentosus* (*278*).

(**409**)

Mycelial cultures of *Xerula melanotricha* (= *Oudemansiella badia*) produce yellow-orange pigments which have been separated chromatographically into nonpolar and polar fractions (*25, 364*). The nonpolar fraction gave a mixture (1:2) of xerulin (**406**) and dihydroxerulin (**407**) which proved inseparable during subsequent application of the usual chromatographic techniques. The structures of both pigments were deduced from the infrared, electronic, and 500 MHz ^{1}H-n.m.r. spectra of the mixture, and from its mass spectrum which exhibited two molecular ions (m/z 262 and 264). A small coupling constant ($J < 1$ Hz) between the protons at C-2 and C-5 in (406) and in (407) is consistent with a (Z)-configuration at the exocyclic double bond. The polar fraction afforded a closely related, highly unsaturated carboxylic acid possessing an identical lactone unit. This third pigment, xerulinic acid, has been tentatively assigned the structure (408) (490).

The mixture of (406) and (407) possesses inhibitory activity against pig liver esterase.

3.7.2. Quinones with Extended Unbranched Side Chains

The sulphur yellow fruit bodies of *Sarcodontia setosa*, a fungus living on old apple trees, owe their colour to the presence of sarcodontic acid (410) (440, 442) together with its dihydro and didehydro derivatives (411) and (412), respectively (439). The pigments are present in the dried fungus in a ratio of 45:45:10 [(410):(411):(412)] and are best separated by high performance liquid chromatography involving their methyl ester derivatives. The geometry of the side chain double bonds in (410) and (412) was clear from the magnitude of the vicinal coupling constant between the respective olefinic protons ($J = 15.2$ Hz).

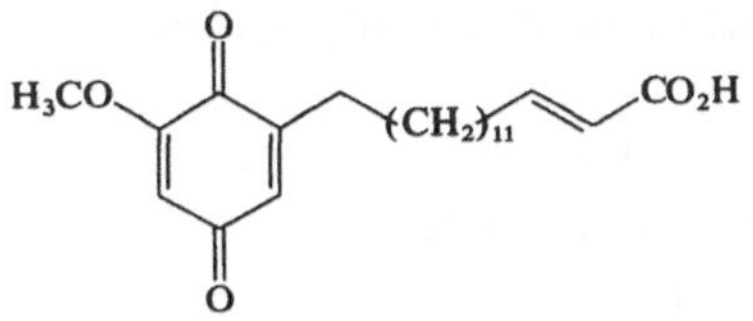

(410) 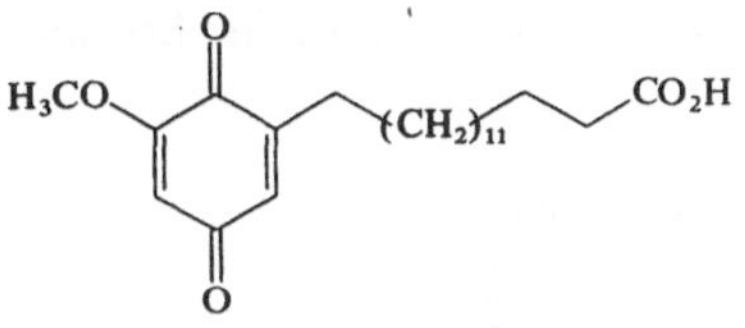   (411)

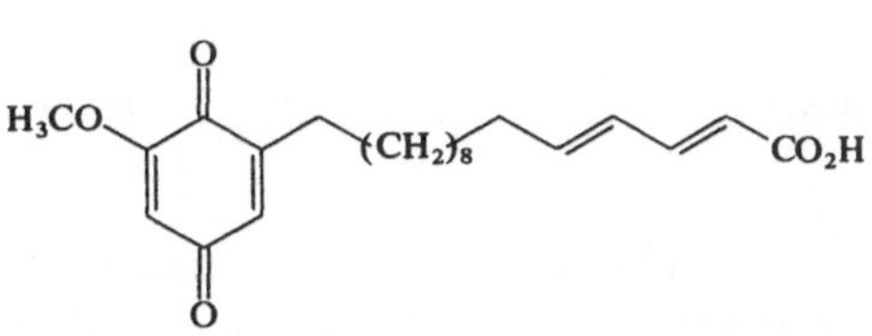

(412)

A colourless compound, betulachrysoquinone hemiketal (413), which must be related biogenetically to the sarcodontic acids, has been isolated from wood of *Betula lutea* inoculated with *Phanerochaete chry-*

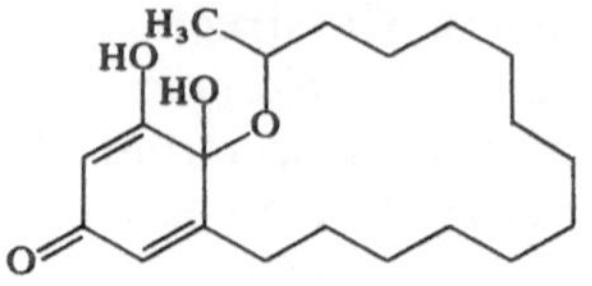

(413)

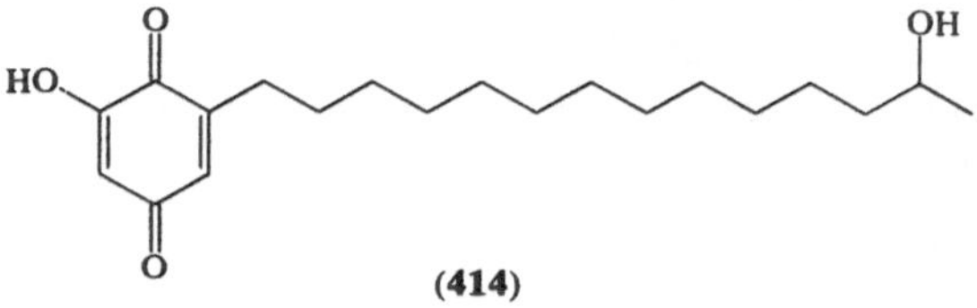

(414)

sosporium (*166*). When the hemiketal (413) was treated with mineral acid it furnished betulachrysoquinone (414).

It is a mixture of the ceriporiones-A (415) and -B (416) which imparts a cherry red colour to the acetone extracts of *Ceriporia viridans* (*155*). The mixture [m.p. 149–150 °C; R_f 0.46, silica gel (benzene:ethyl formate:formic acid = 10:5:3)], which has not been separated into its individual components, exhibited absorption maxima at 303 and 463 nm and formed a green monoanion with ammonia, and a blue dianion with aqueous potassium hydroxide. The structures (415) and (416) were deduced from ^{1}H- and ^{13}C-n.m.r. spectra while bearing in mind biogenetic considerations. The position of the side chain double bonds in these pigments, however, awaits further confirmation.

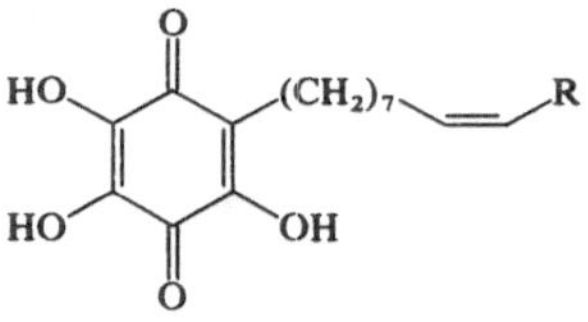

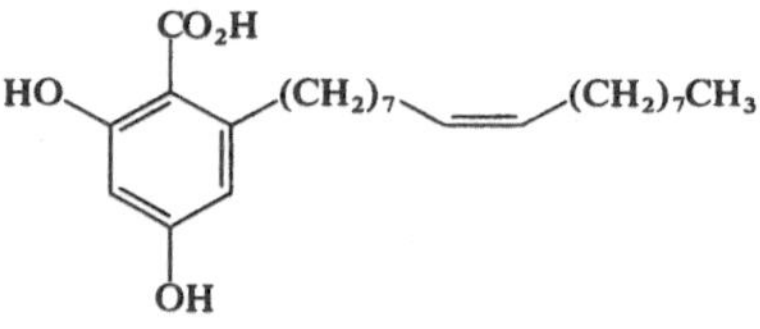

(415) R = (CH$_2$)$_5$CH$_3$
(416) R = C$_8$H$_{13}$

(417)

The ceriporiones are closely related to the antibiotic merulinic acids, e.g. merulinic acid-A (417), which have been isolated from fruit bodies of *Merulius tremellosus* and *Phlebia radiata* (*275*).

The progressive development of functionality in the aromatic rings in the merulinic acids, the sarcodontic acids, and the ceriporiones is a strong indication of the close biogenetic connections between these metabolites of Aphyllophorales (Scheme 67).

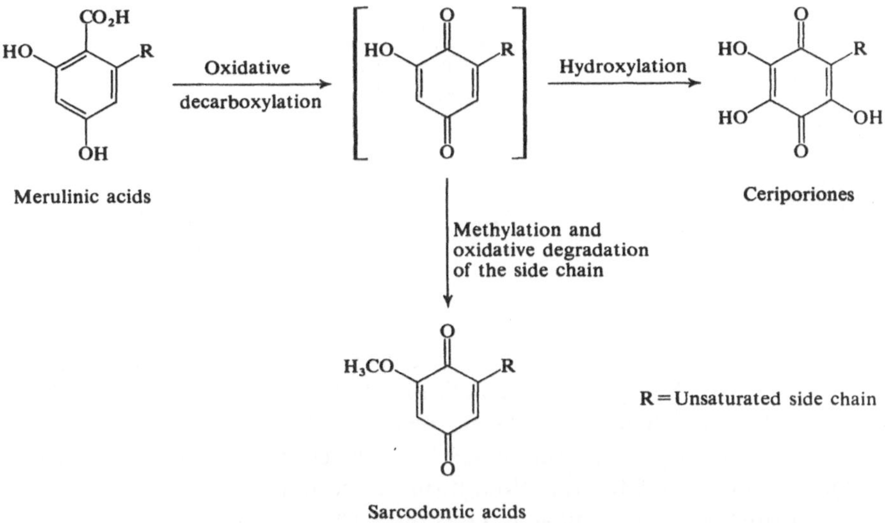

Scheme 67. Biogenetic relationships between polyketides from Aphyllophorales

4. Pigments from the Mevalonate Pathway

Contrary to their wide distribution in the plant and animal kingdoms, the occurrence of pigments deriving from the mevalonate pathway in Macromycetes is restricted to a few families and, in certain instances, even to a single species. Certain *Lactarius* toadstools produce sesquiterpene pigments based on the guaiane skeleton, while benzoquinones derived from cuparane type precursors prevail in cultures of several *Coprinus* species. Carotenoids are characteristic for Ascomycetes from Pezizales and Helotiales, and for Basidiomycetes from Tremellales, Cantharellales, Dacrymycetales and Phallales, but are found only rarely in Agaricales.

4.1. Sesquiterpenoids

Fruit bodies of *Lactarius* species belonging to the Section Dapetes produce a latex which may show orange, red, green, or blue colours (*563*). The genuine pigment of *Lactarius deliciosus* and *L. deterrimus* is the orange-yellow stearate ester (**418**) of 14-hydroxyguai-1,3,5,9,11-pentaene (**419**) (*82, 685*). Thus, the ester (**418**) was the only sesquiter-

(418) R = CH$_2$OCO.C$_{17}$H$_{35}$
(419) R = CH$_2$OH
(420) R = CHO

(421) R = CH$_2$OH
(422) R = CHO
(423) R = CH$_3$
(424) R = CH$_2$OCO.C$_{17}$H$_{35}$

(425)

pene obtained when young, undamaged mushrooms were frozen ($-197\,°C$) at their place of growth and subsequently extracted under an inert atmosphere with hexane below $-10\,°C$ (*82*). On exposure to air and to enzymes present in the latex the ester (**418**) suffers changes which bring about the discolouration of the milk which occurs when the mushrooms are damaged. Thus, all earlier attempts to isolate the pigments from *L. deliciosus* yielded azulenes and related secondary products (Table 41) which may now be regarded as artefacts. Since these by-products arise from the ester (**418**) only in the injured fruit bodies of *L. deliciosus* and *L. deterrimus* and never *in vitro*, where green polymeric materials result, it may be concluded that these secondary processes are enzyme controlled. It should be noted that some of the pigments identified as artefacts from *L. deliciosus* and *L. deterrimus* may be present in the intact fruit bodies of related species.

The presence of a pre-existing azulene, 1-stearoyloxymethylene-4-methyl-7-isopropenylazulene (**424**), has recently been convincingly demonstrated in the magnificent blue latex from the North American mushroom *Lactarius indigo* (*345*). The structure of the deep blue crystalline pigment, m.p. 48.5–49 °C, was deduced from the electronic and ^{1}H-n.m.r. spectra which pointed firmly to an azulene possessing a guaiane skeleton. Alkaline hydrolysis of the substance afforded pure stearic acid and the original location of the ester moiety in the nucleus of (**424**) was ascertained from the ^{1}H-n.m.r. data. The parallel between the blue pigment (**424**) from *L. indigo* and the genuine colouring matter (**418**) from *L. deliciosus* is notable.

The azulene (**421**, isopropyl in place of isopropenyl) isolated from dried specimens of *L. deterrimus* from Kashmir (*441*) is almost certainly yet another artefact.

The lagopodins-A (**426**) and -B (**427**) and hydroxylagopodin-B (**428**) are sesquiterpenoid quinones which have been isolated from cultures of various species of *Coprinus* (Table 42). Their chemistry has been dealt with by THOMSON (*655*).

Table 41. *Dihydroazulenes and Azulenes from* Lactarius *Species*

Pigment	Colour	Occurrence[a]	TLC R_f^b	Ultraviolet/Vis. [λ_{max} (log ε)]
(418)	Orange-yellow	Isolated: *L. deliciosus* (82, 685), *L. deterrimus* (82); TLC: *L. salmonicolor*, *L. sanguifluus*[c], *L. semisanguifluus* (563)	0.85	244 (4.34), 278 (3.76), 424 nm (3.24) [cyclohexane]
(419)	Orange-yellow	Isolated: *L. deliciosus* (685), [A] (82), *L. deterrimus* [A] (82); TLC: *L. salmonicolor* [A], *L. sanguifluus*, *L. semisanguifluus* (563)	0.10	245 (4.30), ca. 270 (3.77), 426 nm (3.01) [cyclohexane]
(420)	Yellow	Isolated: *L. deliciosus* [A], *L. deterrimus* [A] (82)	–	229 (4.06), 265 (3.98), 298 (3.71), 435 nm (3.42) [hexane]
(421)	Blue	Isolated: *L. deliciosus* [A], *L. deterrimus* [A] (82)	–	239 (4.34), 290 (4.67), 370 (3.88), 580 nm (2.67) [ethanol]
(422) (Lactaroviolin)	Wine red	Isolated: *L. deliciosus* (78, 359, 399, 531, 532, 587, 588, 589, 716, 719, 720, 722), [A] (79, 82, 563, 685, 721), *L. indigo* (345); TLC: *L. sanguifluus* [A], *L. semisanguifluus* (563)	0.25	240 (4.40), 290 (4.80), 376 (3.80), 496 (2.68), 514 (sh., 2.76), 541 (2.85), 577 (2.82), 635 nm (2.43) [cyclohexane]
(423) (Lactarazulene)	Blue	Isolated: *L. deliciosus* (78, 587, 717), [A] (79, 82, 563, 685, 721); TLC: *L. deterrimus* [A], *L. semisanguifluus* [A] (563)	0.92	240 (4.70), 310 (4.70), 392 (4.20), 560 (2.53), 584 (2.61), 606 (2.68), 630 (2.63), 658 nm (2.63) [cyclohexane]
(425) (Lactarofulvene)	Orange	Isolated: *L. deliciosus* (USA) (83), *L. deliciosus* (Europe) [A] (685); TLC: *L. deliciosus* (Europe) [A] (563)	0.88	252 (3.98), 290 (3.67), 440 nm (2.76) [cyclohexane]
(424)	Blue	Isolated: *L. indigo* (345)	–	–

[a] The symbol [A] indicates that the compound has been shown to be an artefact of the isolation procedure in that species.
[b] Kieselgel G (cyclohexane:diethyl ether = 9:1) (563).
[c] *Lactarius sanguifluus* is reported to contain as its main genuine pigment a still unknown wine red compund ('lipophiles Lactaroviolin'), $R_f = 0.83$[b] (563).

	R¹	R²
(426)	H	H
(427)	H	OH
(428)	OH	OH

(429)

Table 42. *Occurrence of Lagopodins*[a]

Lagopodin-A (426)	*Coprinus cinereus* (*142*), *C. lagopus* (*111*)
Lagopodin-B (427)	*Coprinus cinereus* (*142*), *C. lagopus* (*111*)
Hydroxylagopodin-B (428)	*Coprinus cinereus* (*142*), *C. cinereus* var. *microsporus*
	(= *C. macrorhizus* var. *microsporus*) [mutant strain] (*114*)

[a] The lagopodins occur only in the cultured mycelium of these fungi.

Interestingly, the lack of a distinct carbonyl stretching band near 1750 cm^{-1} in the infra red spectra of lagopodin-B and hydroxylagopodin-B recorded in potassium bromide has been ascribed to the predominance of a hemiketal form in the solid state (*111, 114*). However, no evidence that lagopodin-B exists in such a form in deuteriochloroform solution could be gleaned from the ^{1}H-n.m.r. spectrum (*142*).

It was predicted by THOMSON (*655*) that the dimeric quinone, lagopodin-C, isolated from *C. lagopus* (*111*) would prove to be an artefact formed during work up of the culture fluids. This has subsequently been shown to be the case (*142*). In fact, both lagopodin-A and lagopodin-B are quite unstable in aqueous solution at neutral or slightly alkaline pH. The quinone (427), for example, has a half-life of only 12 h at pH 6.8–7.6 when incubated in the sterile media used for the cultures of *C. cinereus* (*142*). It can be assumed then that many of the pigments observed in the total extracts of these *Coprinus* species arise by nonenzymic reactions of the parent quinones (426) and (427) in the aerobic aqueous culture media.

The lagopodins are closely related to helicobasidin (429), a pigment from *Helicobasidium mompa*, the biosynthesis of which has been extensively studied (*81, 655*).

Pleurotin (430) (Scheme 68) was first isolated from cultures of *Hohenbuehelia grisea* (= *Pleurotus griseus*) in 1947 (*545*). However, it was not until much later that the structure was elucidated on purely chemical grounds following elegant studies by ARIGONI's group. Some of the many interesting chemical transformations which this versatile molecule and its derivatives can be induced to undergo are summarised in Scheme 68 (pp. 190–191).

Following meticulous analysis of the spectroscopic data of each compound in Scheme 68, and others, the structure and relative stereochemistry of pleurotin could be deduced (559). The structure (430) was subsequently confirmed by X-ray analysis of the derived p-bromobenzoate (431) of leucopleurotin (432) (187), and differs in stereochemical detail from that proposed for pleurotin by others (302).

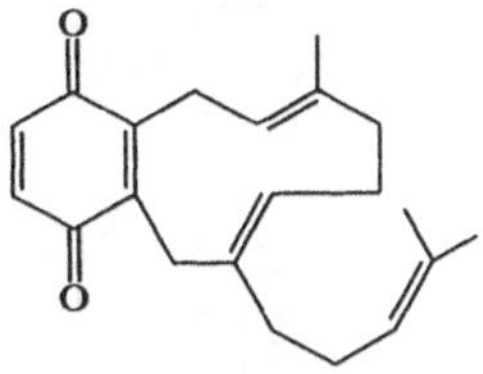

(434)

The structure (430) suggests a biosynthesis of pleurotin from a farnesylhydroquinone such as (433) involving a series of cyclisation, rearrangement and oxidation steps (30). A minor metabolite of *H. grisea*, buehelin (434), possesses a structure in which the initial cyclisation of the farnesyl side chain to the nucleus in (433) has taken place. Structural comparisons suggest that buehelin (434) is generated from the hydroxyfarnesylquinone shown in Scheme 69 and that it plays a central role in the biosynthesis of pleurotin (430) (31).

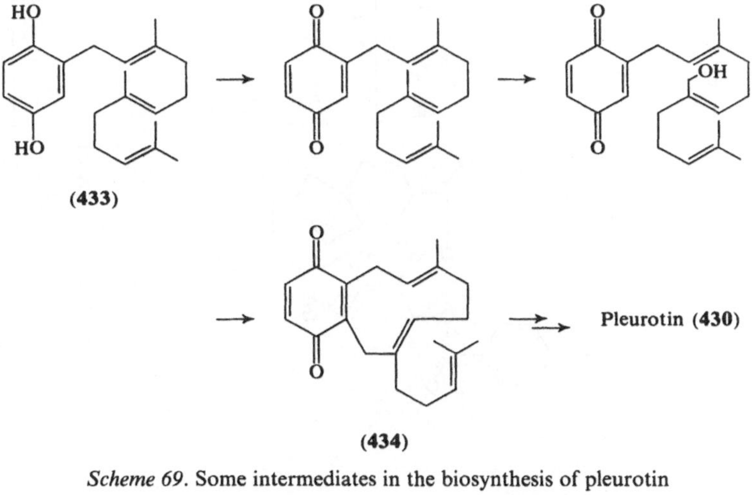

(433)

Pleurotin (430)

(434)

Scheme 69. Some intermediates in the biosynthesis of pleurotin

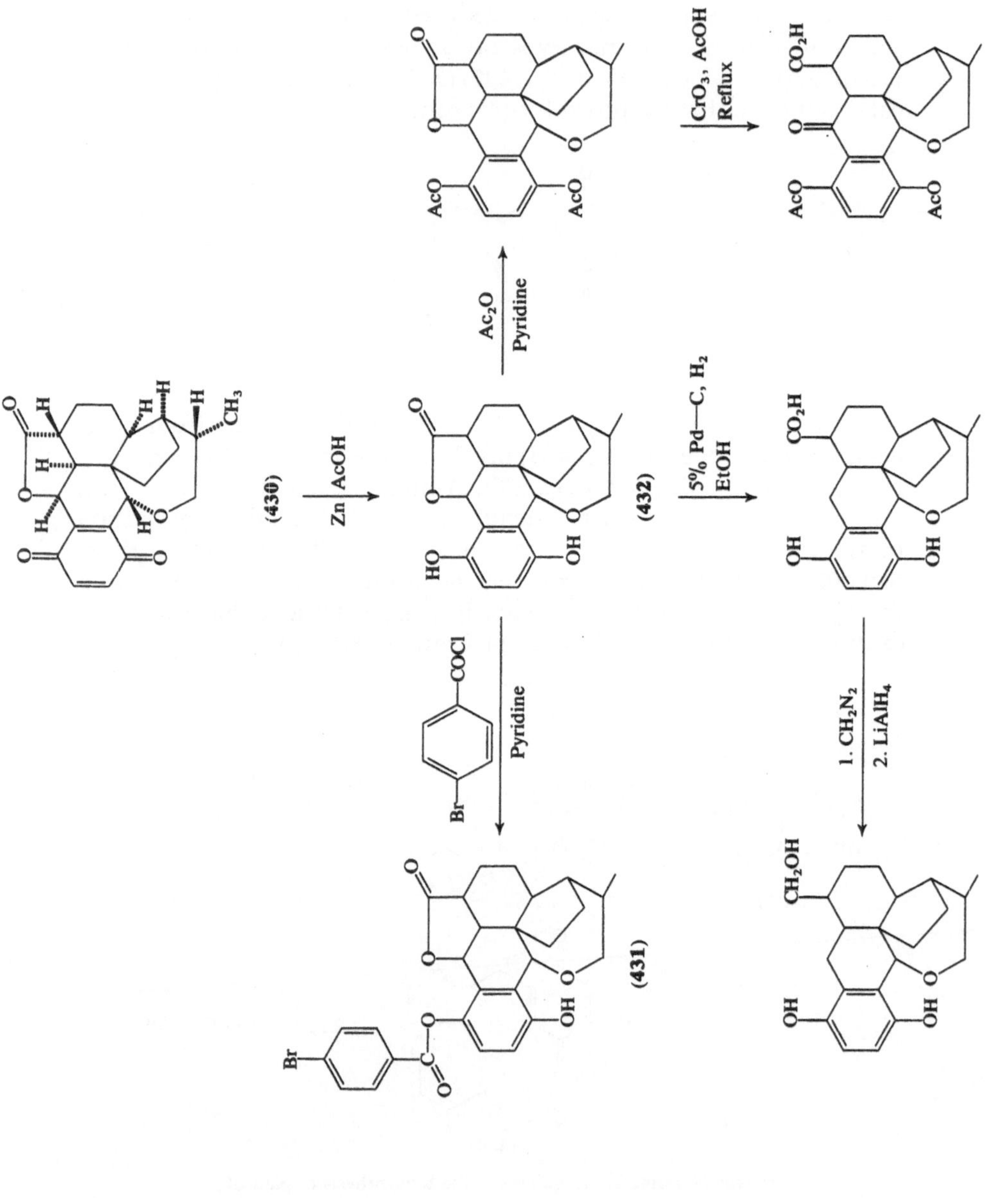

(430)
Zn | AcOH
(432)
Ac2O
Pyridine
CrO3, AcOH
Reflux
5% Pd—C, H2
EtOH
Br—C6H4—COCl
Pyridine
1. CH2N2
2. LiAlH4
(431)
AcO
HO
OH
CO2H
CH2OH
CH3
Br

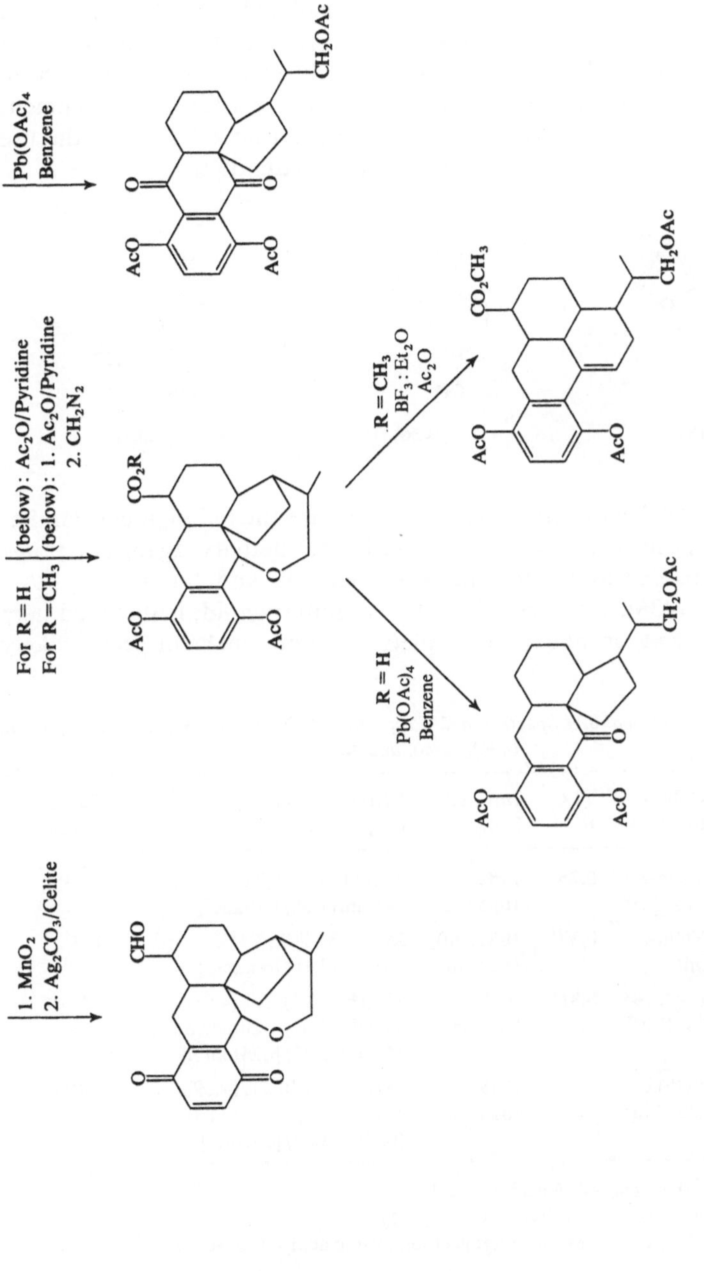

Scheme 68. Some chemistry of pleurotin

The farnesylhydroquinone (433) is reminiscent of the prenylated phenols, e.g. cristatic acid (251) and scutigeral (253), produced by species of *Albatrellus* and discussed in Section 3.1.

'Géogénine', an antibiotic compound isolated from *Hohenbuehelia geogenia* has been assigned, unwittingly, precisely the same structure as pleurotin (*172*). This metabolite is clearly identical with pleurotin, and since the name geogenin had been given earlier to the pigment (219) isolated from *Hydnellum geogenium* (Section 2.3.2) (*611*) the use of the name in the present context becomes redundant.

(435)   (436)   (437)

Cultures of *Lentinellus omphalodes* produce the red pigment omphalone (435) which exhibits marked antibiotic activity against a range of Gram-positive and Gram-negative bacteria and fungi (*490, 491*). Omphalone (435), while clearly not a sesquiterpenoid, is included here since its biogenesis like that of pleurotin and buehelin seems likely

Table 43. *Some Physical and Spectroscopic Properties of Pleurotin* (430) *and the Pigments* (434), (435), *and* (437)

Pigment	Colour M.p.	TLC R_f	Infrared $[\nu_{co}]$	Ultraviolet/Vis. $[\lambda_{max}(\log \varepsilon)]$	References
Pleurotin (430)	Yellow 218–221°	0.35[a]	1782, 1665 cm^{-1}	248(4.08), 320(2.65), 445 nm(1.38) [ethanol]	(*559*)
Buehelin (434)	Yellow oil	0.37[b]	1655(sh.), 1650 cm^{-1}	251(4.28), 295(3.41), 335 nm(3.18) [ethanol]	(*31*)
Omphalone (435)	Dark red 98–100°	0.81[c]	1660, 1640 cm^{-1}	213(sh., 4.12), 258(4.08), 284(sh., 3.87), 320(sh., 3.52), 445 nm(3.48) [methanol]	(*490, 491*)
Frustulosin (437)	Yellow 139–140°	–	1645, 1615 cm^{-1}	237(sh., 4.17), 249(4.29), 273(sh., 3.82), 293(3.83), 402 nm(3.83) [ethanol]	(*505*)

[a] Kieselgel HF (benzene:ethyl acetate = 4:1).
[b] Kieselgel 60 F$_{254}$ (hexane:ethyl acetate = 10:1).
[c] Kieselgel 60 F$_{254}$ (cyclohexane:ethyl acetate:formic acid = 120:40:5).

References, pp. 253–286

to involve a prenylated hydroquinone. A putative precursor (**436**) to omphalone has been found in cultures of *Psatyrella gracilis* (*491*).

Structurally related to omphalone is the yellow aldehyde frustulosin (**437**) which occurs in cultures of *Xylobolus frustulatus* (= *Stereum frustulosum*) (*505*). Frustulosin, which is active against *Staphylococcus aureus* and several other bacteria at very low concentrations (16 p.p.m.), was originally formulated as an unusual benzofulvene but subsequent work resulted in revision to the acetylenic structure (**437**) which has been confirmed by synthesis (*547*).

Mushrooms and related Basidiomycetes are adept in the manufacture of unique sesqui- and diterpenoids. Since the vast majority of these interesting substances are colourless they lie beyond the scope of this article. However, the interested reader is directed to the pertinent review by AYER and BROWNE (*54*).

4.2. Carotenoids

From the early work of BACHMANN (*57, 58*), ZOPF (*732*), KOHL (*435*), HEIM (*360*), and VAN WISSELING (*683*) it became known that a number of fungi contain carotenoids, but it was left to later workers to identify these pigments more accurately. Using a combination of chromatographic and spectroscopic techniques carotenoids have now been detected and characterised from many species of fungi (Table 44). Their occurrence in a relatively small number of families, however, has rendered them a valuable aid to the systematic taxonomy of several otherwise difficult groups of Ascomycetes (*35, 36, 362, 674, 678*) and Basidiomycetes (*34, 230, 241, 361, 362, 674*).

Surveys of the early literature on fungal carotenoids are given in the reviews by HAXO (*355*) and GOODWIN (*298, 300, 301*) and in the introduction to the paper by FIASSON, LEBRETON and ARPIN (*239*). The citation of many review articles dealing with carotenoids, including fungal carotenoids, may be found in the compendium compiled by STRAUB (*633*).

It is unfortunate that much of the early work on identification of fungal carotenoids relied almost entirely on comparison of electronic spectra and chromatographic behavior with those of known carotenoids. Not unexpectedly, the subsequent application of more sophisticated methods of purification and identification has revealed that total reliance on the older criteria had led to misidentification of some pigments (for examples, see ref. *674*). There exists an outstanding need for reinvestigation of many carotenoid producing fungi using state-of-the-art techniques of chromatography and spectroscopy.

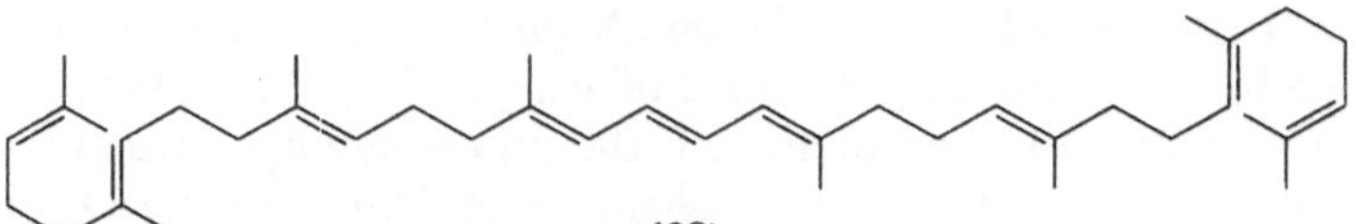

(438)

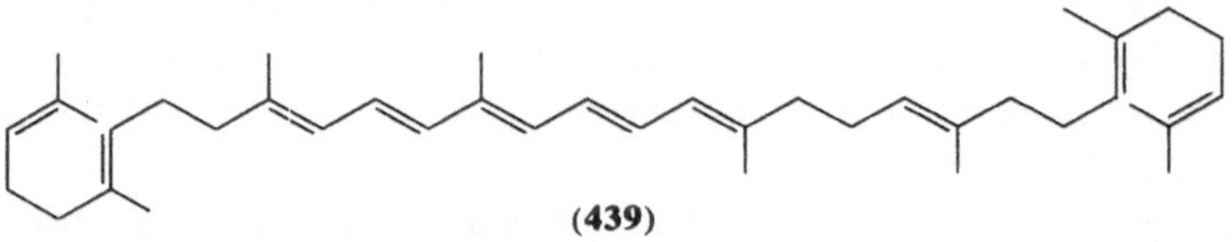

(439)

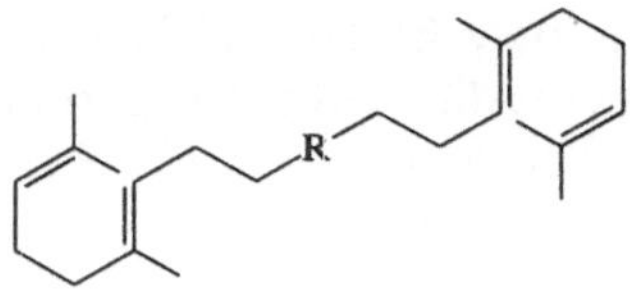

(440)

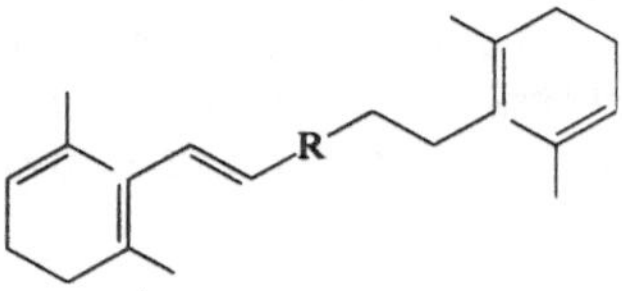

(441)

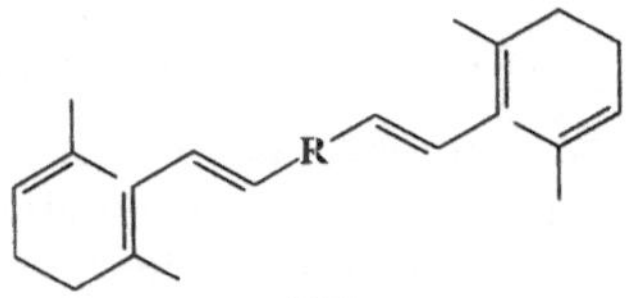

(442)

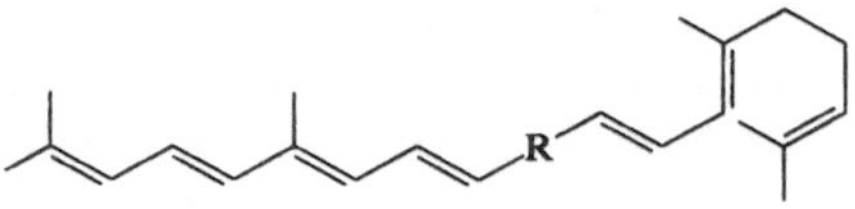

(443)

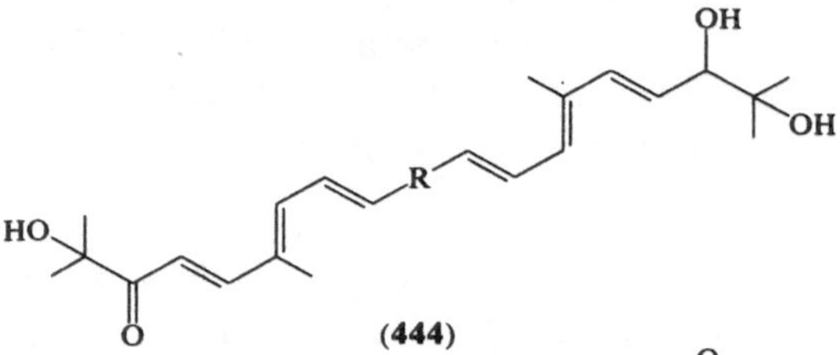

(444)

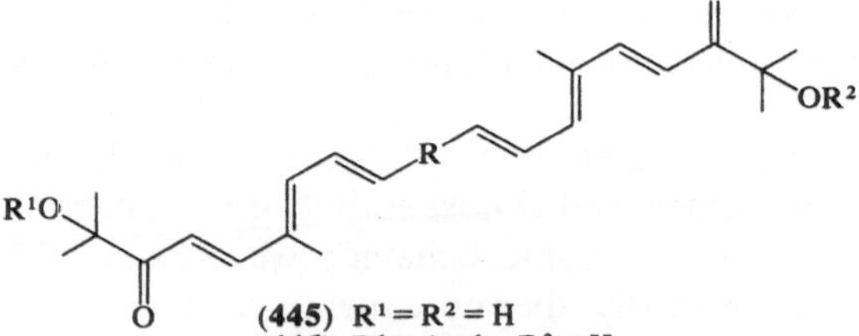

(445) $R^1 = R^2 = H$
(446) $R^1 = Acyl, \quad R^2 = H$
(447) $R^1 = R^2 = Acyl$

In formulae (440)–(447) R represents

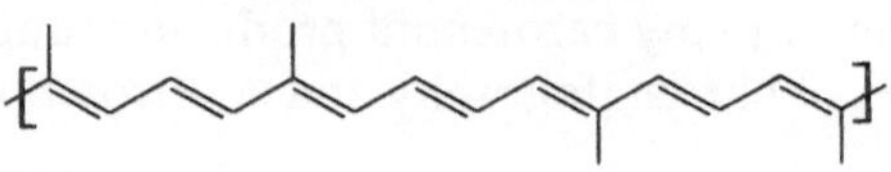

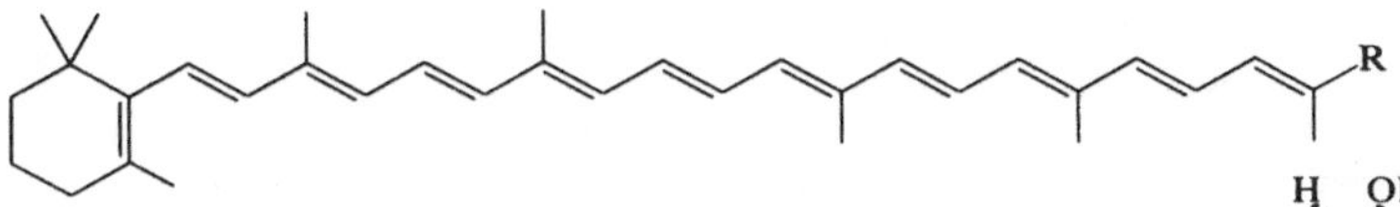

(448) R =

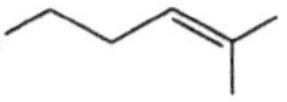

(449) R =

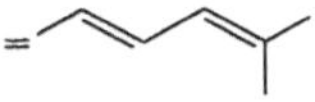

(450) R =

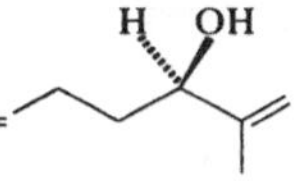

(451) R =

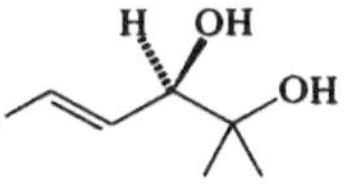

(452) R =

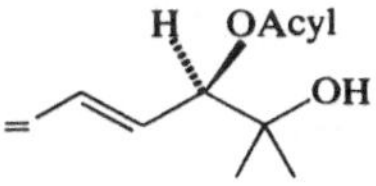

(453) R =

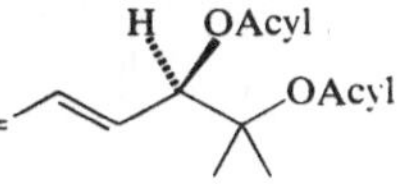

(454) R =

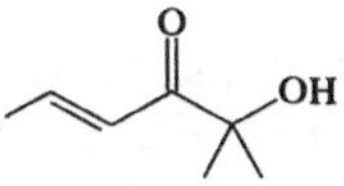

(455) R =

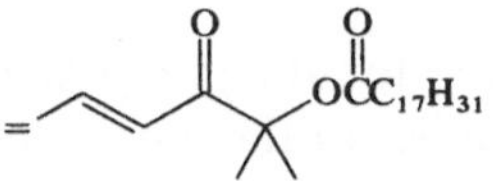

(456) R =

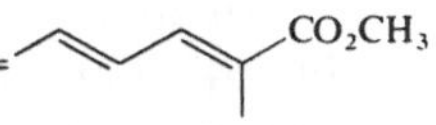

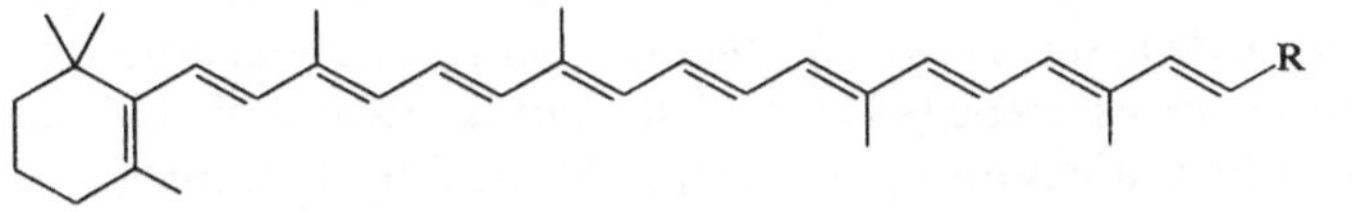

(457) R =

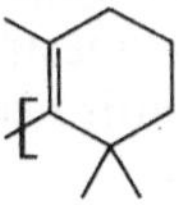

(458) R =

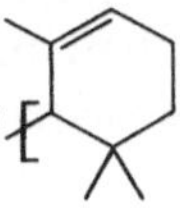

(459) R =

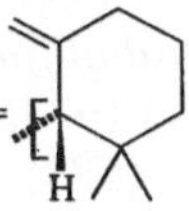

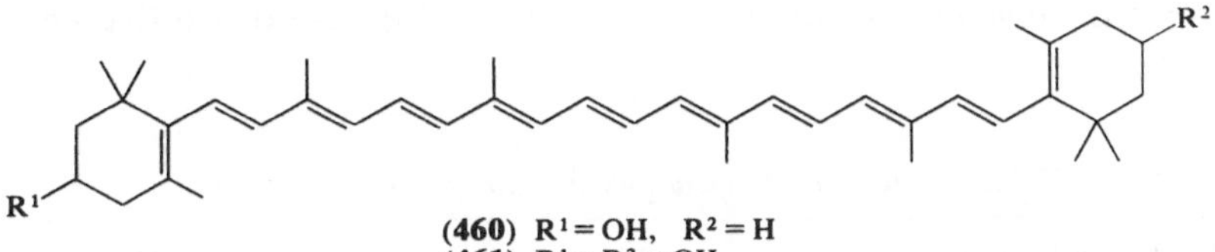

(460) R¹ = OH, R² = H
(461) R¹ = R² = OH

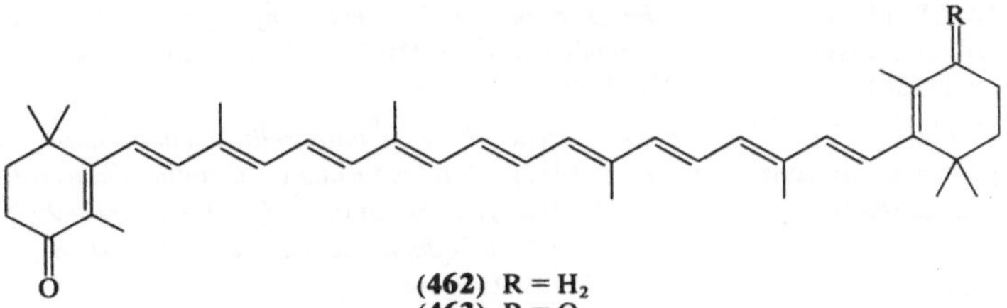

(462) R = H₂
(463) R = O

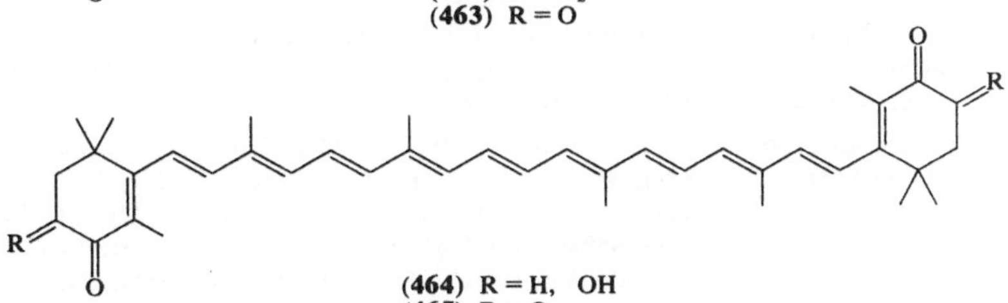

(464) R = H, OH
(465) R = O

(466) R = H
(467) R = CH$_3$

(468)

In many fungi the predominant pigment is β-carotene (457) (Table 44). Thus, the fruit bodies of the 'chanterelle' (*Cantharellus cibarius*) contain β-carotene together with lesser amounts of lycopene (442), and γ- (448) and α-carotene (458). Related species contain lycopene [*Cantharellus melanoxeros* (= *C. ianthinoxanthus*)] or equal amounts of lycopene and neurosporene (441) (*C. lutescens*) as their main colouring matters. The identification by TURIAN (666) and by FIASSON (230) of neurosporene (441) as the major pigment of *Cantharellus tubaeformis* (= *C. infundibuliformis*), which had been seriously questioned on several occasions (674, 677), has subsequently been confirmed by mass spectrometry (678).

Carotenoids present in the orange-red American toadstool, *Cantharellus cinnabarinus*, are composed almost entirely of canthaxanthin (463).

Table 44. *Occurrence of Carotenoid Pigments in Macromycetes*[a, b]

Acyclic Carotenoids

7,8,11,12,7',8',11',12'-Octahydro-ψ,ψ-carotene (Phytoene) (438)	*Anthurus archeri* (177), *Dacrymyces stillatus* [trace] (299), *Iodophanus carneus* [10.5%] (678), *Sphaerobolus stellatus* [10%] (12)
7,8,11,12,7',8'-Hexahydro-ψ,ψ-carotene (Phytofluene) (439)	*Anthurus archeri* (177), *Cantharellus cinnabarinus* (241, 354), *C. friesii* (241), *C. tubaeformis* (= *C. infundibuliformis*) (234, 239, 241), *Dacrymyces capitatus* (= *D. ellisii*) (342), *D. stillatus* [1.2%] (299), *Iodophanus carneus* [0.7%] (678), *Sphaerobolus stellatus* [10%] (12)
7,8,7',8'-Tetrahydro-ψ,ψ-carotene (ζ-Carotene) (440)	*Anthurus archeri* (177), *Cantharellus tubaeformis* [2.8–4.8%] (239, 677), *Dacrymyces capitatus* (342), *D. stillatus* [19%] (299), *Iodophanus carneus* [0.6%] (678)
7,8-Dihydro-ψ,ψ-carotene (Neurosporene) (441)	*Aleuria aurantia* [4.3%] (675), *Anthracobia melaloma* [10%] (36), *Anthurus archeri* (= *A. aseroëformis*) (177, 239), *Caloscypha fulgens* (35), *Cantharellus friesii* [0.7%] (241), *C. lutescens* [45%] (241), (234, 666, 718), *C. melanoxeros* (= *C. ianthinoxanthus*) [26%] (231), *C. tubaeformis* [74%] (241), (234,

Table 44 (*continued*)

	239, 666), *Craterellus cornucopioides* [60%] (*241*), *Iodophanus carneus* [0.8%] (*678*), *Scutellinia umbrarum* [1%] (*36*)
ψ,ψ-Carotene (Lycopene) (**442**)	*Aleuria aurantia* [0.2–1.0%] (*457*), [1.7%] (*675*), *Anthurus archeri* (*177, 213, 665*), *Calocera viscosa* (*666*), *Caloscypha fulgens* (*36*), *Cantharellus cibarius* (*241, 718*), *C. cibarius* var. *pallidifolius* (*241*), *C. lutescens* [50%] (*241*), (*234, 666, 718*), *C. melanoxeros* [74%] (*231*), *C. tubaeformis* [8.8%] (*241*), [5–11%] (*677*), (*234, 239, 666, 718*), *Clathrus ruber* [76%] (*240*), *Cookeina sulcipes* [4%] (*35*), *Craterellus cornucopioides* [40%], *C. fallax* [18.7%], *C. odoratus* [12.4%] (*241*), *Gerronema chrysophyllum* (= *Omphalia chrysophylla*) [0.5%]' (*236*), *Leucoscypha rutilans* [2%] (*36*), *Lycogala epidendron* [18%] (*457*), *Mutinus caninus* (*215*), *Phallus rugulosus* (*343*), *Phillipsia carminea* [7%] (*35, 47*), *P. subpurpurea* [5%] (*35*), *Sphaerobolus stellatus* [15%] (*12*)
3,4-Didehydro-ψ,ψ-carotene (3,4-Dehydrolycopene) (**443**)	*Aleuria aurantia* [1%] (*457*), [0.8%] (*675*), (*137*), *Caloscypha fulgens* (*36*), *Lycogala epidendron* [3%] (*457*)
Dihydro-2′-phillipsiaxanthin (**444**)	*Phillipsia carminea* [in the form of di- and triacyl derivatives: diacyl, 2%: triacyl, 3%] (*47*)
Phillipsiaxanthin (**445**)	*Cookeina sulcipes* (*35, 48*)
Phillipsiaxanthin-1-*O*-acyl derivative (**446**)	*Cookeina sulcipes* (*35*), *Phillipsia carminea* [6%] (*35, 47*), *P. carnicolor* [30%] (*35*)
Phillipsiaxanthin-1,1′-di-*O*-acyl derivative (**447**)	*Cookeina sulcipes* [75%] (*48*), [60%] (*35*), *C. tricholoma* (*35*), *Phillipsia carminea* [74%] (*47*), *P. carnicolor* [70%], *P. subpurpurea* [80%] (*35*)
β,ψ-Carotene (γ-Carotene) (**448**)	*Aleuria aurantia* [39–50%] (*457*), [36%] (*36, 37*), [27.6%] (*673*), [15.5%] (*675*), (*137, 450*), *A. rhenana* [37%], *Anthracobia melaloma* [8%], *Caloscypha fulgens* [8%] (*36*), *Cantharellus cibarius* (*241, 718*), *C. cibarius* var. *pallidifolius* [7%], *C. friesii* [1.7%] (*241*), *Cheilymenia crucipila* [98%], *C. theleboloides* [90%] (*36*), *Clavulinopsis aurantio-cinnabarina* [1.5–5%] (*241*), *Coprobia granulata* [73%] (*36*), *Craterellus fallax* [21.3%], *C. odoratus* [12.3%] (*241*), *Dacrymyces stillatus* [0.1%] (*299*), *Dasyscyphus bicolor* (*35*), *Geopyxis maialis* (*36*), *Gerronema chrysophyllum* [11%] (*236*), *Haasiella venustissima* (= *Clitocybe venustissima*) (*34*), *Helotium subcorticale* [6%] (*35*), *Iodophanus carneus* [2.2%] (*678*), *Leotia lubrica* [12%] (*35*), *Leucoscypha rutilans* [35%] (*36*), *Lycogala epidendron* [27%] (*457*), *Melastiza chateri* [21%], *M. greletii* [25%] (*36, 37*), *Microglossum olivaceum* [73%], *Mitrula paludosa* [1%] (*35*), *Mutinus caninus* (*215*), *Nectria cinnabarina*, perfect form [1.2%], imperfect form [0.8%] (*237*), *Octospora leucoloma* [17%], *O. rubricosa* [17%] (*36*), *Peniophora aurantiaca* [2%] (*46*), *Phyllotopsis nidulans* [29%] (*230a*), *Pulvinula constellatio* [12%], *Scutellinia ampullaceae* [95%], *S. ar-*

Table 44 (*continued*)

	enosa [90%] (*36*), *S. scutellata* [92%] (*37*), [91%] (*457*), *S. scutellata* var. *cervorum* [94%], *S. setosa* [20%], *S. superba* [94%], *S. trechispora* [93%], *S. umbrarum* [84%], *Sowerbyella radiculata* [4%], *S. unicolor* [6%] (*36*), *Sphaerobolus stellatus* [15%] (*12*), (*564*), *Stereum hirsutum* [62%] (*50*)
3′,4′-Didehydro-β,ψ-carotene (Torulene) (**449**)	*Aleuria aurantia* [trace], *A. rhenana* [3%], *Caloscypha fulgens* (*36*), *Clavulinopsis aurantio-cinnabarina* (*241*), *Dacrymyces stillatus* [9.7%] (*299*), *Geopyxis carbonaria* [50%], *G. maialis* (*36*), *Gerronema chrysophyllum* [1.5%] (*236*), *Lycogala epidendron* [13%] (*457*), (*450*), *Scutellinia ampullaceae* [3%], *S. arenosa* [8%] (*36*), *S. scutellata* [6%] (*37, 457*), *S. scutellata* var. *cervorum* [6%], *S. superba* [3%], *S. trechispora* [6%], *S. umbrarum* [7%], *Sowerbyella unicolor* [2%] (*36*)
Aleuriaxanthin (**450**)[c]	*Aleuria aurantia* [22–25%] (*457*), [21.5%] (*675*), [20%] (*36, 37*), *Melastiza chateri* [47%] (*37*), *M. greletii* [47%] (*36, 37*)
Plectaniaxanthin (**451**)	*Leucoscypha rutilans* [7%], *Octospora leucoloma* [54%], *O. rubricosa* [58%] (*36*), *Pithya vulgaris* (*35*), *Sarcoscypha coccinea* (= *Plectania coccinea*) [6%] (*49*), [3%] (*35*)
Plectaniaxanthin-2′-O-acyl derivative (**452**)	*Sarcoscypha coccinea* [6%] (*35, 49*)
Plectaniaxanthin-1′,2′-di-O-acyl derivative (**453**)	*Pithya vulgaris* (*35*), *Pulvinula constellatio* [6%] (*36*), *Sarcoscypha coccinea* [47%] (*35, 49*)
2′-Dehydro-plectaniaxanthin (**454**)	*Aleuria aurantia* [0.5%] (*36*), (*37, 137*), *Clavaria helicoides* var. *robusta* (*241*)
2′-Dehydro-plectaniaxanthin-1′-O-linoleate (**455**)	*Aleuria aurantia* [4%] (*36, 37*), (*137*), *Melastiza chateri* [10%] (*36, 37*), *M. greletii* [3%] (*37*), [2%] (*36*), *Phillipsia carminea* [9%] (*35, 47*), *Sarcoscypha coccinea* [20%] (*35, 49*)
Torularhodin methyl ester (**456**)	*Cookeina sulcipes* [25%] (*48*), [20%] (*35*), *Phillipsia subpurpurea* [3%] (*35*)
Bicyclic Carotenoids	
β,β-Carotene (β-Carotene) (**457**)	*Aleuria aurantia* [43.9%] (*675*), [38%] (*36, 37*), [26–34%] (*457*), (*450*), *A. rhenana* [53%], *Anthracobia melaloma* [80%] (*36*), *Bisporella citrina* (= *Calycella citrina*) [95%] (*35*), (*666*), *Calocera viscosa* (*666*), *Caloscypha fulgens* [70%] (*36*), (*41*), *Cantharellus cibarius* (*241, 718*), *C. cibarius* var. *pallidifolius* [67.1%], *C. cinnabarinus* [2.3%], *C. friesii* [43.9%], *C. minor* [57.6%] (*241*), *C. tubaeformis* [0.6–1.7%] (*677*), *Clathrus ruber* [24%] (*240*), *Clavaria helicoides* var. *robusta* [80%], *Clavulinopsis aurantio-cinnabarina* [84.5–95%] (*241*), *Coprobia granulata* [18%] (*36*), *Craterellus fallax* [49.5%], *C. odoratus* [60.2%] (*241*), *Dacrymyces capitatus* (*342*), *D. stillatus* [39.5%] (*299*), *Dasyscyphus bicolor* (*35*), *Geopyxis maialis* (*36*), *Gerronema chrysophyllum* [87%] (*236*), *Haasiella venustissima* (*34*), *Helotium subcorticale* [88–92%], *Lachnellula calycina* (= *Trichoscyphella calycina*) [100%], *L. gallica* (= *Trichoscyphella gallica*) [88%], *Leotia lubrica* [81%] (*35*),

Table 44 (*continued*)

	Leucoscypha rutilans [51%] (*36*), *Lycogala epidendron* [39%] (*457*), (*450*), *Melastiza chateri* [20%], *M. greletii* [23%] (*36, 37*), *Microglossum olivaceum* [23%], *Mitrula paludosa* [96%] (*35*), *Mutinus caninus* (*215*), *Nectria cinnabarina*, perfect form [18.2%], imperfect form [5.2%] (*237*), *Octospora leucoloma* [28%], *O. rubricosa* [24%] (*36*), *Peniophora aurantiaca* [26%] (*46*), *Phallus rugulosus* (*343*), *Phyllotopsis nidulans* [58%] (*230a*), *Pithya vulgaris* (*35*), *Pulvinula constellatio* [58–66%] (*36*), *Sarcoscypha coccinea* [24%] (*35, 49*), *Scutellinia setosa* [80%], *Sowerbyella radiculata* [62%], *S. unicolor* [73%] (*36*), *Sphaerobolus stellatus* [40%] (*12*), (*13, 564*), *Stereum hirsutum* [33%] (*50*), *Tremella mesenterica* (*450*)
β,ε-Carotene (α-Carotene) (**458**)	*Aleuria aurantia* [0.3%] (*675*), *Cantharellus cibarius* (*718*), *Dacrymyces stillatus* [7%] (*299*)
β,γ-Carotene (**459**)	*Caloscypha fulgens* [8%] (*36, 41*), *Cantharellus cibarius* var. *pallidifolius* [5.2%], *C. minor* [22.7%] (*241*), *Mitrula paludosa* [3%] (*35*), *Sowerbyella radiculata* [11%], *S. unicolor* [14%] (*36*)
Cryptoxanthin (**460**)	*Calocera viscosa* (*666*), *Dacrymyces stillatus* [11.3%] (*299*), *Sphaerobolus stellatus* [5%] (*12*)
Zeaxanthin (**461**)	*Dacrymyces stillatus* [12.3%] (*299*)
Echinenone (**462**)	*Cantharellus cinnabarinus, C. friesii* (*241*), *Peniophora aurantiaca* (*46, 239*), *Phyllotopsis nidulans* [8%] (*230a*)
Canthaxanthin (**463**)	*Cantharellus cinnabarinus* [91%] (*241*), (*354*), *C. friesii* [9.1%] (*241*), *Sphaerobolus stellatus* [5%] (*12*)
Astaxanthin (**464**)	*Peniophora aurantiaca* (*46, 239*), *Phyllotopsis nidulans* [3.5%] (*230a*)
Astacin (**465**)	*Peniophora aurantiaca* (*46, 239*)
Apo-carotenoids	
Neurosporaxanthin (**466**)	*Iodophanus carneus* [33.2%] (*678*), *Nectria cinnabarina*, perfect form [37%], imperfect form [65%] (*237*)
Neurosporaxanthin methyl ester (**467**)	*Iodophanus carneus* [47%, plus 2.2% of a *cis*-isomer] (*678*), *Nectria cinnabarina*, perfect form [43.6%], imperfect form [29%] (*237*)

[a] The identification of carotenoids has been based, in some cases, on chromatographic or limited spectroscopic comparison with data for known pigments. When these criteria alone have been applied, then the presence or absence of a particular carotenoid must be regarded as not rigorously proved.

[b] Figures in parentheses refer to the proportion of the total carotenoids extracted.

[c] Present in the most part in the form of esters incorporating several saturated and unsaturated fatty acids (*44, 457*).

In the apothecia of the 'orange peel' fungus *Aleuria aurantia* β-
and γ-carotene are accompanied as major pigments by aleuriaxanthin
(**450**), a xanthophyll possessing an end group which is unique among
naturally occurring carotenoids (*457*). Aleuriaxanthin occurs in *A. aurantia* as a mixture of esters formed with C_{25}-saturated hydroxy, C_{21}-
monounsaturated, C_{20}-triunsaturated, C_{18}-diunsaturated, and C_{16}-sat-
urated fatty acids (*44*). The structure of the free xanthophyll (**450**)
followed principally from the electronic and ^{1}H-n.m.r. spectra of the
pigment and its acetate which identified the β,ψ-carotene chromophore
and located the hydroxy group at C-2′ in the end group. The gross
structure (**450**) so derived was confirmed by total synthesis of racemic
aleuriaxanthin commencing with a novel photosensitised auto-oxida-
tion of linalool (Scheme 70) (*413*).

The (2 *R*)-absolute configuration of natural aleuriaxanthin was first
deduced by application of a modified Horeau technique (*137*) and has
subsequently been corroborated by a stereospecific total synthesis of
the pigment in both enantiomeric modifications (*220*). The natural enantiomer (**450**) was produced from the optically active epoxygeraniol
derivative (**469**), itself obtained from geraniol acetate by allylic hydroxylation followed by Sharpless asymmetric epoxidation (Scheme 71)
(*220*).

The presence of rubixanthin (**470**) among the minor carotenoids
of *Aleuria aurantia* has been described (*450, 673*) and apparently confirmed, at least in British specimens of the fungus (*675*). However,
the identity of this pigment with rubixanthin has been seriously chal-
lenged (*36, 457*) and, in the absence of firm spectroscopic data, its
structure must remain in doubt.

(**470**)

High performance liquid chromatography has revealed the presence
in the mixture of carotenoids produced by *Aleuria aurantia* of a predominant (*Z*)-aleuriaxanthin of as yet undetermined structure (*220*).

Lycopene (**442**) is responsible for the intense red colour of several
visually striking fungi belonging to the Phallales, e.g. *Clathrus ruber,
Mutinus caninus, Phallus rugulosus* and *Anthurus archeri* (= *A. aseroëformis*). In the developing fruit bodies of *A. archeri* lycopene is accompanied by its biosynthetic precursors of the Porter-Lincoln series [phy-
toene (**438**), phytofluene (**439**), ζ-carotene (**440**), and neurosporene

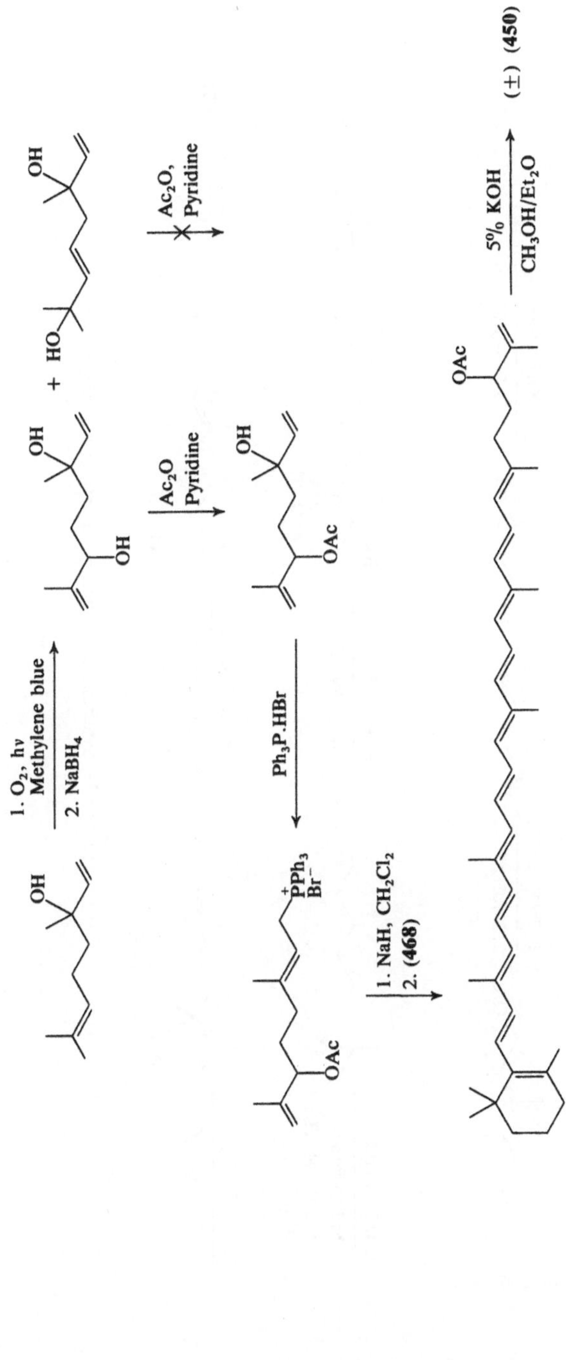

Scheme 70. Synthesis of (±)-aleuriaxanthin from linalool

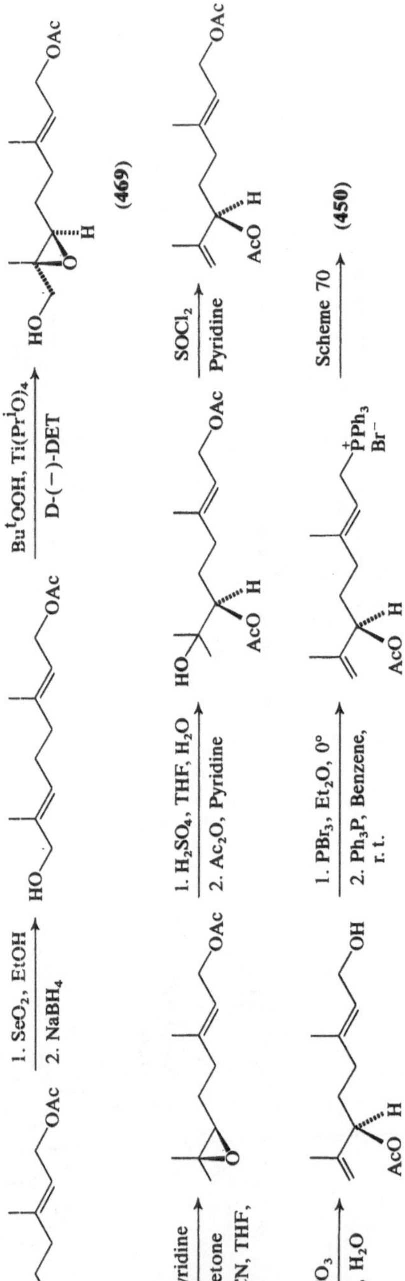

Scheme 71. Synthesis of the natural enantiomer of aleuriaxanthin

(441)] in proportions which vary with the maturity of the sporophores (*177*).

The beautiful colours of some cup fungi are partially caused by more unusual carotenoids. Thus, plectaniaxanthin (**451**) and 2'-dehydroplectaniaxanthin (**454**) are present in the brilliant red cups of *Sarcoscypha coccinea* (= *Plectania coccinea*) (*35, 49*), where they occur mainly in esterified form, e.g. as the linoleates (*672*). The chemical structures of these carotenoids and their esters were established by chemical and physical methods. During the degradative studies on the naturally occurring diacyl derivatives of plectaniaxanthin it was noted that alkaline hydrolysis of the secondary and tertiary ester groups in (**453**) proceeded with almost equal facility, thereby allowing no accumulation of a monoacylated intermediate. This interesting observation has been rationalised by assuming a base catalysed intramolecular acyl group migration as depicted in Scheme 72.

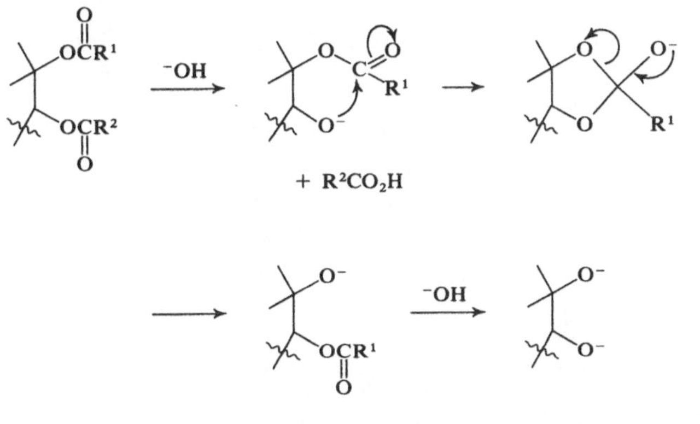

Scheme 72. Neighbouring group participation in alkaline hydrolysis of plectaniaxanthin-1',2'-di-*O*-acyl derivatives

Chiroptical comparison of plectaniaxanthin acetonide with (2' *S*)-16',17'-dinorplectaniaxanthin acetonide (**472**) was correctly taken as proof of (2' *R*) chirality for natural plectaniaxanthin (**451**) and its esters (**452**) and (**453**). The synthesis of the chiral model compound (**472**), which in the event proved to be enantiomeric with the acetonide of the natural product, was effected from D-mannitol *via* 2,3-*O*-isopropylidene-D-glyceraldehyde (**471**) as the key synthon (Scheme 73) (*548*).

The optically active aldehyde (**473**), available from L-serine, has been used for the synthesis of (2' *S*)-plectaniaxanthin (*197*). The strategy followed very closely that shown in Scheme 73 and the CD spectrum

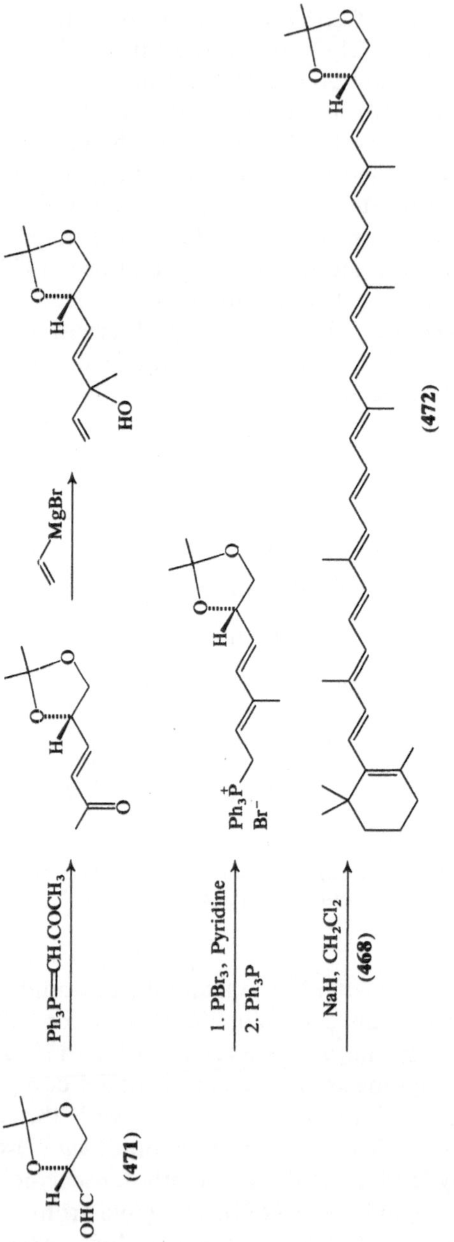

Scheme 73. Synthesis of chiral reference compound (472)

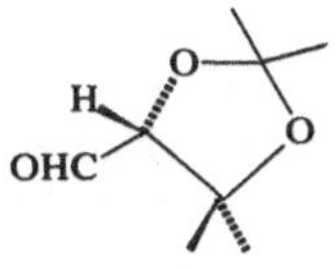

(473)

of the final product confirmed the absolute configuration of the natural pigment as (2′ R).

β,γ-Carotene **(459)**, the first naturally occurring carotenoid possessing a terminal methylene group, was isolated as a minor component from the yellow-orange cups of *Caloscypha fulgens* (41). The novel structure was deduced from the spectroscopic data and confirmed by synthesis of the pigment **(459)** in racemic form starting from (±)-γ-ionone (Scheme 74) (20, 21). When (+)-γ-ionone, partially resolved *via* its menthydrazone derivative, was used in the synthetic transformations depicted in Scheme 75, β,γ-carotene enriched in the (6′ R)-enantiomer **(474)** was obtained (340). By CD correlation with β,γ-carotene from *Caloscypha fulgens*, the (6′ S)-stereochemistry could be demonstrated for the natural product. Interestingly, the chiralities of the fungal carotenoids **(451)** and **(459)** are opposite to those encountered in analogous bacterial and algal carotenoids (459).

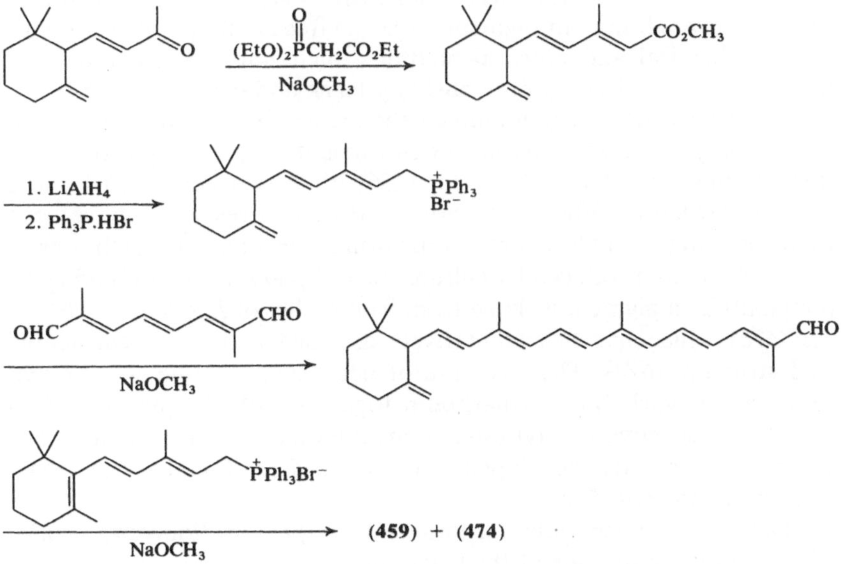

Scheme 74. Synthesis of racemic β,γ-carotene

β,γ-Carotene (459) exhibits very similar absorption properties and an electronic spectrum identical with that of α-carotene (458). It is probable, therefore, that the two pigments have been confused in the older literature (674). The pigment P444, detected in several fungi by ARPIN (35, 241), is identical with β,γ-carotene (459) (41).

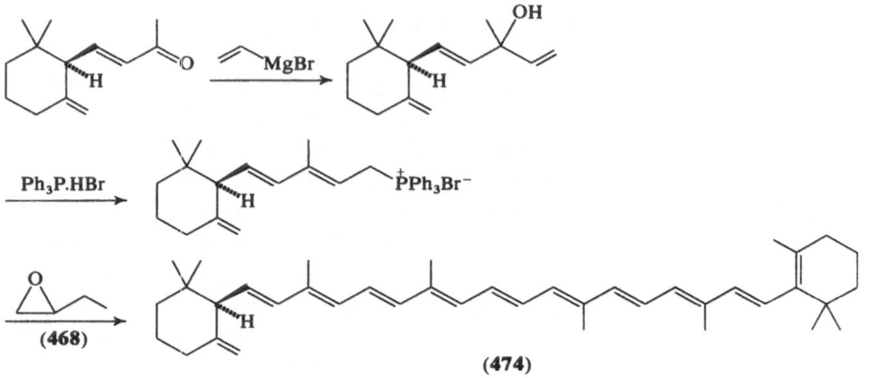

Scheme 75. Synthesis of optically active (6' R)-β,γ-carotene

Red fruit bodies of Phillipsia carminea collected in the Central African Republic contain the diester derivative (447) of phillipsiaxanthin (445) as the predominant pigment (47). The free xanthophyll was identified by chemical and physical methods, including direct comparison with material synthesised previously by ISLER and coworkers (572).

The diester (447) also dominates the pigmentation in a second African discomycete, Cookeina sulcipes, in which it occurs along with significant amounts of the methyl ester (456) of the carotenoid carboxylic acid torularhodin. Other carotenoid acid derivatives which have been found in fungi include neurosporaxanthin (466) and its methyl ester (467), which are produced by cultures of Iodophanus carneus, and laetiporaxanthin, a pigment isolated from fruit bodies of Laetiporus sulphureus (676) which has been tentatively assigned a β-apo-8'-carotenoic acid structure (674). The structure of neurosporaxanthin was known from earlier work (1). Torularhodin together with the pigment (466) and their respective methyl ester derivatives have all been synthesised using what has now developed into standard methodology by ISLER and his group (379, 552).

The orange resupinate carpophores of the Basidiomycete Peniophora aurantiaca are one of the few known sources of ketocarotenoids. The main pigment is astaxanthin (464), but echinenone (462) and astacin (**465**) are also present along with β-carotene. The presence of canthaxanthin (**463**) in the red American toadstool *Cantharellus cinnabarinus* has already been mentioned. This fungus too contains echinenone (**462**), and both of the ketocarotenoids (**462**) and (**463**) are present in the closely related *Cantharellus friesii*.

Carotenoids have been rarely found among Agaricales. β-Carotene (**457**) and γ-carotene (**448**) are responsible for the remarkable orange colour of the fruit bodies of *Haasiella venustissima* (= *Clitocybe venustissima*), while the same two pigments, together with lycopene (**442**) and torulene (**449**), impart the brilliant golden-yellow colour to carpophores of *Gerronema chrysophyllum* (= *Omphalia chrysophylla*). The carotenes (**448**) and (**457**), together with several ketocarotenoids, have been isolated from the wood-destroying agaric *Phyllotopsis nidulans* (*230a*). Reports that carotenoids are to be found extensively among members of such genera as *Amanita* and *Suillus* (*175, 176*), among others, must be regarded with the utmost caution until independent studies have been undertaken.

The carotenoid composition of the slime mould *Lycogala epidendron* has been reinvestigated in detail and a minor constituent, originally identified (*450*) as spirilloxanthin (**476**), has been shown to be 3,4-dehydrolycopene (**443**) (*457*). It seems likely that spirilloxanthin and other methoxylated carotenoids are not produced by fungi but are restricted in their distribution to photosynthetic purple bacteria (*457*).

(**476**)

Aspects of the early and later stages of carotenoid biosynthesis have been reviewed (*136, 181*) and a comprehensive account of the stereochemistry of naturally occurring carotenoids, including fungal pigments, has been published (*458*).

Essays in the older literature (*42, 43*) dealing with the purification and characterisation of carotenoids have been superseded by recent accounts of the application of HPLC to carotenoid analysis and separation (*551*) and of modern techniques of high field n.m.r. spectroscopy to elucidation of their structures (*216*). The mass spectra of several of the fungal carotenoids discussed above have been studied in detail by LIAAEN-JENSEN (*217*).

5. Nitrogen Heterocycles

5.1. Phenoxazin-3-ones

Substituted phenoxazin-3-one pigments impart the very stable bright orange-red colour to the conspicuous bracket shaped fruit bodies of the wood-rotting fungi *Pycnoporus cinnabarinus*, *P. sanguineus*, and *P. coccineus* (Table 45).

The elucidation of the structure of cinnabarin (**477**) involved an interesting series of classical experiments carried out contemporaneously in the laboratories of GRIPENBERG in Finland and CAVILL in Australia. After a considerable amount of initial speculation (*159, 163, 308, 326*) regarding the nature and location of the substituents in the distinctive phenoxazin-3-one nucleus, the correct structure (**477**) was eventually formulated (*160, 161, 309*) and subsequently confirmed by chemical methods (*161, 162, 309, 312, 328*).

Cinnabarinic acid (**478**) and tramesanguin (**479**), which have been isolated only from specimens of *Pycnoporus sanguineus* collected in Central Africa (*314, 327*), clearly share with cinnabarin (**477**) a biogenesis which involves oxidative coupling of an anthranilic acid precursor, a process for which there is well known laboratory analogy. Thus, cinnabarin (**477**) has been synthesised by oxidative coupling between 2-amino-3-hydroxybenzyl alcohol and 3-hydroxyanthranilic acid using both *p*-benzoquinone (*709a*) and air (with the reagents adsorbed on silica gel) (*274a*) as oxidants. Cinnabarinic acid (**478**), a minor product of both these processes, has been obtained upon oxidative dimerisation of 3-hydroxyanthranilic acid (*146a*). 3-Hydroxyanthranilic acid is a degradation product of tryptophan.

The suggestion, tentatively made, that the phenoxazinones may ultimately prove of value to the taxonomy of *Pycnoporus* (*314*) has not been borne out by subsequent studies. Thus, the pigments (**477**), (**478**), and (**479**), along with several unidentified phenoxazinones, are present

	R^1	R^2
(**477**)	CH_2OH	CO_2H
(**478**)	CO_2H	CO_2H
(**479**)	CO_2H	CHO
(**480**)	H	H

(**481**)

Table 45. *Occurrence and Physical Properties of Phenoxazin-3-one Pigments*

Pigment	Occurrence[a]	Ultraviolet/Vis. [λ_{max} (log ε)]	TLC R_f[b]	Methyl ester M.p.
Cinnabarin (**477**) (Polystictin)	*Pycnoporus cinnabarinus* [synonym: *Trametes cinnabarina* (*305, 314*)], [TLC] (*639*), [cultures, TLC] (*470*), *P. coccineus* [TLC] (*639*), [c., TLC] (*470*), *P. sanguineus* [synonyms: *Coriolus sanguineus* (*152, 163, 451*), *Polyporus sanguineus* (*163*), *Polystictus cinnabarinus* (*451*), *P. sanguineus* (*327*), *Trametes cinnabarina* var. *sanguinea* (*314*)], [TLC] (*639*), [c., TLC] (*470*)	450 nm (br., 4.45) [pyridine]	0.47	200–202° (dec.)
Cinnabarinic acid (**478**)	*Pycnoporus cinnabarinus* [TLC] (*639*), *P. coccineus* [TLC] (*639*), *P. sanguineus* [synonym: *Polystictus sanguineus* (*327*)], [TLC] (*639*)	450 nm (br., 4.35) [pyridine]	0.35	224–226° (dec.)
Tramesanguin (**479**)	*Pycnoporus cinnabarinus* [TLC] (*639*), *P. coccineus* [TLC] (*639*), *P. sanguineus* [synonym: *Trametes cinnabarina* var. *sanguinea* (*314*)], [TLC] (*639*)	238 (4.55), 430 (4.41), 450 nm (infl., 4.38) [dioxan]	0.67	245–250°
2-Amino-phenoxazin-3-one (**480**)	*Calocybe gambosa* [c] (*560*)	425 nm (4.37) [methanol]	0.57	–
Phenoxazin-3-one (**481**)	*Calocybe gambosa* [c] (*560*)	244, 341, 445 nm (4.06) [methanol]	–	–

[a] We have followed the subdivision of *Pycnoporus* given by NOBLES and FREW (*513*).
[b] Silica gel G (benzene:ethyl acetate:glacial acetic acid:formic acid = 12:6:1:1) (*639*).

in similar concentrations in both the fruit bodies (*639*) and cultured mycelium (*470*) of *P. cinnabarinus*, *P. sanguineus*, and *P. coccineus* obtained from a wide range of geographical locations.

The relationship, if any, between the known phenoxazinone pigments, ZOPF's 'xanthotrametin' (*731*) and LEMBERG's 'polystictinin' (*451*) remains obscure.

The orange phenoxazin-3-one (**481**) and its yellow-brown amino derivative (**480**) have recently been isolated from cultures of the agaric, *Calocybe gambosa* (*560*).

5.2. Riboflavin and Russupteridines

Vitamin-B$_2$ and pteridine derivatives which are closely connected with the biosynthesis of this primary metabolite are responsible for the pigmentation and the intense fluorescence under UV light of many fungi belonging to *Lyophyllum* and *Russula*. Riboflavin (**482**) is now recognised as an important yellow pigment in a great many species within these, and closely related, genera (Table 46). In fruit bodies of *Panellus serotinus*, riboflavin (**482**) is accompanied by larger amounts (0.04% dry weight) of its *N*-methyl derivative (**483**) (*630*).

The aldehyde and the carboxylic acid corresponding to riboflavin have been isolated from cultures of *Schizophyllum commune* and named schizoflavin-F$_1$ and -F$_2$, respectively (*646*).

(**482**) R = H
(**483**) R = CH$_3$

Table 46. *Occurrence of Riboflavin in Macromycetes*[a]

Calocybe chrysenteron (= *Lyophyllum chrysenteron*) [1.05%], *C. fallax* (= *Lyophyllum fallax*) [0.75%], *C. gambosa* (= *Lyophyllum georgii*) [5.8 × 10⁻³%], *C. ionides* (= *Lyophyllum ionides*) [1.1 × 10⁻²%], *C. onychina* (= *Lyophyllum onychinum*) [0.45%], *Lyophyllum buxeum* [6.9 × 10⁻²%] (*289*), *L. decastes* (= *L. aggregatum*) [5 × 10⁻³%] (*294*), *L. favrei* [0.14%], *L. hypoxanthum* [5.8 × 10⁻²%] (*289*), *L. multiforme* [8 × 10⁻³%] (*294*), *Panellus serotinus* [7 × 10⁻³%, isolated] (*630*), *Russula adulterina*, *R. adusta*, *R. aeruginea* [1.5 × 10⁻³%], *R. badia*, *R. bresadoliana* [3.6 × 10⁻³%], *R. cavipes*, *R. coerulea* [2.9 × 10⁻³%], *R. decolorans*, *R. densifolia* [4.8 × 10⁻³%], *R. emetica* [6 × 10⁻³%], *R. emetica* var. *longipes* [2.6 × 10⁻³%], *R. fragrantissima* [1.9 × 10⁻³%], *R. grisea*, *R. illota* [1.5 × 10⁻⁴%], *R. maculata*, *R. ochroleuca* [6 × 10⁻³%], *R. paludosa*, *R. pulchella*, *R. queletii* [4.4 × 10⁻³%], *R. sardonia* [7.7 × 10⁻³%] [208 mg from 200 kg toadstools], *R. variata*, *R. ventricosipes* [4.3 × 10⁻³%], *R. vesca*, *R. virescens* [2.2 × 10⁻³%], *R. viscida* [7.4 × 10⁻³%], *R. xerampelina* [9.2 × 10⁻³%] (*382*), *Tephrocybe rancida* (= *Lyophyllum rancidum*) (*294*)

[a] Yields refer to the proportion of riboflavin in the dried fungus. Concentrations in *Calocybe* and *Lyophyllum* were determined by UV/vis. spectrophotometry, those in *Russula* by fluorimetry.

References, pp. 253–286

The cap skin pigments and fluorescent principles present in fungi belonging to the large and difficult genus *Russula* have long been recognised as valuable taxonomic characters. The early literature on *Russula* colouring matters has been concisely documented by WATSON (*704*) and is given more extensive coverage in the review by EUGSTER (*222*).

Interest in these substances both by mycologists and chemists has led to the development of increasingly sophisticated techniques for their separation and analysis. Thus, electrophoresis (*62*), paper and cellulose chromatography (*62, 288, 290, 292*), thin layer chromatography (*290, 704*), isoelectric focusing (*224*) and gel permeation chromatography (*224, 382*) have revealed the presence of colourless, yellow, red and blue-violet compounds, all of which show strong fluorescence under UV light. In careful investigations requiring colossal quantities of fungal material EUGSTER and coworkers have succeeded in separating and purifying several of these intriguing substances in quantities which have permitted structure elucidation. Thus, from 200 kg of *Russula sardonia*, 36 kg of *R. paludosa*, 25 kg of *R. emetica* and 6 kg of *R. ochroleuca*

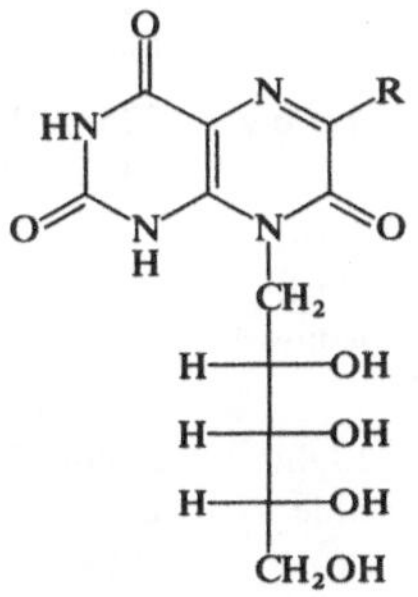

(484) R = CH₃
(485) R = H

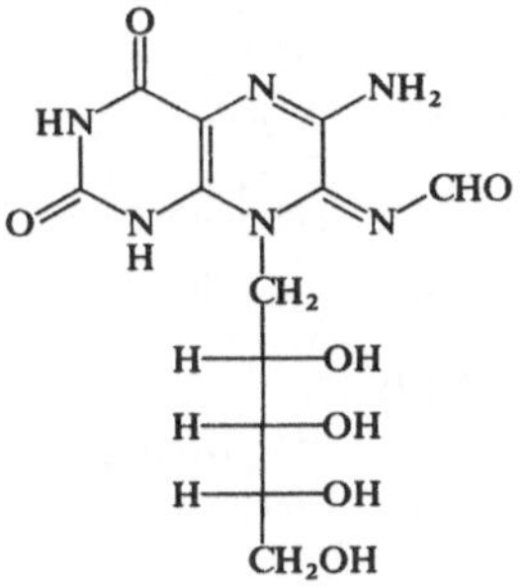

(486)

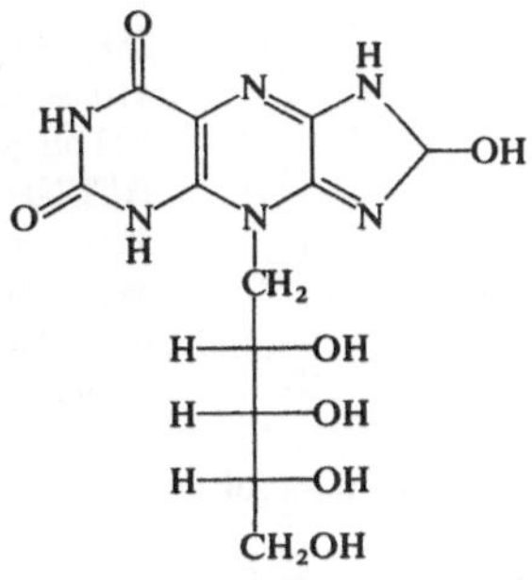

(487)

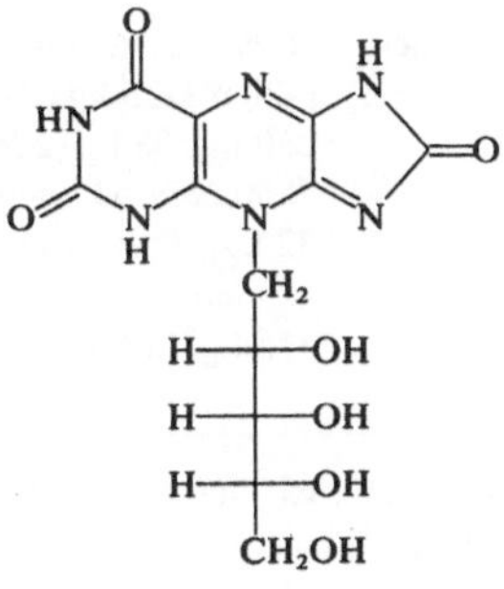

(488)

these chemists have isolated following extensive chromatography on cellulose and various Sephadex gels the colourless russupteridines (**484**) and (**485**), which exhibit blue fluorescence under UV, and five yellow

Table 47. *Occurrence and Physical Properties of Russupteridines*

Compound	Occurrence	Colour	TLC $R_f{}^a$	Fluorescence[b]	Ultraviolet/Vis. $[\lambda_{max}(\log \varepsilon)]$
6-Methyl-7-oxo-*N*(8)-(D-ribityl)-7,8-dihydro-lumazine (**484**)	*Russula emetica, R. ochroleuca, R. paludosa, R. sardonia, R. vinosa* (= *R. obscura*) (*382*)	Colourless	0.61	Violet	287, 343 nm [water]
7-Oxo-*N*(8)-(D-ribityl)-7,8-dihydrolumazine (**485**)	as (**484**) (*382*)	Colourless	0.56	Violet	286, 346 nm [water]
Russupteridine-yellow I (**486**)	*R. paludosa* [155 mg/36 kg fresh fungus], *R. sardonia* [110 mg/200 kg] (*382*)	Yellow-brown	0.31	Blue	261 (4.12), 306 (3.50), 412.5 nm (4.04) [water]
Russupteridine-yellow II	*R. paludosa* [40 mg/36 kg], *R. sardonia* [72 mg/200 kg] (*382*)	Dark yellow	0.22	Yellow	226, 287, 446 nm [water]
Russupteridine-yellow III ≡ Riboflavin (**482**)	see Table 46	Yellow	0.79	Dark yellow	–
Russupteridine-yellow IV (**488**)	*R. sardonia* [25 mg/300 kg] (*382*)	Pale yellow	0.22	Light blue	219 (4.52), 277 (4.24), 296 (sh., 3.95), 389 nm (4.32) [water]
Russupteridine-yellow V	*R. ochroleuca* [7 mg/6 kg] (*382*)	Yellow	0.40	Green-yellow	247, 290, 375, 454 nm [water]
Russupteridine-s_{III} (**496**)	*R. badia, R. emetica* [420 mg/35 kg (*380*), 250 mg/38 kg (*224*)], *R. paludosa* [206 mg/32 kg], *R. sardonia, R. vinosa* [933 mg/155 kg] (*224*)	Deep red	–	–	288, 390, 410, ~497, 523 nm [0.1 M HCl/NaCl]; 268, 391, ~411, 509, 541 nm [0.2 M phosphate buffer]

[a] Cellulose layer CEL 300-UV$_{254}$ (Macherey and Nagel) (n-butanol:acetic acid:water = 5:3:5) (*381, 382*).
[b] Colour of compound under UV light ($\lambda_{max} = 360$ nm) after chromatography as in footnote a.

pigments[1], the russupteridines-yellow I–V, which fluoresce intensely yellow-green (Table 47).

The novel 6,7-diaminolumazine structure (486) of russupteridine-yellow I was deduced from a detailed analysis of the spectroscopic data and from its distinctive hydrolytic cleavage to the dioxolumazine (489) *via* the aminolumazine (490) (Scheme 76) (*225, 381, 382*). The lumazines (489) and (490) have been prepared from 1-deoxy-1-(2,4,7-trioxo-1,2,3,4,7,8-hexahydropteridin-8-yl)-D-ribitol (491) *via* the azo compound (492) and its reduction with tin and formic acid (Scheme 77) (*471*).

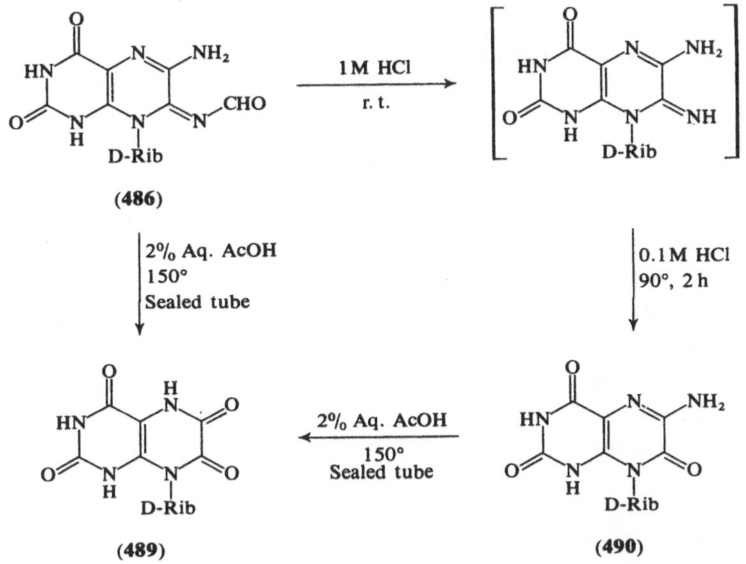

[D-Rib = a C-1 substituted 1-deoxy-D-ribitol]

Scheme 76. Hydrolysis of russupteridine-yellow I

With silver oxide in alkaline solution russupteridine-yellow I was oxidised to russupteridine-yellow IV which is present in very small amounts in *Russula sardonia*. The electronic spectrum of russupteridine-yellow IV proved to be similar to those of several 6,7,8-substituted lumazines and, together with the infrared and n.m.r. data, allowed structure (488) to be proposed for this pigment. The structure (488) is in accord with its derivation by oxidation of russupteridine-yellow I

---

[1] Yellow pigments from *Russula* are referred to elsewhere as russulaxanthines (*290*).

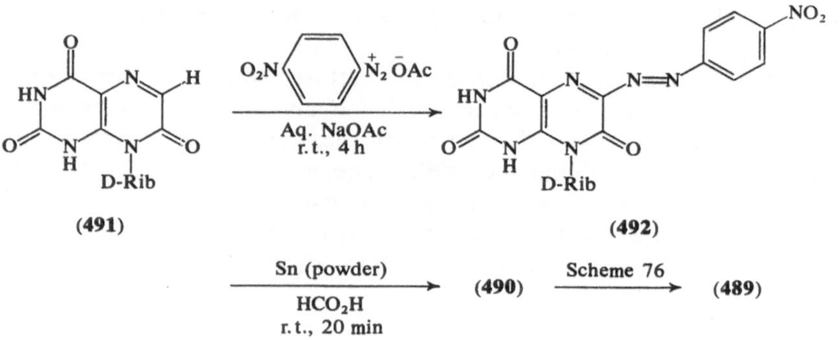

[D-Rib = a C-1 substituted 1-deoxy-D-ribitol]

Scheme 77. Synthesis of the lumazines (**489**) and (**490**)

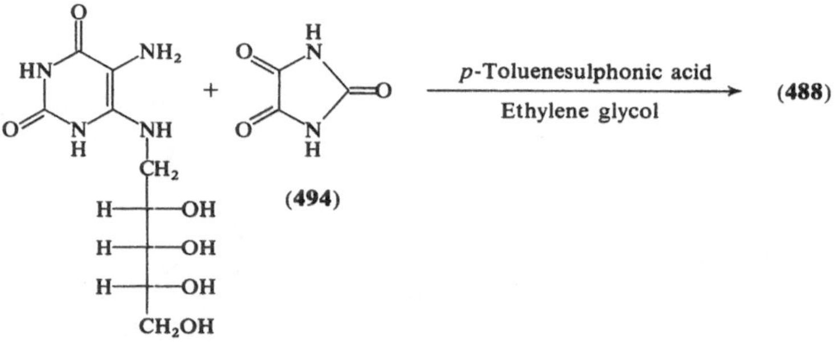

(**493**)

Scheme 78. Synthesis of russupteridine-yellow IV

[reacting in the aminal form (**487**)] and has been confirmed by unambiguous synthesis, albeit in extremely poor yield, from the uracil derivative (**493**) and parabanic acid (**494**) (Scheme 78) (*381*, *382*, *471*).

It is interesting that the uracil derivative (**493**), an intermediate in riboflavin biosynthesis, has also been detected in extracts of *Russula paludosa* and *R. sardonia* (*382*).

Russupteridine-yellow II closely resembles russupteridine-yellow I (**486**) in that it is hydrolysed to the dioxolumazine (**489**) on treatment with a degassed 2% aqueous solution of acetic acid in a sealed tube at 150 °C, but no more is known about its detailed structure. Russupteridine-yellow III is identical with riboflavin (**482**) (*223*, *381*, *382*).

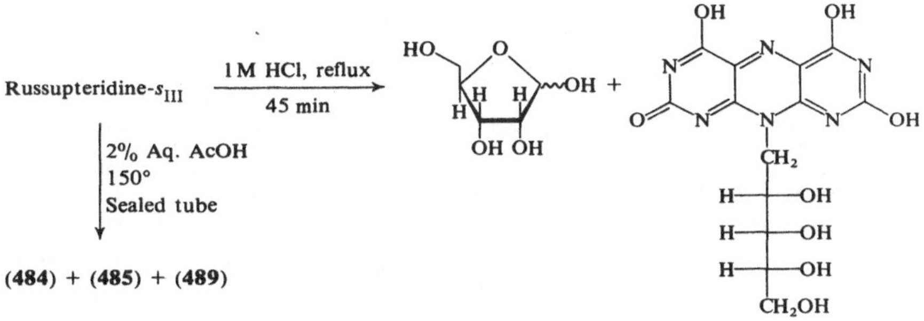

Scheme 79. Degradation of russupteridine-s_{III}

The colourless russupteridines (484) and (485) have been obtained together with ribose, the dioxolumazine (489) and the heterocycle (495) as degradation products (Scheme 79) of the more complex red russupteridines[1], e.g. russupteridine-s_{III}, for which dimeric structures represented by partial formula (496) have been suggested (222, 223, 380). The possibility, however, that imino functions are present which have been transformed into oxo groups during the hydrolysis experiments cannot be excluded (382). Interestingly, these red pigments contain a ribose residue bound as an N-riboside.

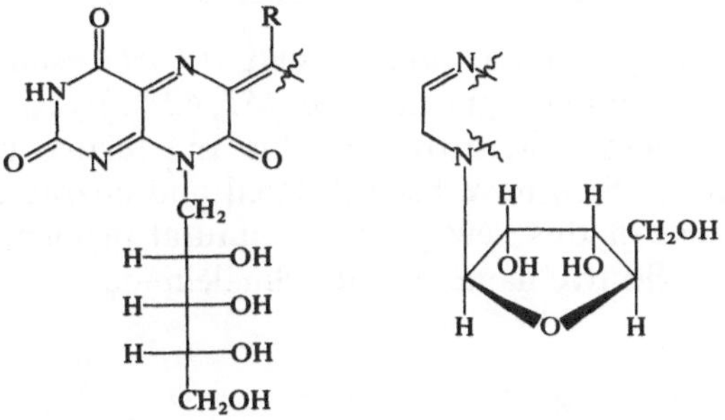

(496)

The structures of the violet and blue pigments of *Russula* [russulacyanins (290)] remain completely unknown at the present time.

---

[1] The red pigments of *Russula* are referred to elsewhere as rubéine (363), ruberine (526, 726), and as the russularhodins (62, 288, 290, 292).

5.3. Indole Pigments

5.3.1. Simple Indoles

Indigo (**497**) has been isolated from mutant strains of *Schizophyllum commune (484, 643)* and *Agaricus campester* (= *A. campestris*) (*227*) and from cultures of *Auriculariopsis ampla* (*23*) and identified by UV and infrared comparison with synthetic material. In *Schizophyllum commune* and *Agaricus campester* the blue pigment is accompanied by the red indirubin (**498**) (*218, 227*), and a yellow compound tentatively identified as isatin (**499**) (*218*).

(**497**)   (**498**)   (**499**)

It is of interest to note in passing the occurrence in various *Chaetomium* species (*133, 394*), and elsewhere among the lower fungi (*724*), of analogues of the terphenylquinone (Section 2.1.2) and pulvinic acid pigments (Section 2.1.3) which derive not from phenylalanine or from tyrosine but from tryptophan.

5.3.2. Bisindolylmaleimides

The red sporangia of the slime mould *Arcyria denudata* contain a number of unique indole pigments (*156, 157, 623*). By careful chromatography of the methanolic extract of this and related myxomycetes the pigments (**500**)–(**513**) have been isolated and characterised. Some physical properties of this new group of natural pigments, for which we propose the collective name bisindolylmaleimides, are given in Table 48.

(**500**)

	R¹	R²
(**501**)	H	H
(**502**)	OH	H
(**503**)	OH	OH

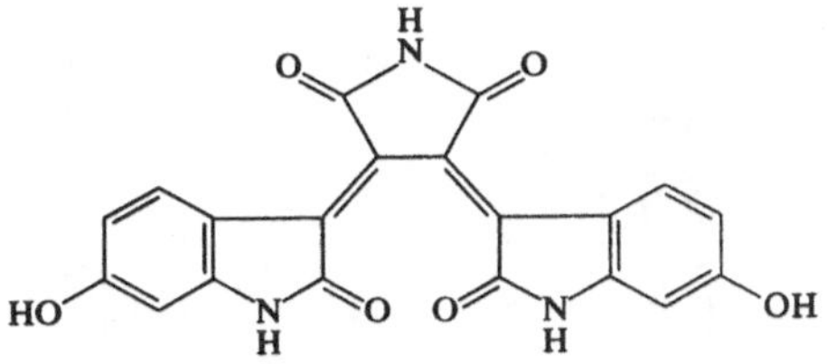

(504)

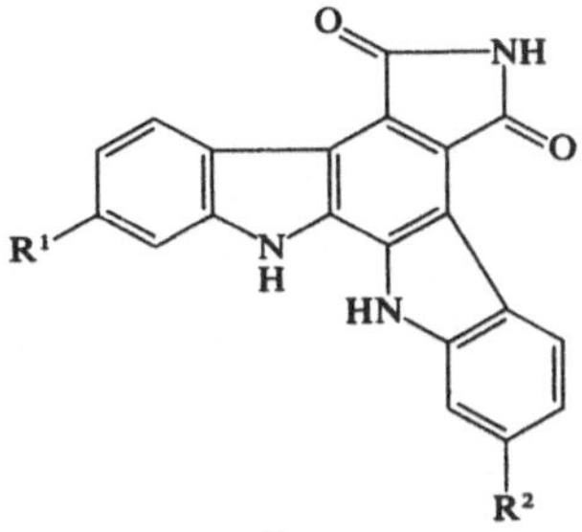

	R¹	R²
(505)	H	H
(506)	OH	H
(507)	OH	OH

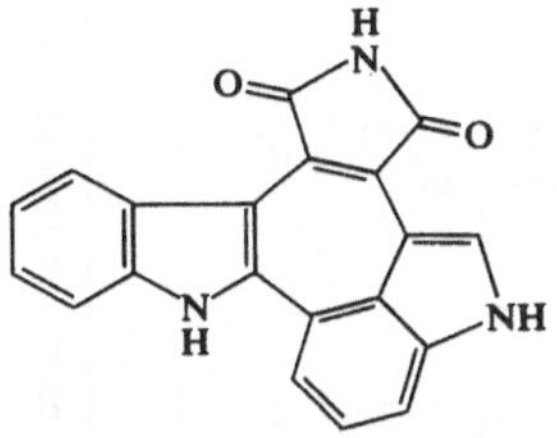

(508)

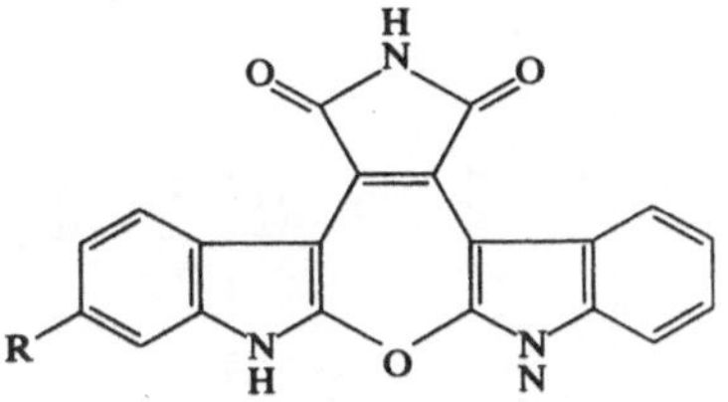

(509) R = H
(510) R = OH

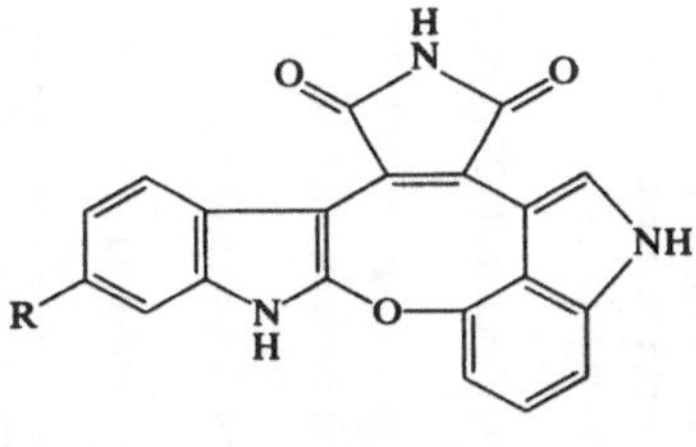

(511) R = H
(512) R = OH

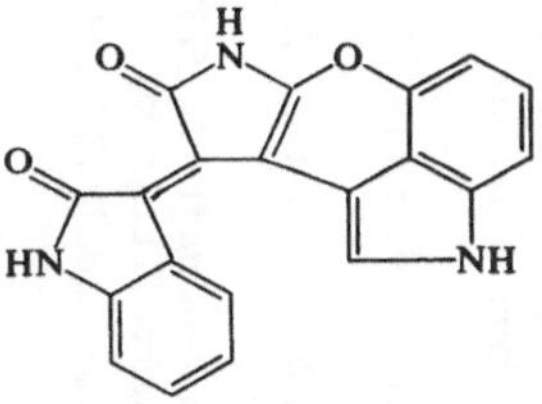

(513)

Dihydroarcyriarubin-B (500) is a colourless compound which gives a red spot on silica gel plates which gradually turns to dark green on exposure to air. In the infrared spectrum (KBr) of (500) two absorptions, at 1770 and 1710 cm⁻¹, were indicative of the dihydromaleimide moiety, and in the ¹H-n.m.r. spectrum the components of an AB-quartet centred, respectively, at δ 4.48 and 4.53 ($J = 6$ Hz) could be assigned to the protons of the maleimide ring. The structure of dihydroarcyriarubin-B has been confirmed by hydrogenation of a second

Table 48. Occurrence and Physical Properties of Bisindolylmaleimides

Pigment[a]	Occurrence[b]	Colour M.p.	TLC R_f^c	Ultraviolet/Vis. [λ_{max} (log ε), methanol]
Dihydroarcyriarubin-B (500)	Arcyria denudata [0.1%] (156)	Colourless 250° (dec.)	0.31	218 (3.98), 271 (3.34), 288 (3.27), 306 nm (sh., 2.85)
Arcyriarubin-A (501)	Arcyria denudata [5 × 10⁻²%] (156)	Red 281°	0.44	248 (sh., 4.23), 276 (4.14), 284 (sh., 4.10), 371 (3.74), 465 nm (3.90)
Arcyriarubin-B (502)	Arcyria denudata [0.4%] (623)	Red 154–155°	0.33	281 (3.93), 392 (sh., 3.57), 465 nm (3.77)
Arcyriarubin-C (503)	Arcyria denudata [1.5%] (623), A. ferruginea [TLC] (156)	Red 205–206°	0.21	283 (3.93), 474 nm (3.76)
Arcyriaverdin-C (504)	Arcyria denudata [0.1%] (156)	Green >300°	0.21	256 (sh., 3.91), 336 (3.51), 433 (3.69), 634 nm (2.82)
Arcyriaflavin-A (505)	Arcyria nutans (595)	Pale yellow >300°	0.44	235 (4.80), 257 (4.42), 272 (4.45), 281 (4.55), 300 (sh., 4.64), 314 (4.86), 402 nm (3.81)
Arcyriaflavin-B (506)	Arcyria denudata [5 × 10⁻²%] (623), Metatrichia vesparium (438)	Pale yellow 350°	0.33	229 (4.07), 271 (3.70), 280 (3.75), 323 (4.11), 414 nm (3.19)
Arcyriaflavin-C (507)	Arcyria denudata [5 × 10⁻²%] (623), Metatrichia vesparium (438)	Pale yellow 350°	0.21	229 (4.14), 255 (sh., 3.77), 270 (3.66), 280 (3.70), 318 (sh., 4.09), 330.5 (4.29), 422 nm (3.34)
Arcyriacyanin-A (508)	Arcyria nutans (595)	Green-blue >300°	0.42	224 (4.06), 243 (4.15), 264 (sh., 3.88), 360 (3.72), 625 nm (2.49)
Arcyroxepin-A (509)	Arcyria denudata [0.3%] (623)	Red 268–270°	0.42	226 (4.44), 273 (3.82), 283 (sh., 3.83), 362 (3.51), 471 nm (3.68)
Arcyroxepin-B (510)	Arcyria denudata [0.1%] (595)	Violet	0.32	274, 388, 484 nm
Arcyroxocin-A (511)	Arcyria denudata [0.2%] (156)	Red >300°	0.42	276 (4.15), 285 (4.12), 364 (3.76), 465 nm (3.91)
Arcyroxocin-B (512)	Arcyria denudata [0.4%] (156)	Violet 235° (dec.)	0.32	274 (4.13), 388 (3.80), 484 nm (3.93)
Arcyroxindol-A (513)	Arcyria denudata [5 × 10⁻²%] (156)	Orange 240° (dec.)	0.30	247 (sh., 4.08), 280 (sh., 3.89), 341 (sh., 3.43), 493 nm (3.45)

[a] The suffix A, B or C in the names of the bisindolylmaleimides indicates the number of hydroxy groups in the 6- and 6'-positions of the indole rings.
[b] Isolated yields are based on the weight of fresh sporangia.   [c] Silica gel Merck 60 F₂₅₄ (benzene:ethyl formate:formic acid = 10:5:3).

pigment, arcyriarubin-B (502), using catalytic hydrogen transfer (Pd-C) to (502) from cyclohexene in boiling xylene. Only one diastereoisomer was formed under these conditions and this proved to be identical with the natural product (500). Because the thermodynamically more stable *trans*-isomer should prevail under these forcing conditions, the same configuration has been tentatively assigned to (500) (*156*).

The arcyriarubins-B (502) and -C (503) are the main red pigments of *Arcyria denudata*. Arcyriarubin-A (501) is present in this organism in only minor amounts. Chromatography has shown that arcyriarubin-C (503) is also the major red pigment of *Arcyria ferruginea*. The structure of the arcyriarubin pigments was deduced from spectroscopic data, especially the ^{1}H- and ^{13}C-n.m.r. spectra which are summarised concisely in Table 49.

Arcyriarubin-A has been synthesised starting from indolylmagnesium bromide and 3,4-dibromo-*N*-methylmaleimide (Scheme 80) (*623*). By heating these compounds together in benzene a mixture of the indole derivatives (514) and (515) was formed which could be conveniently separated by chromatography. Hydrolysis of (515) with aqueous sodium hydroxide followed by heating with ammonium acetate afforded

Table 49. ^{1}H- and ^{13}C-N.m.r. Data for Arcyriarubin-B (502)
(δ values, with TMS as internal standard)

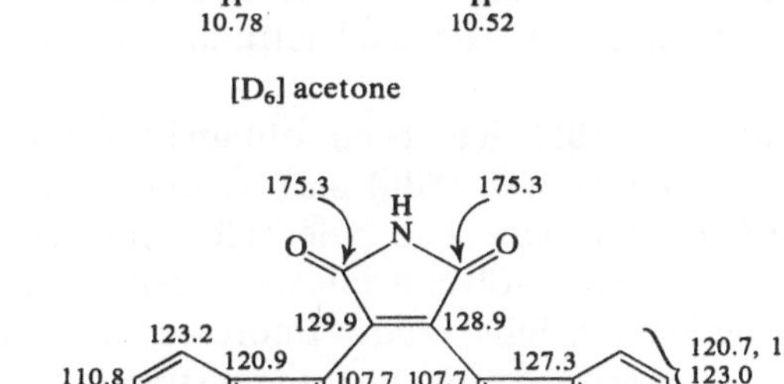

[D$_6$] acetone

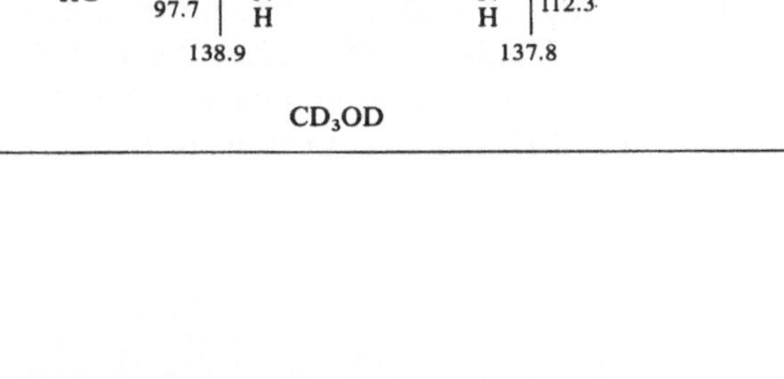

CD$_3$OD

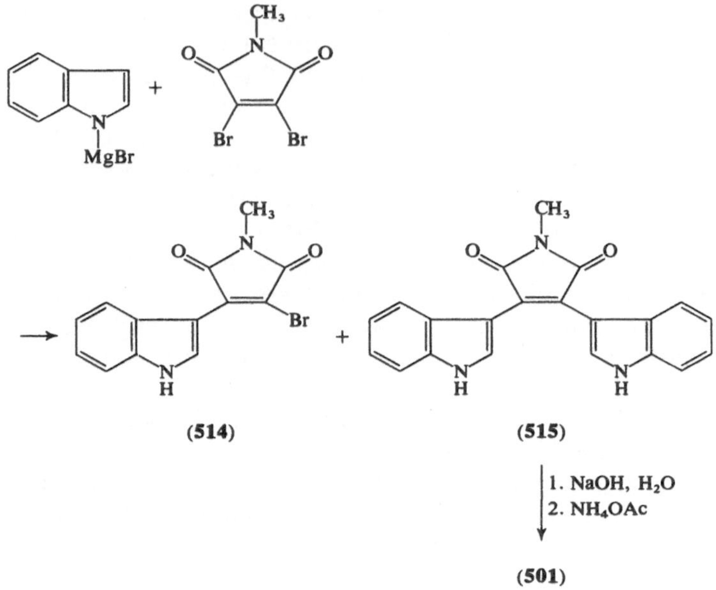

Scheme 80. Synthesis of arcyriarubin-A

arcyriarubin-A (**501**). The arcyriarubins-B and -C exhibit moderate antibiotic activity against *Bacillus brevis* and *B. subtilis*.

A green pigment, arcyriaverdin-C, exhibited spectra in accord with the symmetrical structure (**504**) (*156*). Due to the deshielding effect of the maleimide carbonyl groups the signals of the protons at C-4 and C-4′ in (**504**) resonate at δ 8.65 in the ^{1}H-n.m.r. spectrum, a downfield shift when compared to the corresponding protons in (**503**). Arcyriaverdin-C has been chemically correlated with arcyriarubin-C (**503**) by oxidation of the latter with lead tetra-acetate in chloroform (*156, 593*).

Arcyriaflavin-A (**505**) has been obtained from *Arcyria nutans*, whereas the arcyriaflavins-B (**506**) and -C (**507**) occur in *A. denudata* and *Metatrichia vesparium*. The pale yellow arcyriaflavins are only slightly soluble in organic solvents but were easily detected during chromatography by their bright yellow fluorescence under UV light. In the mass spectra of these compounds the distinctive loss of a C_2HNO_2 fragment from the molecular ion is diagnostic for the presence of a maleimide moiety. The ^{1}H- and some ^{13}C-n.m.r. data for arcyriaflavin-B (**506**) are given in Table 50 (*156, 623*). As was the situation with arcyriaverdin-C, the proximity of the maleimide carbonyl groups leads to deshielding of the 4-H and 4′-H protons in this rigid system.

Table 50. *¹H- and Some ¹³C-N.m.r. Data for Arcyriaflavin-B* **(506)**
(δ values, with TMS as internal standard)

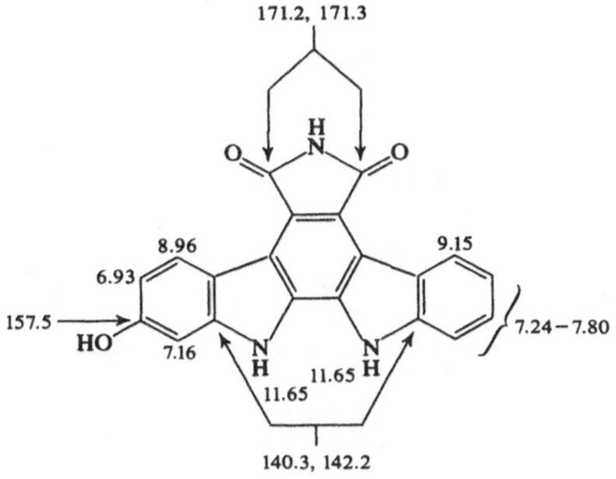

171.2, 171.3

8.96   9.15
6.93
157.5
HO   7.16
7.24–7.80
11.65   11.65
11.65

140.3, 142.2

[D₆] acetone

The structure of arcyriaflavin-C **(507)** was confirmed by its formation from arcyriarubin-C **(503)** on brief exposure of the latter pigment to warm, concentrated sulphuric acid *(623)*. Arcyriaflavin-B **(506)** has been synthesised in excellent overall yield by RAPHAEL (Scheme 81) *(376)*.

The same chromophore as is found in the arcyriaflavins occurs also in rebeccamycin **(516)** and in its dechloro derivative, two antitumour antibiotics isolated from *Nocardia aerocoligenes* *(510)*. Also closely related to these compounds are staurosporin **(517)**, an antibiotic from *Streptomyces staurosporeus* which exhibits pronounced antihypotensive activity *(258)*, and other carbohydrate bridged bisindoles *(575)*. Due to the small amounts of *Arcyria* metabolites available from the

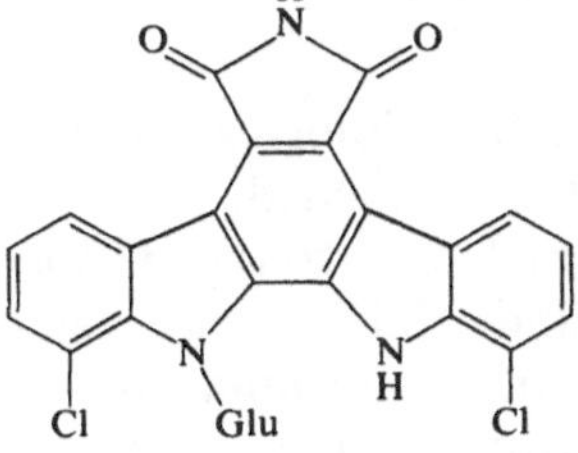

(516)

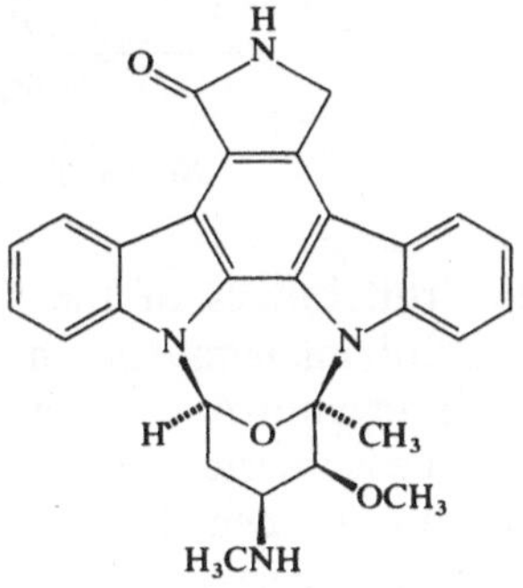

(517)

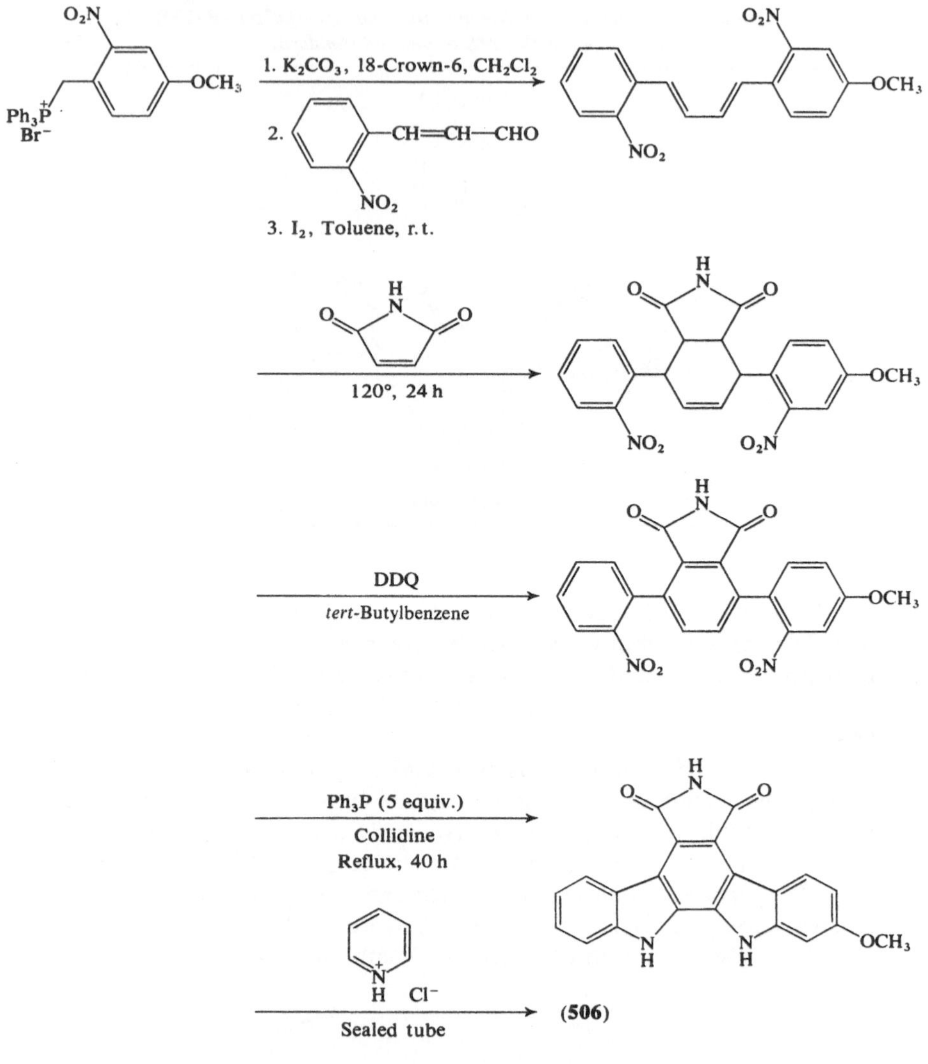

Scheme 81. Synthesis of arcyriaflavin-B

minute fruit bodies of the slime moulds, the biological activities of the bisindolylmaleimides remain to be fully investigated.

Arcyriacyanin-A, a green-blue pigment from *Arcyria nutans* was found to be isomeric with arcyriaflavin-A and exhibited mass and infrared spectra in accord with the presence of a maleimide moiety. In the ^{1}H-n.m.r. spectrum the lack of signals from protons at the positions 2- and 4'- in the indole rings was taken to indicate bond formation between these carbons thus leading to the structure (**508**). As expected, the protons 4-H and 2′-H experience paramagnetic shifts due to the anisotropy of the neighbouring carbonyl groups (δ_{4-H} 8.48, $\delta_{2'-H}$ 7.95) and the structure was further confirmed by ^{13}C-n.m.r. measurements (*595*).

The red pigments, arcyroxepin-A (**509**) and arcyroxepin-B (**510**), from *Arcyria denudata* may be easily recognised during thin layer chromatography by their colour change from red to violet on exposure to ammonia vapour. The mass and infrared spectra of these pigments proved consistent with the presence of a maleimide group and the appearance of only half of a full set of signals in the ^{1}H-n.m.r. spectrum of arcyroxepin-A indicated a symmetrical structure. The absence from the ^{1}H-n.m.r. spectrum of arcyroxepin-A of signals emanating from 2-H and 2′-H together with the necessity to incorporate an additional oxygen atom in order to satisfy the molecular formula, led to the oxepin

Table 51. ^{1}H- and ^{13}C-n.m.r. Data for Arcyroxocin-B (**512**)
(δ values, with TMS as internal standard)

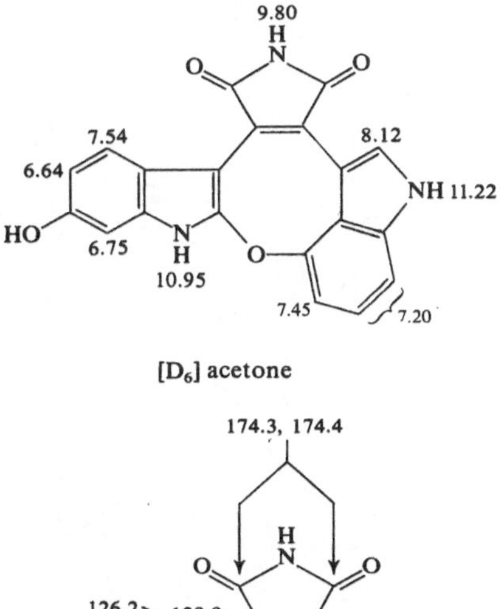

[D$_6$] acetone

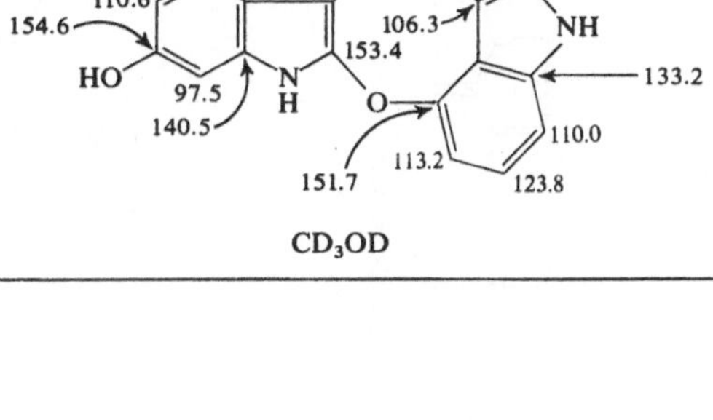

CD$_3$OD

224

References, pp. 253–286

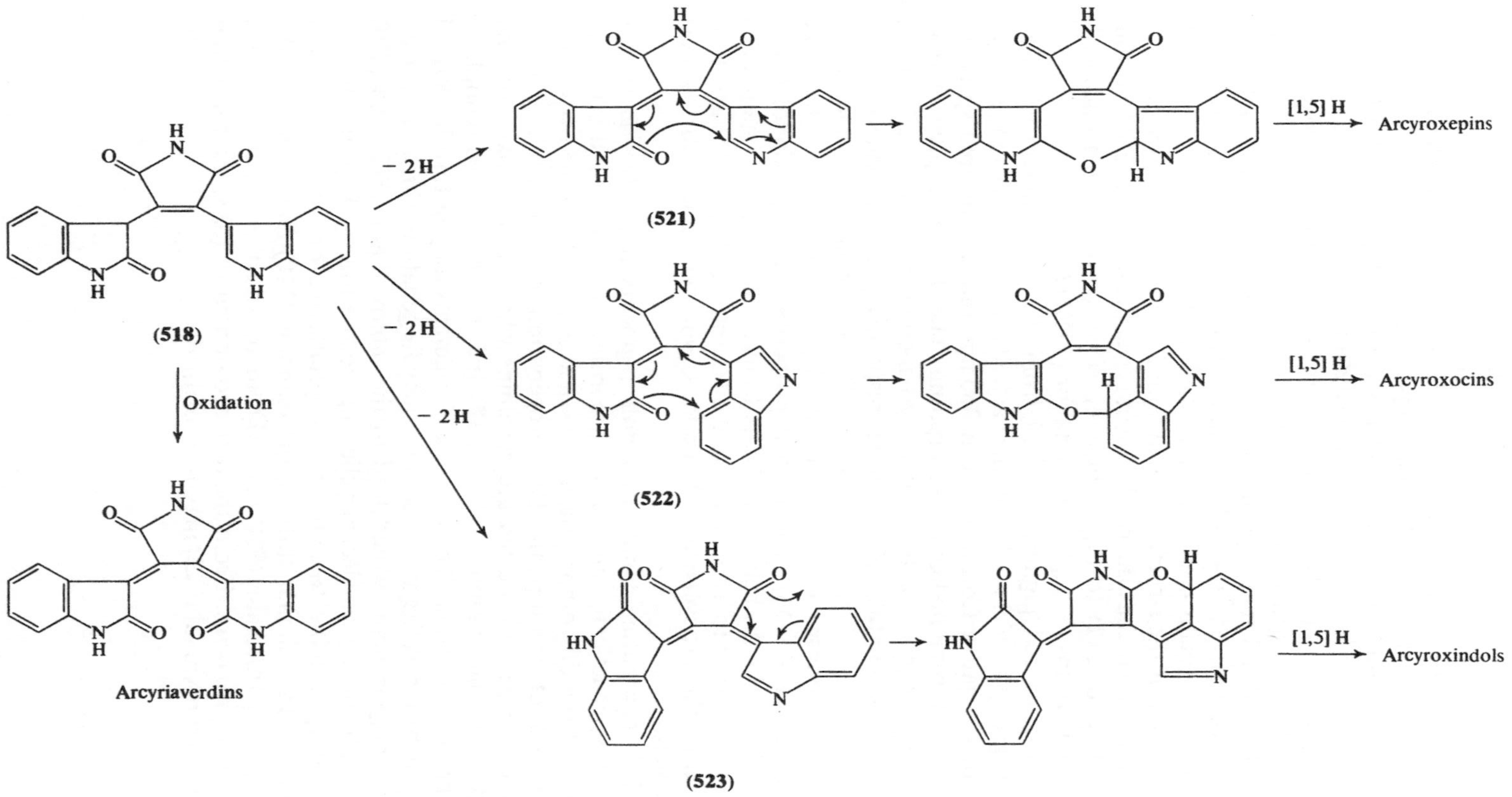

Scheme 82. Hypothetical biogenetic relationships of the bisindolylmaleimides

structure (**509**) for this pigment. The ^{1}H-n.m.r. spectra of the arcyroxe-pins show broad signals at room temperature.

Two red pigments, arcyroxocin-A (**511**) and arcyroxocin-B (**512**), have been isolated from *Arcyria denudata*. Both compounds possess chromatographic properties very similar to those of the arcyroxepins-A and -B from which they were separated only with difficulty. Again, the infrared and mass spectra indicated the presence of a maleimide ring and from the ^{1}H- and ^{13}C-n.m.r. data (Table 51) the structures of the arcyroxocins could be established. In the case of arcyroxocin-B (**512**) proof of the structure was obtained from an X-ray analysis from which it is clear that the rigid oxocin system exposes the 2′-H protons in (**511**) and (**512**) to the deshielding effect of one of the maleimide carbonyl groups [$\delta_{2'-H}$ 8.12 *vs.* 7.82 for (**502**)] (*157*).

A further red pigment isolated from *Arcyria denudata* is unique in this group by virtue of its optical activity. Its CD spectrum exhibited a strong negative Cotton effect at 360 nm and the infrared spectrum lacked the typical maleimide absorptions. The ^{1}H-n.m.r. spectrum indi-cated the presence of two indole systems which must be substituted in the 2,3- and in the 3′,4′-positions, respectively. From a careful analy-sis of the spectral data, including n.O.e. experiments, formula (**513**) has been proposed for this pigment which has been given the name arcyroxindol-A (*156*). The optical properties of arcyroxindol-A can be explained by the helicity of the molecule, which could be maintained by a high barrier to inversion caused by the sterically demanding substi-tuents at the exocyclic double bond. Computer simulations support this picture.

This fascinating family of indole pigments from *Arcyria* species can be biogenetically inter-related according to Scheme 82. It is reason-able that the dihydroarcyriarubin system is initially formed from two molecules of tryptophan. Dehydrogenation of dihydroarcyriarubin could then lead to the red arcyriarubins which may be oxidised first to the hypothetical intermediate (**518**) and thence to the chromophore of the arcyriaverdins. The ring closed pigments may in turn be derived from their respective precursors by dehydrogenation which in the case of the arcyriarubins would lead to the intermediates (**519**) and (**520**) and subsequently *via* electrocyclic ring closure followed by sigmatropic 1,5-hydrogen shifts to either the arcyriaflavins or the arcyriacyanin pigments. In a similar fashion, the oxindole (**518**) may cyclise *via* the dehydro intermediates (**521**) and (**522**) to afford either the arcyroxepins or the arcyroxocins. From the dehydro derivative (**523**) the formation of the arcyroxindol (**513**) may be explained.

5.4. Necatorone

The dark olive brown fruit bodies of *Lactarius necator* ($=$ *L. turpis*) change to a deep purple colour when exposed to ammonia vapour. This reaction was noted as early as 1896 (*344*) at which time it was ascribed to the presence of substances related to polyporic acid. More recent studies, however, have established the unusual 5,10-dihydroxy-6*H*-pyrido[4,3,2-*k,l*]acridin-6-one structure (**524**) for a metabolite responsible, in part, for this colour change (*250, 253*). The alkaloid (**524**), necatorone, forms red needles [UV/vis. (methanol): λ_{max} (log ε) $=$ 212 (sh., 4.38), 233 (4.60), 265 (sh., 4.13), 293 (3.88), 310 (sh., 3.85), 431 nm (4.13)] which dissolve in dimethyl sulphoxide to produce a green solution showing strong green-yellow fluorescence. With aqueous ammonia, successive deprotonations of (**524**) produce blue and purple anions. The proton coupled ^{13}C-n.m.r. spectrum of necatorone in deuteriodimethyl sulphoxide clearly demonstrated the presence in solution of a single tautomer having the structure (**524**) (*250*).

(**524**)   (**525**)

Necatorone has been synthesised from 2-(3,4-dimethoxyphenyl)ethylamine according to the sequence of reactions shown in Scheme 83 (*368*). The conversion of the aminophenol (**526**) into necatorone (**524**) occurs *via* the quinone imine intermediate which is formed by oxidative dehydrogenation of (**526**). The alternative mode of cyclisation involving intramolecular addition of the amino group in (**526**) to an incipient *ortho*-quinone does not operate; the intact *p*-aminophenol moiety being essential for the cyclisation to occur.

A second constituent of *L. necator* which gives a purple colour with ammonia vapour has also been isolated and characterised (*416*).

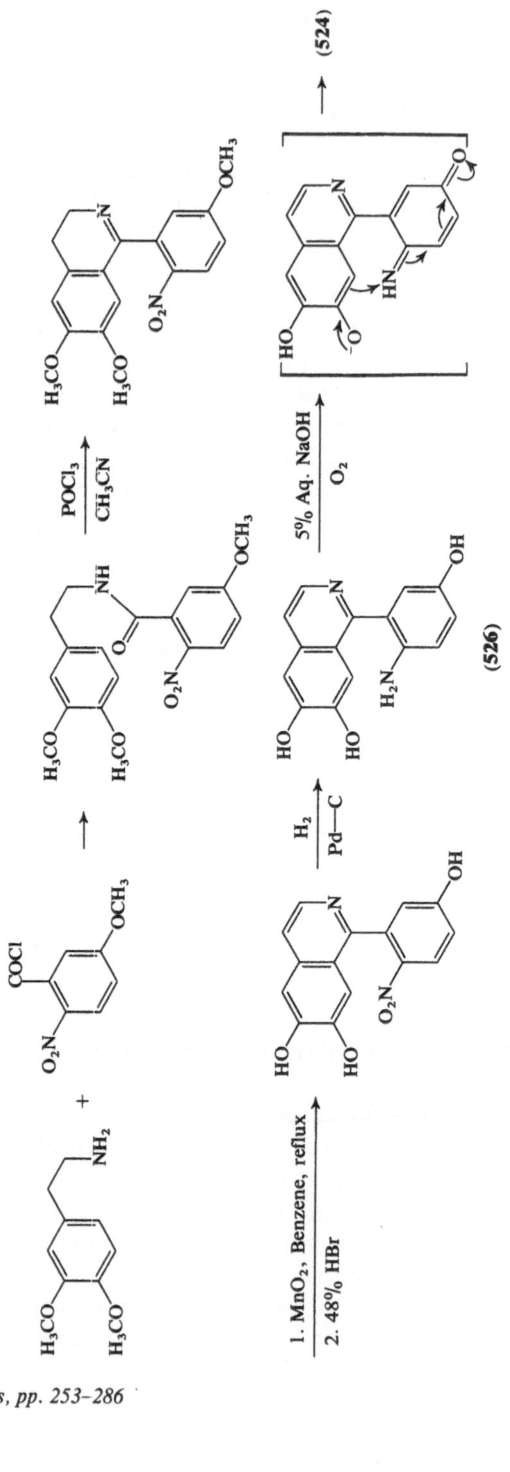

Scheme 83. Synthesis of necatorone

This compound possesses the structure (525) composed of two necator-one subunits linked at the 4,4'-positions. This dark brown pigment proved extremely insoluble in organic solvents but could be character-ised spectroscopically in the form of its pertrimethylsilyl derivative. Interestingly the dimer (525) has been produced *in vitro* by the action of horse-radish peroxidase on necatorone.

Interest in the considerable mutagenicity of extracts of *L. necator* led to the isolation of 'necatorin', a red crystalline compound exhibiting high mutagenic activity in the Ames *Salmonella* assay (*641, 642*). 'Neca-torin', for which an alternative structure to (524) was originally pro-posed (*642*), has been shown by direct comparison to be identical with necatorone (*368*).

5.5. Miscellaneous N-Heterocyclic Pigments

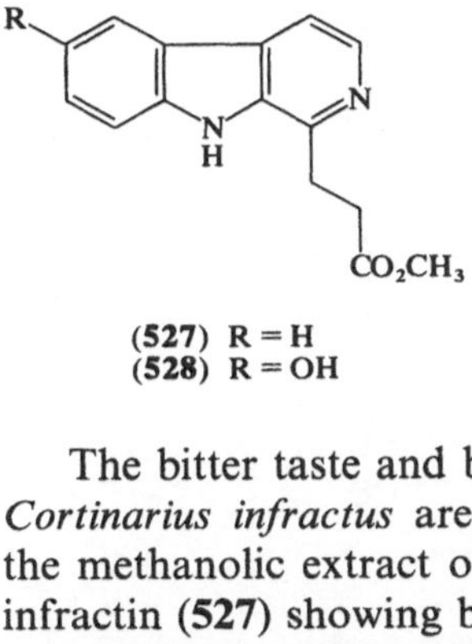

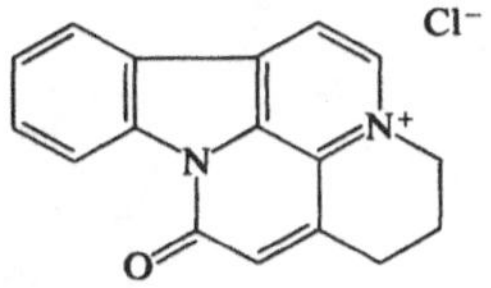

(527) R = H
(528) R = OH

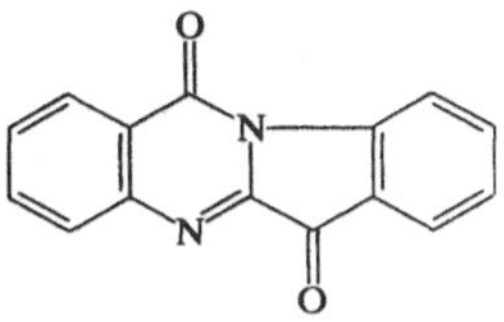

(529)

(530)

The bitter taste and blue fluorescence of the flesh of the toadstool *Cortinarius infractus* are due to β-carboline derivatives (*613*). From the methanolic extract of freeze dried fruit bodies have been isolated infractin (527) showing bright blue fluorescence under UV light, 6-hy-droxyinfractin (528) which fluoresces green-yellow, and a bitter princi-ple, infractopicrin (529). The yellow pigment (528) [UV/vis. (methanol): $\lambda_{max}(\log \varepsilon) = 214$ (4.36), 231 (4.53), 246 (4.37), 258 (4.24), 290 (sh., 4.14), 296 (4.33), 360 (3.72)] has been synthesised according to the reactions depicted in Scheme 84 (*567*).

Yellow crystals isolated from the culture fluids of *Leucopaxillus cerealis* var. *piceina* have been identified as tryptanthrene (530) and the 300 MHz ^{1}H-n.m.r. spectrum of this compound has been fully as-signed by reference to the spectra of its 2- and 8-chloro derivatives (*393*).

A minor constituent of each of the acutely toxic mushrooms *Corti-narius orellanus* and *C. speciosissimus* is the yellow bipyridyl orelline (531) (*26, 27*). The mono- and bis-*N*-oxides of orelline, named orellinine

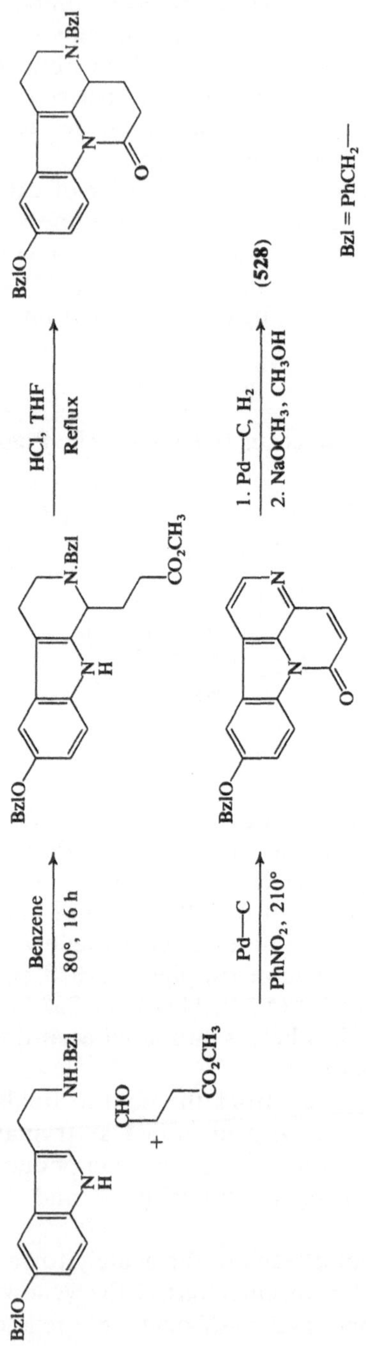

Scheme 84. Synthesis of 6-hydroxyinfractin

Bzl = PhCH$_2$—

(532) and orellanine (533), respectively, have been held responsible for the toxicity of these mushrooms, the ingestion of which has caused numerous poisonings and several deaths in the recent past (129, 337). Historical aspects of the isolation and properties of the toxins of C. orellanus have been nicely brought together by MOSER (446), ANTKOWIAK (27) and HØILAND (570).

(531)   (532)   (533)

The unusual structures of these substances coupled with some dispute as to the chemical nature of the toxic constituents of C. orellanus and C. speciossisimus [for alternative views, see (651)] have stimulated synthetic activity in this area. Two similar syntheses of orelline (531) and the major toxin, orellanine (533), have recently appeared (Scheme 85) (182, 656).

The nephrotoxic substance orellanine is transformed into the nontoxic orelline (531) on slow heating at 150 °C with elimination of molecular oxygen. This elimination takes place explosively if orellanine is heated rapidly above 267 °C (26). Sunlight converts orellanine to a mixture of orellinine (532) and orelline (531) (27).

Orellanine (533) has also been detected chromatographically in extracts from Cortinarius orellanoides and C. rainierensis (407).

(534)   (535) R = H
                             (536) R = CH₃

The tender yellow-green stalks of the agaric Entoloma incanum (= Rhodophyllus incanus) turn blue-green on bruising, a phenomenon which was explained by the presence of leuco compounds as early as 1917 (697). On extraction of the fresh toadstools with water or

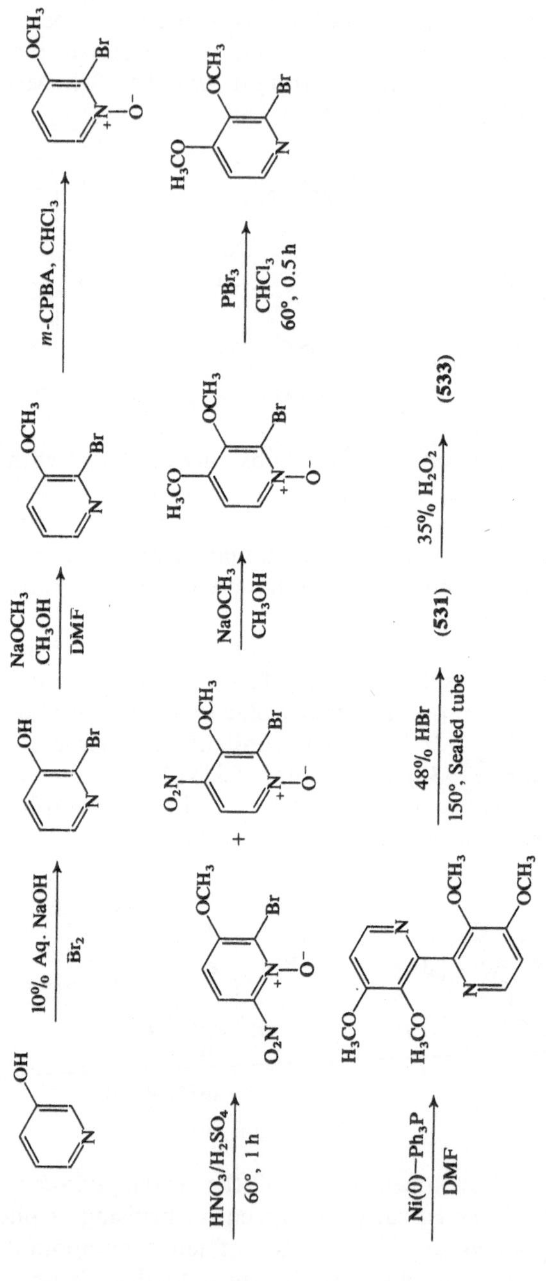

Scheme 85. Synthesis of orelline and orellanine

methanol dark green solutions result which have been separated into red, blue, and yellow fractions by chromatography on columns of Sephadex G-10 (*60, 61*). A yellow, fluorescent compound, incaflavin [UV/vis. (methanol): λ_{max} (log ε) = 246.5 (3.76), 415 (3.44)] which shows only a single resonance (δ 1.90) in the ^{1}H-n.m.r. spectrum recorded in deuteriomethanol has been assigned the azaquinone structure (**534**). On electron impact incaflavin suffers two diagnostic retro-Diels-Alder fissions which have been interpreted in terms of the fragmentations shown in Scheme 86.

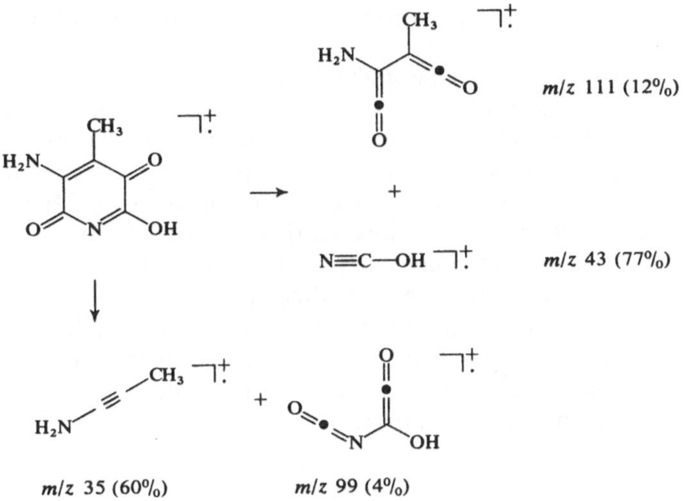

Scheme 86. Fragmentation of incaflavin on electron impact

The structure of incaflavin was confirmed by synthesis as depicted in Scheme 87 (*60*). Thus, condensation between the enol silyl ether

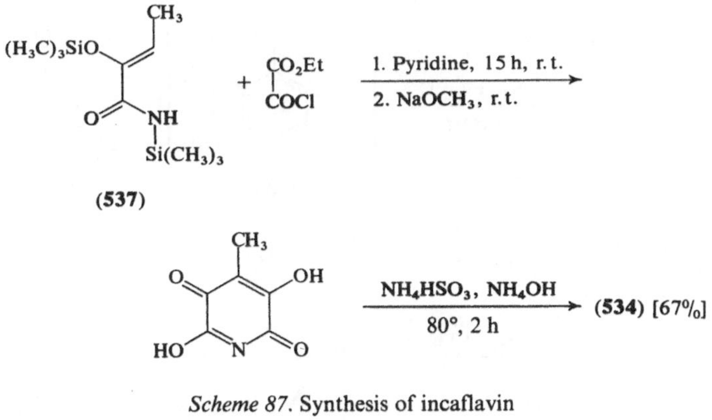

Scheme 87. Synthesis of incaflavin

(537) of α-ketobutyramide and ethyl oxalyl chloride afforded an aza-quinone derivative in low yield (17%) which in turn was converted to incaflavin (534) using the Bucherer reaction.

A blue pigment from extracts of *E. incanum* was oxidised to incaflavin (534) merely on standing in aqueous solution. This observation, together with the similarity between the electronic spectrum of this blue substance and that of the bacterial pigment indigoidin (535) (*417*), prompted the proposal that the blue fungal compound possessed the dimeric structure (536). This suggestion was supported by the presence of a molecular ion of appropriate mass in the field desorption mass spectrum.

In a series of model experiments 3-amino-2,6-dihydroxy-4-methyl-pyridine (538) was prepared as a putative precursor to the pigment (536) (Scheme 88) (*61*). A colourless solution of (538), obtained by hydrogenolysis of an azo precursor, immediately turned blue on exposure to air and after standing over night incaflavin (534) could be isolated from the solution in low yield. It is possible, therefore, that *Entoloma incanum* fruit bodies contain either the substituted pyridine (538) itself or a leuco form of the dimer (536), both of which would be expected to oxidise rapidly to the blue pigment (536) when the toadstool is damaged. Chemistry close to that proposed here for *E. incanum* has been established for the blueing of *Mercurialis annua* and *M. perennis* (Euphorbiaceae) (*644*).

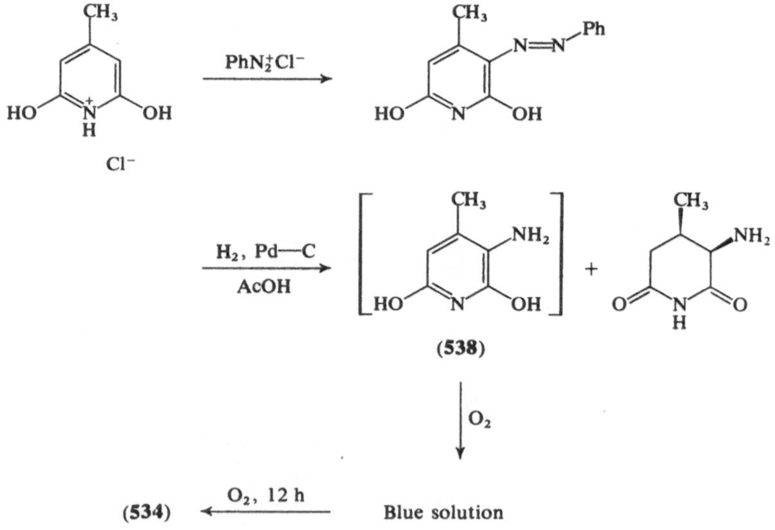

Scheme 88. Production of incaflavin *via* the chromogenic pyridine derivative (538)

The chemical nature of the yellow plasmodial pigments of *Fuligo* and *Physarum* species (Myxomycetes) has intrigued chemists for a long time because of the role which these substances may play in phototaxis and in the induction of sporulation in these slime moulds. The structure of the main pigment of *Fuligo septica* var. *flava* has been elucidated (*156, 158*). The pigments are present in the plasmodia and in the aethalia of *F. septica* in the form of their yellow calcium salts which after addition of mineral acid have been extracted into methanol to yield orange-red solutions of the free pigments. By careful chromatography one major orange-red pigment, fuligorubin-A, has been obtained in pure form. The compound exhibited a molecular ion corresponding to the formula $C_{20}H_{23}NO_5$ which fragmented to prominent ions at m/z 212 ($C_9H_{10}NO_5$), 173 ($C_{12}H_{13}O$), and 145 ($C_{11}H_{13}$). From these data and the ultraviolet [λ_{max}(methanol) = 243 and 425 nm] and ^{1}H-n.m.r. spectra, the presence in fuligorubin-A of a conjugated pentaenone chromophore was deduced. Further spectroscopic studies using two dimensional n.m.r. experiments established structure (**539**) for the pigment.

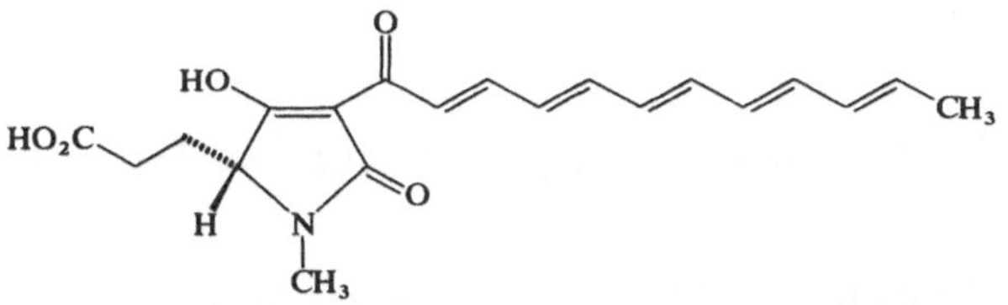

(**539**)

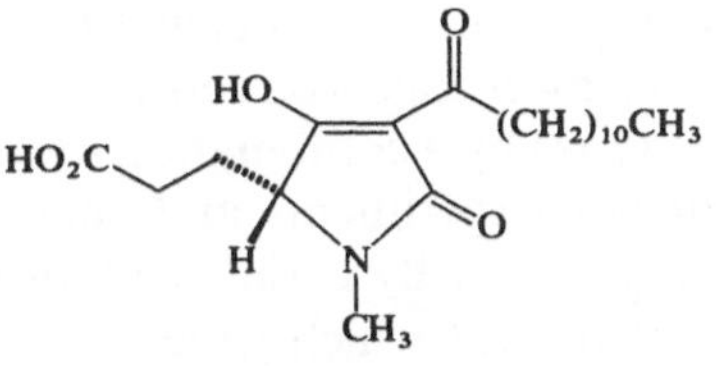

(**540**)

The (*R*)-absolute configuration of the chiral centre in fuligorubin-A was established by comparison of the chiroptical properties of the perhydro derivative (**540**) of fuligorubin-A with those of the model compound (**541**). The compound (**541**) was synthesised from dimethyl (*S*)-glutamate according to the reactions depicted in Scheme 89 and exhibited an opposite Cotton effect in the CD spectrum to that displayed by the perhydro derivative (**540**) of the natural product.

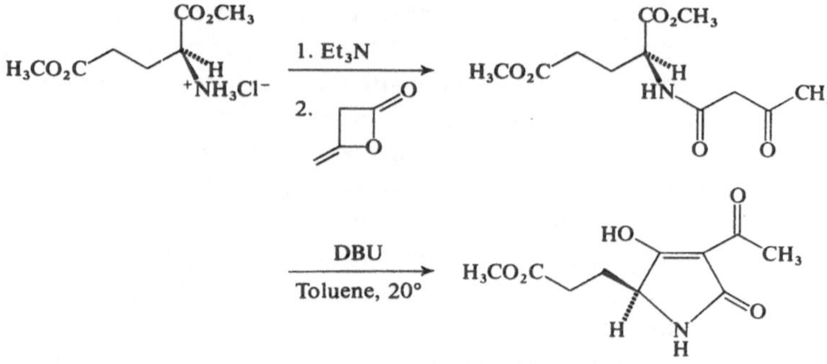

(541)

Scheme 89. Synthesis of the chiroptical model (**541**)

Fuligorubin-A is structurally closely related to several tetramic acids which are known as biologically active metabolites of microorganisms (*668*). The biological properties of fuligorubin-A have not yet been determined.

6. Further Pigments Containing Nitrogen

The red pigment, '490-quinone' found in the sporulating gill tissue of the common commercial mushroom *Agaricus bisporus* is a potent inhibitor of a number of enzymes containing thiol groups at their active sites (*705*). '490-Quinone', initially so called because of its absorption maximum at that wavelength, has been shown to be 2-hydroxy-4-imino-2,5-cyclohexadienone (**542**) by a combination of biosynthetic considerations, cyclic voltammetry experiments and chemistry (*489*). Thus, the structure (**542**) was confirmed by aerial oxidation of 4-aminocatechol in aqueous solution at pH 7.8 which gave a quinone imine exhibiting

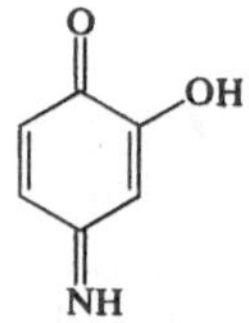

(**542**)

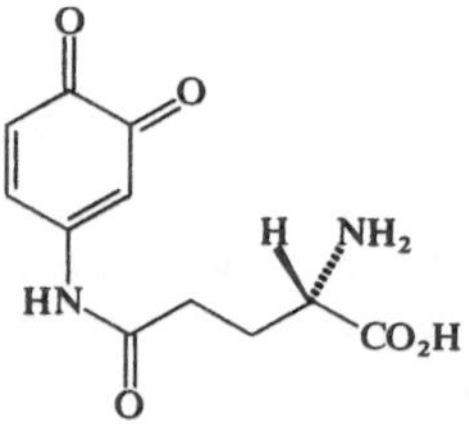

(**543**)

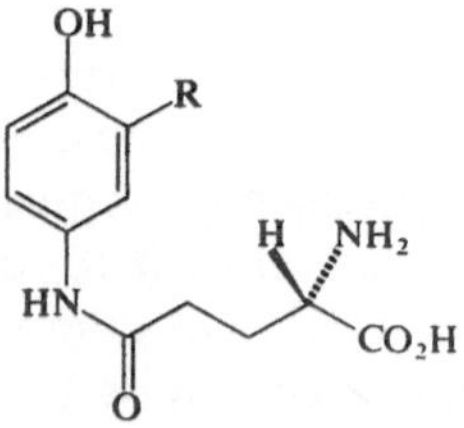

(**544**) R = H
(**545**) R = OH

References, pp. 253–286

the same cyclic voltammogram and electronic spectrum as the natural quinone. Furthermore, the quinone from *A. bisporus* could be converted to 3,4-diacetoxyacetanilide by reduction with sodium borohydride and subsequent acetylation.

The quinone imine (542) is formed *in vitro* and also presumably in the mushroom from γ-L-glutaminyl-3,4-benzoquinone (543) according to the mechanism shown in Scheme 90 (*489*). γ-L-Glutaminyl-3,4-benzoquinone (543) is itself produced by the action on γ-L-glutaminyl-4-hydroxybenzene (544) of an enzyme with tyrosinase activity which has been purified from extracts of *A. bisporus* (*109, 706, 707*). γ-L-Glutaminyl-4-hydroxybenzene (544) constitutes 1–2% of the dry matter of the gills of *A. bisporus* (*635, 708*) and is present also in the closely related *A. hortensis* (*383*).

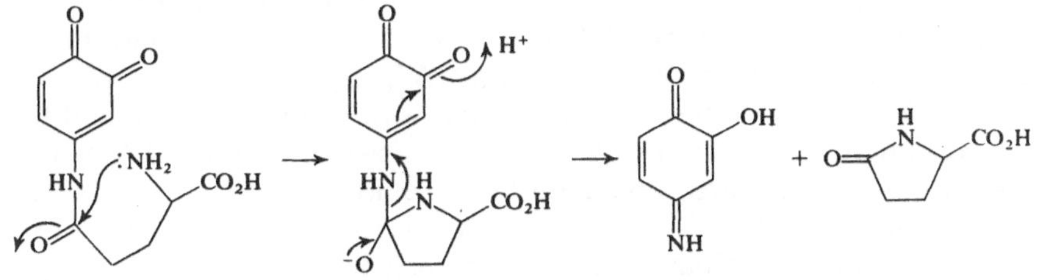

(543)   (542)

Scheme 90. Formation of '490-quinone'

In *Agaricus campester* (= *A. campestris*) an intermediate in the oxidative sequence (544) → (543) → (542) occurs in the form of the amino acid, agaridoxin (545) (*645*). The structure of agaridoxin was deduced principally from the ¹H-n.m.r. spectrum of the compound and from the mass spectrum of its pertrimethylsilyl ether derivative, and was confirmed by synthesis. Agaridoxin (545) and L-DOPA have been isolated from *A. bisporus* (*663*).

Labelling experiments have confirmed that the aromatic moiety of γ-L-glutaminyl-4-hydroxybenzene (544) is formed *via* shikimic acid in *A. bisporus* (*635, 664*). It is probably a metabolite of *p*-aminobenzoic acid as is the closely related compound agaritine (553) (*668*). *p*-Aminobenzoic acid may be oxidatively decarboxylated to yield *p*-aminophenol and thence (544).

The role of these aromatic metabolites in the fruit bodies is not known with certainty but they have been implicated in the control of sporulation (*109, 538, 706*) and as precursors in the formation of spore wall melanin (*358, 635*).

Several *Agaricus* species suffer reddening of the flesh on bruising. This phenomenon may be ascribed to the presence of the red quinones

(542) or (543) arising by tyrosinase action on the phenols (544) and (545). Reddening due to the action of tyrosinase on tyrosine as the chromogen has been observed with *Russula nigricans* (*115*) and with *Daedaleopsis confragosa* (*599*).

Several mushrooms belonging to *Agaricus* develop characteristic yellow zones when damaged. Fruit bodies of *A. xanthoderma* develop bright chrome yellow stains and exude a strong carbolic odour on handling. Underlying these changes is an intriguing series of transformations, the early details of which have only recently come to light. From ethanol extracts of the mascerated fruit bodies of *A. xanthoderma* have been isolated phenol in quantities sufficient to explain the aroma of the flesh and the unpleasant symptoms which accompany its ingestion and a minor yellow pigment identified as 4,4'-dihydroxyazobenzene (546). The nature of these compounds and their isolation together with quinol and 4,4'-dihydroxybiphenyl suggested a common biogenesis from 4-hydroxybenzenediazonium ion (*284*). By careful extraction of *A. xanthoderma* with methanol saturated with sulphur dioxide and by performing purification steps under argon at 0–3 °C the 4-hydroxybenzenediazosulphonate (547) could be obtained (*367*). The diazo compound (547) is an artefact formed from 4-diazo-2,5-cyclohexadien-1-one (548), which is present as such in the fruit bodies (*195, 367*). The diazo compound (548) has been detected chromatographically in methanolic extracts of the fungus obtained at low temperature in darkness and may be effectively trapped as the corresponding azo dye by coupling with resorcinol and with β-naphthol (*367*). Presumably, the azobenzene (546) arises by coupling of the diazo compound (548) with phenol produced during the original masceration procedure.

(546)   (547)   (548)   (549)

When fruit bodies of *A. xanthoderma* were extracted with cold ethyl acetate and the yellow extracts were concentrated and chromatographed at low temperature, the red crystalline pigment agaricone (549)

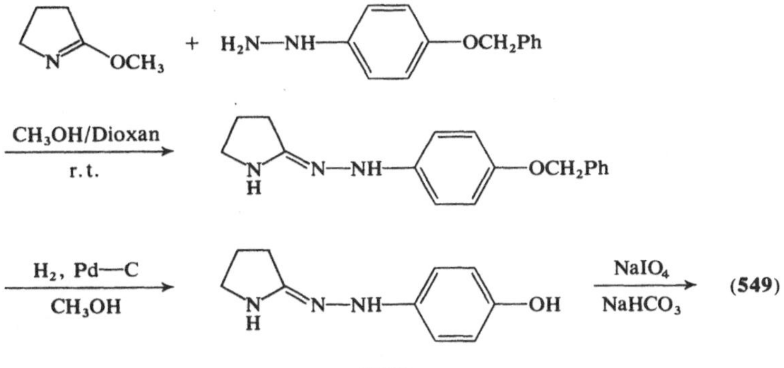

Scheme 91. Synthesis of agaricone

[UV/vis. (methanol): $\lambda_{max}(\log \varepsilon) = 242$ (3.74), 427 nm (4.19)] was obtained. The structure of agaricone followed from the spectroscopic data and was confirmed by synthesis *via* the colourless precursor, leucoagaricone **(550)** (Scheme 91) (*367*).

The major chromogenic principle present in *A. xanthoderma*, xanthodermin, has been isolated by extraction of the fungus using the methanol-sulphur dioxide technique and identified as the acylhydrazine **(551)** by spectroscopic methods and total synthesis (Scheme 92) (*367, 411*).

On treatment of xanthodermin [UV/vis. (water): $\lambda_{max}(\log \varepsilon) = 228$ (3.34), 292 nm (3.90)] with alkaline potassium ferricyanide solution or with sodium hydroxide solution an intense yellow colour [UV/vis.: $\lambda_{max} = 242$, 348, 445 nm] was produced. The same phenomenon is observed with an aqueous extract of the fungus and may be ascribed to the delocalised anion **(552)** of the corresponding acylazo compound (*337*).

The yellow discolouration of the flesh of *A. xanthoderma* may be attributed then to the production by oxidase enzymes in the fungus of a mixture of agaricone **(549)** and the anion **(552)** of dehydroxanthodermin from colourless precursors present in the intact mushroom. Although agaricone **(549)** could arise directly from leucoagaricone **(550)**, the latter compound has not been detected in *A. xanthoderma* and an alternative pathway to agaricone from xanthodermin **(551)** may in fact prevail (Scheme 93).

Agaricone **(549)**, xanthodermin **(551)** and the diazo compounds **(547)** and **(548)** are antibiotically active against *Bacillus subtilis* and *B. brevis* (*367*). Especially active against both bacteria and fungi are the two diazo compounds the minimum inhibitory concentrations of which are comparable with those of established antibiotics (*195, 367*).

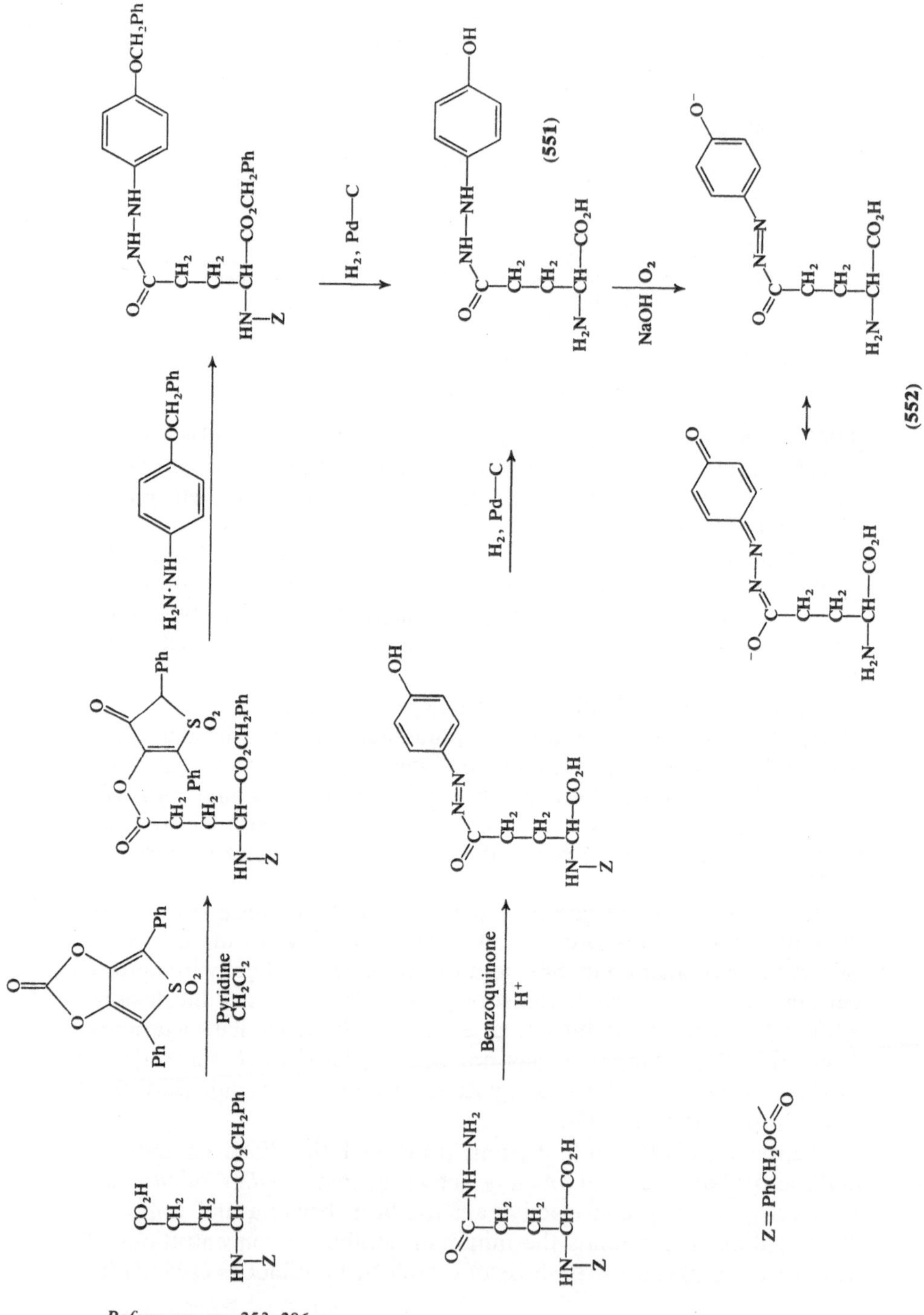

Scheme 92. Synthesis of xanthodermin

(551)

(552)

Z = PhCH₂OC⟩=O

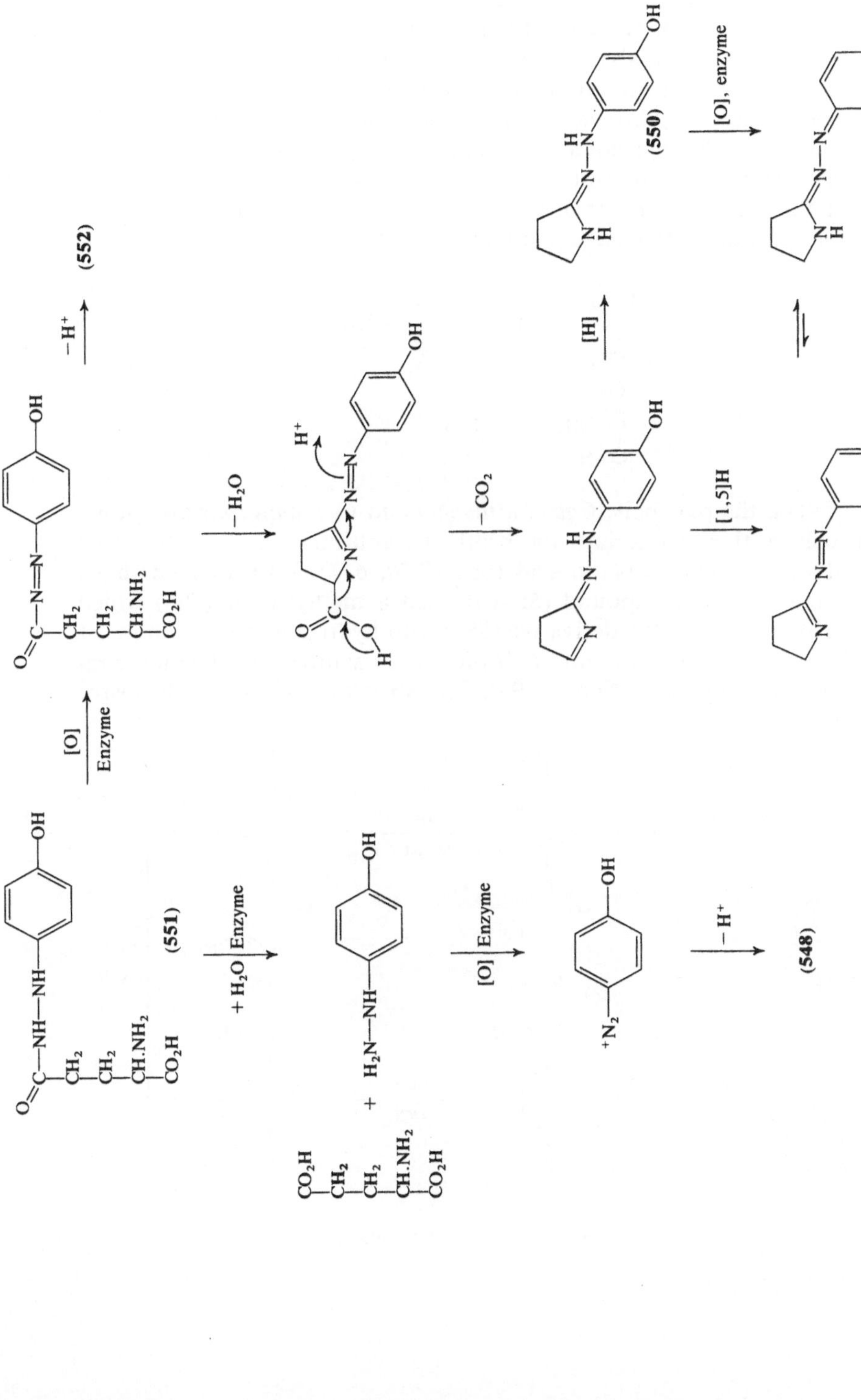

Scheme 93. Possible modes of pigment production in *Agaricus xanthoderma*

Xanthodermin (**551**) bears an obvious biogenetic relationship to agaritine (**553**), a metabolite of several *Agaricus* species (*668*) including *A. xanthoderma* (*454*). Interestingly, *A. bisporus* is known to produce enzymes that are capable of cleaving agaritine (**553**) to 4-(hydroxymethyl)phenylhydrazine and glutamate and thereafter of oxidising the hydrazine to the 4-(hydroxymethyl)phenyldiazonium ion (*549*). This diazonium ion, a known carcinogen in mice (*658*), is present at appreciable levels (0.6 p.p.m.) in fruit bodies of *A. bisporus* (*549*). It is possible that the same or similar enzymes in *A. xanthoderma* are responsible for generation of the diazo compound (**548**) (Scheme 93).

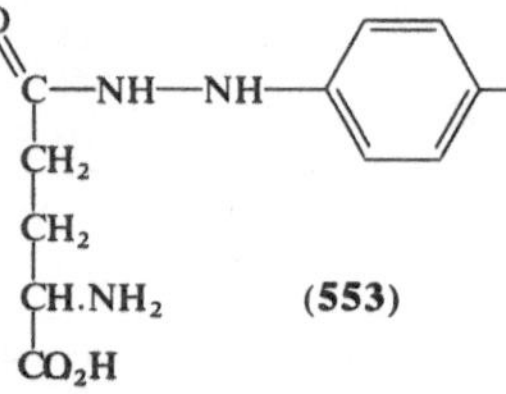

When the puff balls *Calvatia lilacina* and *C. craniformis* are grown in culture they exude into the broth the antifungal and antibacterial arylazoxycyanide, calvatic acid (**554**) (*270, 671*). With diazomethane the pale yellow compound (**554**) formed a methyl ester (**555**) which was hydrolysed to the derivative (**556**) with hydrogen chloride in moist ether. The azoxy compound (**556**) has been synthesised starting from *p*-aminobenzoic acid (Scheme 94). The structure and stereochemistry

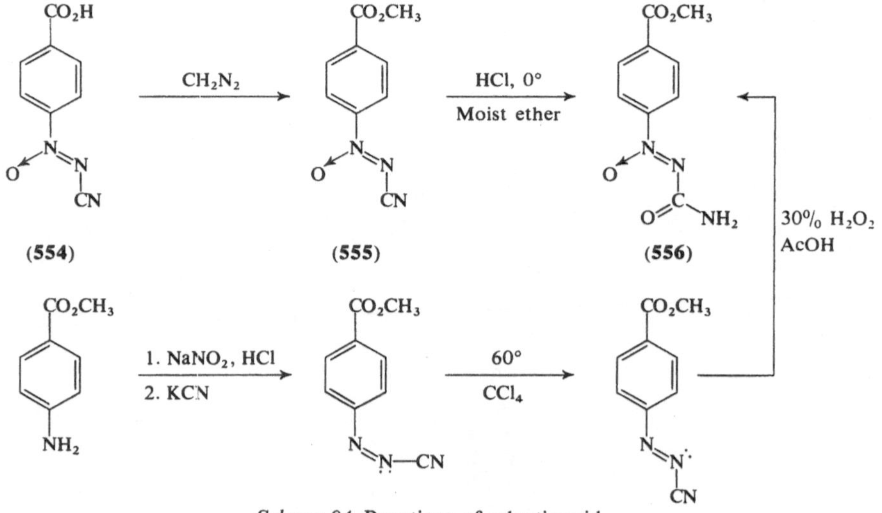

Scheme 94. Reactions of calvatic acid

of calvatic acid have been confirmed by an X-ray crystallographic analysis (*684*).

The white fruit bodies of the North American puff ball *Calvatia rubro-flava* turn golden yellow when the toadstools are touched or dried. The dried fruit bodies contain a unique group of benzoquinone semicarbazone pigments (*250*). These compounds may be isolated from the fungus by extraction with methanol and,have been purified by column chromatography on Sephadex LH-20 followed by preparative thin layer chromatography on silica gel.

The orange-red main pigment, rubroflavin (**557**) constitutes *ca.* 1% of the dry weight of the fruit bodies. From the F.A.B. mass spectrum the molecular formula $C_9H_{11}N_3S_2O_3$ was assigned which was in agreement with elemental analysis. From the ^{1}H-n.m.r. (Table 52) and electronic spectra (Table 53) the benzoquinone nucleus was recognised and the very high optical activity of the pigment, $[\alpha]_D = -2180°$, suggested that a chiral methanesulphinyl substituent was conjugated with the quinonoid chromophore.

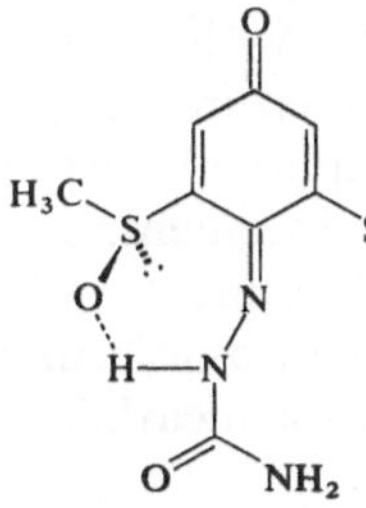

(**557**) 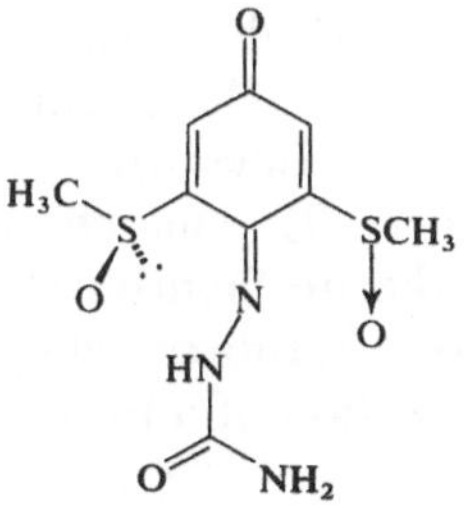   (**558**)

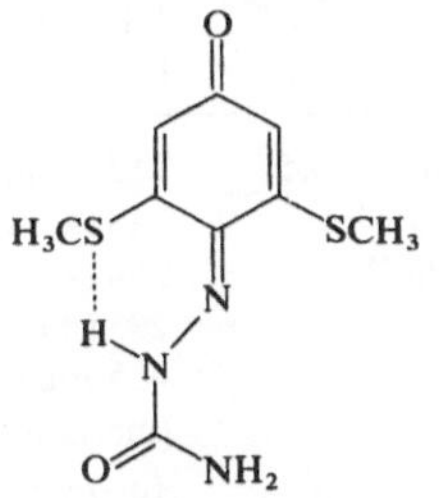

(**559**) 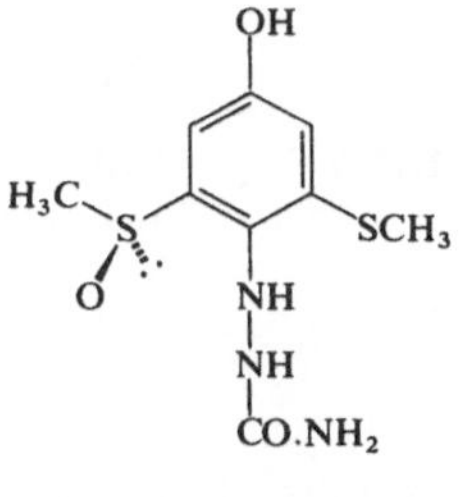   (**560**)

The *para*-quinone semicarbazone structure (**557**) for rubroflavin was deduced from the proton coupled ^{13}C-n.m.r. spectrum of the pigment and those of several model *ortho*- and *para*-quinone semicarbazones. These experiments allowed assignment of carbon resonances to rubroflavin as shown in Table 52.

Table 52. ^1H- and ^{13}C-N.m.r. Data for Rubroflavin (557)
(δ values, with TMS as internal standard)

[D₆] DMSO	CD₃OD
-6.57 d, 2.3 Hz; 5.97 d, 2.3 Hz; 2.64 s H₃C; 2.25 s CH₃; S; O; S; N; 6.42 br. H—N; O NH₂; 6.53 br.	180.8; 115.4; 116.9; 44.7 H₃C; 154.0; 148.9; 15.0 CH₃; 132.2; N; H—N; 166.4; O NH₂

In the mass spectrum, rubroflavin (557) showed no molecular ion but underwent instead a facile elimination of the elements of HNCO and N_2 to produce a base ion with the composition $C_8H_{10}S_2O_2$. Collapse of the molecular ion promoted by the expulsion of HNCO and N_2 was also observed on electron impact with oxyrubroflavin (558) and leucorubroflavin (560), but was not seen in the spectra of benzoquinone semicarbazones which lack a sulphoxide substituent, e.g. deoxyrubroflavin (559). This characteristic fragmentation has been rationalised by the mechanism shown in Scheme 95, which entails the sulphoxide group in rubroflavin providing intramolecular base catalysis for initial abstraction of a proton from the semicarbazone side chain.

Table 53. Physical Properties of the Rubroflavin Pigments

Pigment	Yield[a]	Colour M.p.	$[\alpha]_D^{22}$	TLC R_f[b]	Ultraviolet/Vis. $[\lambda_{max}(\log \varepsilon)]$
Rubroflavin (557)	1.14%	Red 184–185°	−2180° [c=0.07, methanol]	0.57	213 (4.44), 238 (4.36), 243 (4.38), 248 (4.36), 254 (4.28), 260 (4.18), 295 (4.04), 438 (4.48), 455 nm (sh., 4.44) [methanol]
Oxyrubroflavin (558)	7.6 × 10⁻²%	Orange-red 210° (dec.)	−860° [c=0.1, methanol]	0.24	190 (3.98), 304 (3.41), 345 (sh., 3.16), 434 (3.96), 455 nm (sh., 3.87) [methanol]
Deoxyrubroflavin (559)	1.5 × 10⁻³%	Orange –	±0°	0.77	347, 415 nm (sh.) [chloroform]

[a] Isolated yield as a proportion of the dried fungus.
[b] Kieselgel 60 (chloroform:methanol = 2:1) (250).

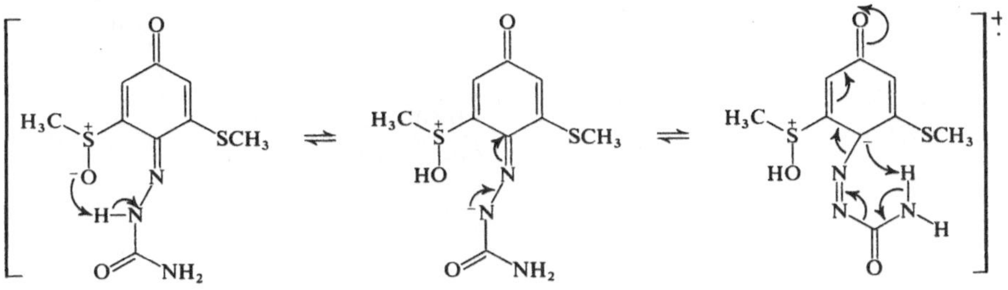

(557) M⁺ absent

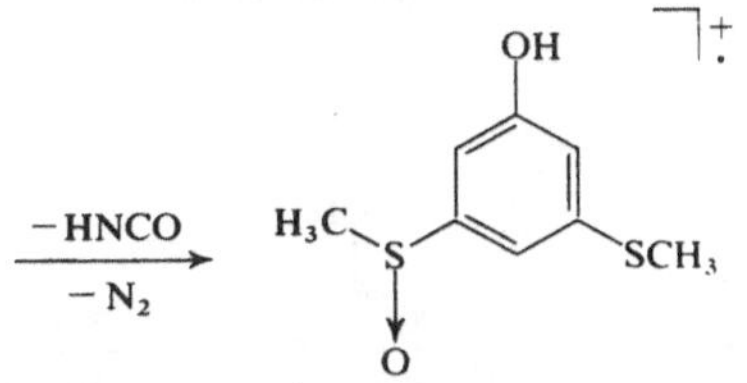

$-HNCO \atop -N_2$

$C_8H_{10}S_2O_2$
100%

Scheme 95. Fragmentation of rubroflavin on electron impact

The fragmentation depicted in Scheme 95 has also been thermally induced and has provided thereby a convenient method for determining the absolute configuration at the chiral sulphur atom in rubraflavin. Thus, brief thermolysis of rubroflavin (557) followed by preparative thin layer chromatography afforded the chiral sulphoxide (561) (Scheme 96). The CD spectrum of (561) exhibited a strong negative Cotton effect ($\Delta\varepsilon = -13.6$) at 241 nm which by comparison with those of aromatic-aliphatic sulphoxides of established stereochemistry (488) enabled the (S) -configuration to be assigned to (561) and hence to rubroflavin.

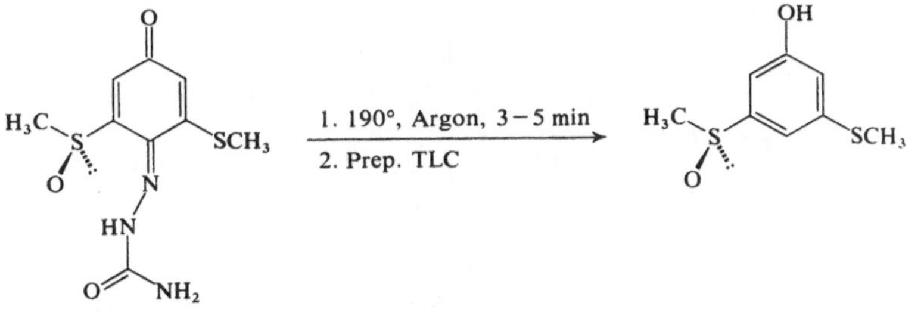

1. 190°, Argon, 3−5 min
2. Prep. TLC

(557)   (561)

Scheme 96. Thermolysis of rubroflavin

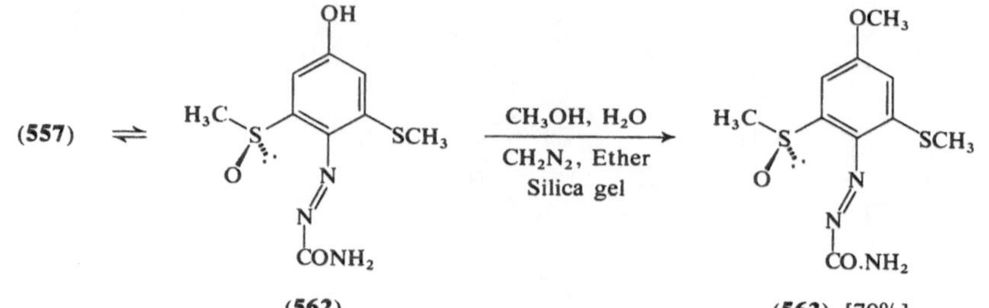

Scheme 97. Methylation of rubroflavin

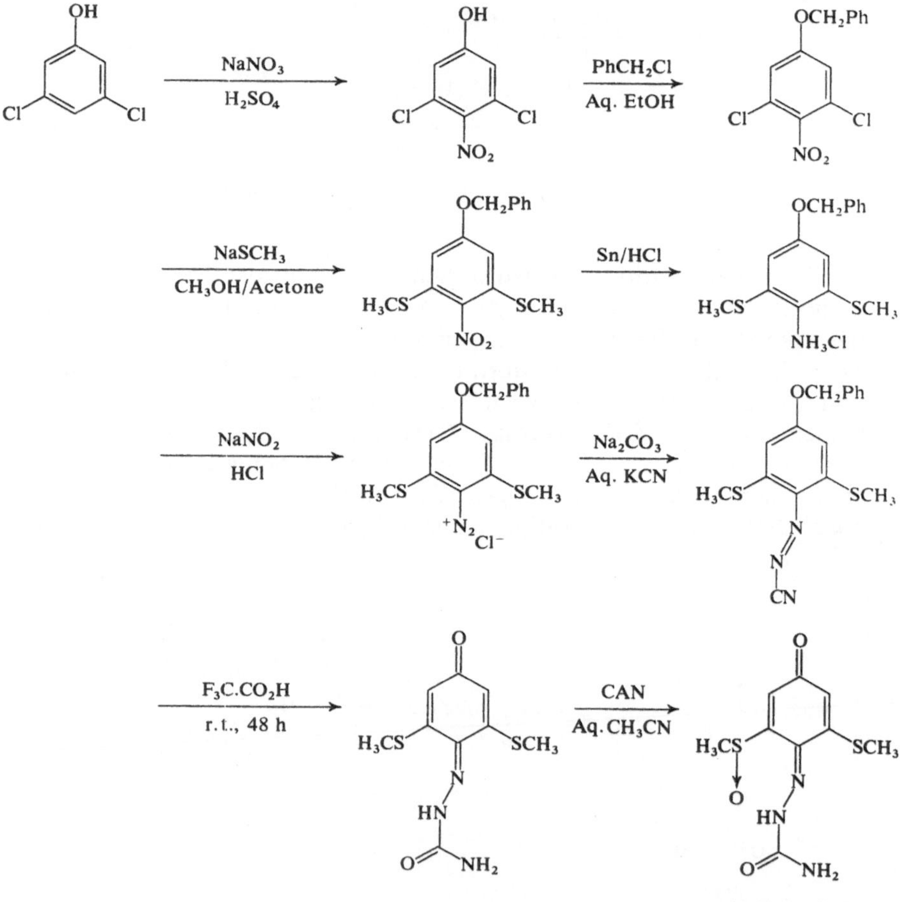

Scheme 98. Synthesis of (±)-rubroflavin

With diazomethane in a heterogenous mixture (Scheme 97) rubro-flavin was trapped in high yield as the methyl ether (563) of its phenolic tautomer (562). The ether (563) like rubroflavin itself is powerfully laevorotatory and its structure is in full accord with the spectroscopic data.

Racemic rubroflavin has been synthesised from 3,5-dichlorophenol according to the reactions shown in Scheme 98 (33).

An orange pigment, oxyrubroflavin (558) is more polar than rubro-flavin itself and showed stronger absorption in the infrared spectrum due to the presence of sulphoxide groups. In the ^{1}H-n.m.r. spectrum of oxyrubroflavin the presence of singlets at δ 2.82 and 2.87, and the coincidence at δ 7.31 of the signals from the protons of the quinonoid ring implied a degree of symmetry in the molecule and led to the bis-methanesulphinyl formula (558) for this compound. Significantly, oxidation of rubroflavin (557) using hydrogen peroxide in acetic acid produced oxyrubroflavin which was indistinguishable in its n.m.r. and CD spectra from the naturally derived pigment. This provided a strong indication that oxyrubroflavin, while possessing one methanesulphinyl group with (S)-stereochemistry, is racemic at the other chiral centre. This was proved to be the case by thermolysis of oxyrubroflavin which afforded the phenol (564). In the ^{1}H-n.m.r. spectrum of (564) the signal due to 4-H appeared as an apparent 'doublet' of triplets ($J_t = 1.7$ Hz) due to superimposition of the individual triplet resonances arising from the (R, S) (i.e. meso)- and (S, S)-diastereoisomers.

(564)

Deoxyrubroflavin (559) has been isolated in only trace amounts. The pigment is optically inactive and in the ^{1}H-n.m.r. spectrum exhib-ited singlets at δ 2.27 and 2.49 indicative of two methylmercapto substi-tuents.

The pigments (557), (558), and (559) are accompanied in dried fruit bodies of Calvatia rubro-flava by small quantities (0.038%) of leucoru-broflavin (560) in which form the pigment (557) is stored in the fresh fruit body. Leucorubroflavin identical with the natural product has been obtained by reduction of rubroflavin (557) with zinc in acetic acid. The oxidation of the leuco compound (560) to the pigments (557) and (**558**) took place rapidly on silica gel on exposure to air. The reactions were followed by two dimensional thin layer chromatography and they demonstrated that rubroflavin and oxyrubroflavin can arise from leucorubroflavin by nonenzymatic processes.

A plausible pathway for the biosynthesis from *p*-aminobenzoic acid of pigments in *Calvatia* is presented in Scheme 99 (*250*).

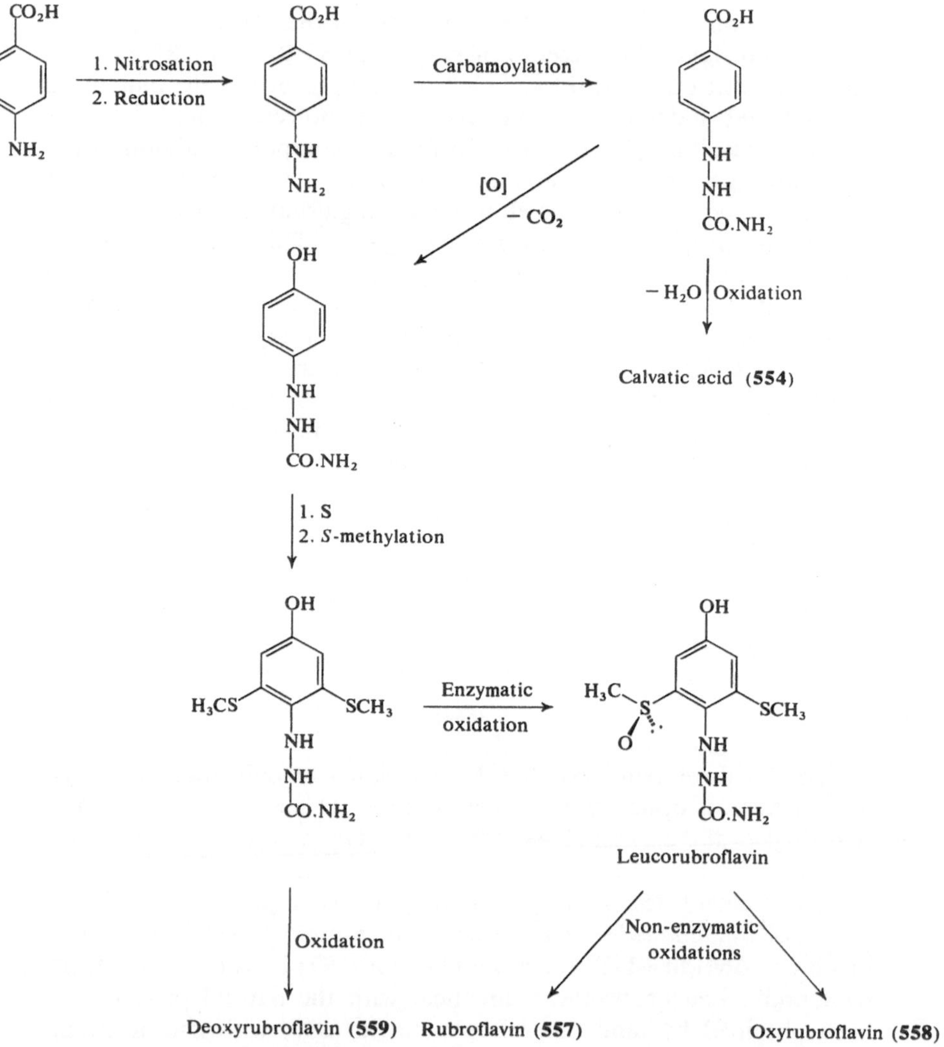

Scheme 99. Hypothetical biosynthetic relationships between *Calvatia* metabolites

7. Compounds Responsible for Colour Reactions

The reactivity of certain chemicals, including atmospheric oxygen, towards fungal tissue has been known since the earliest times (565). Thus, in addition to using the colours present in the intact fruit body as taxonomic characters numerous investigators have included the results of chemical spot tests in their descriptions of mushrooms and toadstools. Unfortunately much of this information remains widely dispersed in the mycological literature although several attempts have been made to bring much of it together (68, 482, 550, 557, 703).

Despite the importance of chemical tests to fungal systematics (64, 336, 346, 582, 583, 703) comparatively little has been published to-date about the nature of the fungal metabolites involved. In this final section we have endeavoured to collect together those few colour reactions for which a firm molecular basis has already been established. In doing so we necessarily preclude reference to numerous species of fungi which are known to contain chromogenic principles but on which little or no structural studies have as yet been undertaken.

In many cases the constituents which are responsible for colour changes are themselves colouring matters of the intact fungus and consequently are cited elsewhere in this review. In such cases further details of the structures and the chemistry of the substances involved may be found by reference to the appropriate section in the preceding text.

The reaction between constituents within the toadstool and atmospheric oxygen mediated by oxidase enzymes brings about the types of colour changes which are observed when some fungi are damaged by contact with man, animals or insects. It is noteworthy that a great many fungi contain high concentrations of oxidation sensitive polyphenolics which are capable of survival only in the reducing medium of the cell. The blueing of members of the Boletaceae and the red and yellow stains shown by several ubiquitous *Agaricus* species are familiar examples of such oxidative processes and the chemistry involved is now well understood. On the other hand the blueing of some other toadstools, e.g. *Inocybe aeruginascens*, various *Psilocybe* species, and *Panaeolus cyanescens*, is caused by the conversion of psilocin (**565**) into a pigment of still unknown constitution (77, 106, 269, 455).

The structures of several metabolites which react with atmospheric oxygen in the presence of an appropriate enzyme are now known and are given along with a brief summary of their respective distributions in Table 54. Within this category could also be included species such as *Cortinarius rufoolivaceus* (Section 3.5.3.G) and *Entoloma incanum*

(Section 5.5) which contain leuco compounds susceptable to purely chemical oxidation on exposure to air.

Oxidation reactions which result in the production of new pigments are not confined in their occurrence to the accidental damage of fungal fruit bodies. That such transformations may also take place under natural conditions is exemplified by the formation in many boletes of red colours due to the generation of variegatorubin (**74**) from variegatic acid (**73**). The red pigment variegatorubin arises in the external parts of the toadstool where the precursor is more exposed to oxygen.

The application of specific chemical reagents such as alkalis and their effect on fungal tissue was reported as early as 1872 by MÜLLER (*500*). He observed that when aqueous potassium hydroxide solution is applied to the fruit body of *Hapalopilus rutilans* (= *Polyporus rutilans*) an intense purple colour is produced. This and similar reactions, which are now known to be due to the presence of hydroxylated quinones

Table 54. *Colour Changes by Oxidation*

Species	Colour Change	Chemical Change
Daedaleopsis confragosa, *Russula nigricans*	Orange-red → grey → black	Tyrosine → L-DOPA (**172**) → Melanin (Scheme 31)
Rhodocybe mundula, *Hygrocybe conica*, *H. ovina*, *Strobilomyces floccopus*	Red → grey → black	L-DOPA (**172**) → Melanin (Scheme 31)
Boletus spp., *Xerocomus* spp., *Suillus variegatus* and other boletes (Table 8)	Yellow → blue	Variegatic acid (**73**) and/or Xerocomic acid (**72**) → Anions of the type (**77**) (Scheme 10)
Gyroporus cyanescens, *Leccinum* spp. (Table 13)	Pale yellow → bright blue	Gyrocyanin (**142**) → Anion (**147**) (Scheme 24)
Chamonixia caespitosa, *Gyrodon lividus*	Colourless → bright blue	Chamonixin (**144**) → Gyrocyanin anion (**147**)
Paxillus involutus	Colourless → brown	Oxidation of involutin (**145**)
Gomphidius glutinosus, *G. maculatus*	Colourless → red → black	Benzene-1,2,4-triol (**235**) → polymer
Agaricus bisporus and other *Agaricus* spp.	Colourless → red	(**544**) → (**545**) → (**543**) → (**542**)
Agaricus xanthoderma	Colourless → yellow	Xanthodermin (**551**) → Dehydroxanthodermin anion (**552**) (Scheme 92), and Leucoagaricone (**550**) → Agaricone (**549**)
Inocybe aeruginascens, *Psilocybe* spp., *Panaeolus cyanescens*	Colourless → blue	Oxidation of psilocin (**565**)

Table 55. *Colour Changes by Alkali*

Species	Colour Change	Chemical Change
		Ionisation of:
Hapalopilus rutilans	Brown → purple	Polyporic acid (**11**)
Paxillus atrotomentosus	Brown → purple	Atromentin (**13**)
Anthracophyllum spp.	Brown → green	Cycloleucomelone (**14**)
Thelephoraceae etc. (Table 3)	→ blue	Thelephoric acid (**16**)
Trichia spp.	Orange → violet	Hydroxynaphthoquinones e.g. (**280**)
Cortinarius, Dermocybe	Orange → red, violet	Hydroxyanthraquinones (Section 3.5.1)
Cortinarius, especially Subgenus *Phlegmacium*	Yellow → green, red, violet	Pre-anthraquinones (Section 3.5.3)
Aphyllophorales	Yellow → brown	Styrylpyrones (Section 2.3.2)
Fomes fomentarius	Brown → blood red	Fomentariol (**193**)
Albatrellus ovinus, A. subrubescens	Colourless → yellow	Scutigeral (**249**)
Chroogomphus rutilus, Ch. helveticus, Suillus bovinus	Orange → red	Boviquinone-3 (**221**) and Boviquinone-4 (**222**)

(Section 2.1.2), are also representative of many other fungi which contain quinonoid pigments capable of ionisation in the presence of base (Table 55). In many cases these quinones are themselves formed from still other molecules by the combined action of alkali and oxygen. Such changes apparently provide the basis for the taxonomically useful colour reactions exhibited by several *Cortinarius* species (*516, 617, 627*). Darkening caused by the addition of alkali to the flesh or to extracts of the fruit bodies of many fungi belonging to Aphyllophorales may be ascribed in general to the presence of phenolic constituents of the styrylpyrone type (Section 2.3.2) (*518*).

Because many fungal metabolites contain phenolic hydroxy groups the colour reactions shown by numerous fungi on application of ferric chloride solution is not entirely surprising. The presence in *Clavariadelphus pistillaris* of the bitter compound pistillarin (**566**) (*624*) is responsible for the intense green colour produced when the flesh of this fungus is treated with ferric chloride. The reaction is of interest for the taxonomy of some clavaroid fungi. Fruit bodies of *Lyophyllum connatum* contain the hydroxamic acid derivatives N,N-dimethyl-N'-hydroxyurea

(568) and connatin (567) which are jointly responsible for the violet colour which results when *L. connatum* is brought into contact with ferric chloride (*254*). Several compounds to which a major chromogenic role in this context has been recognised are identified in Table 56.

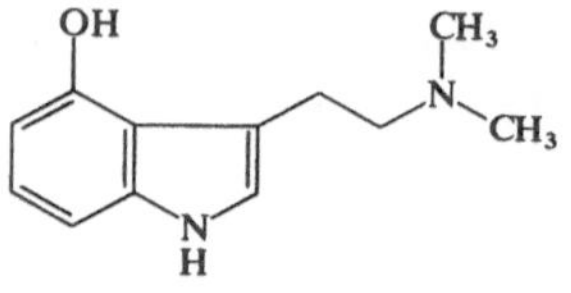

(565)

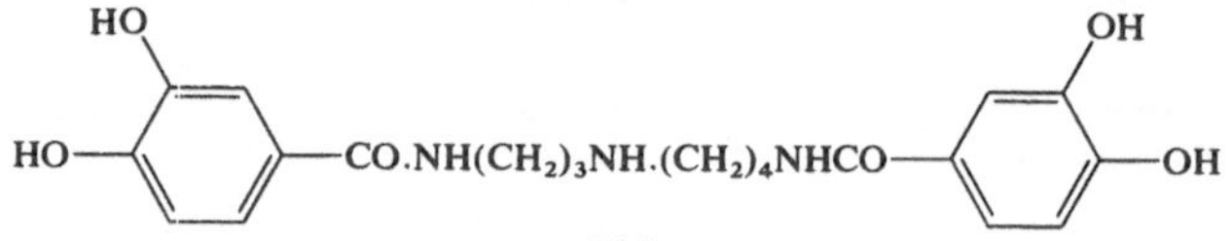

(566)

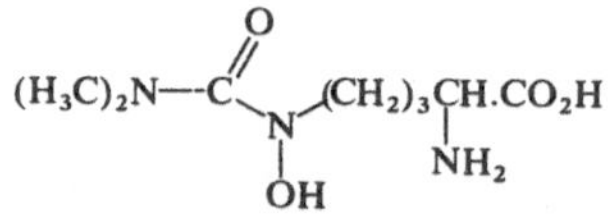

(567)

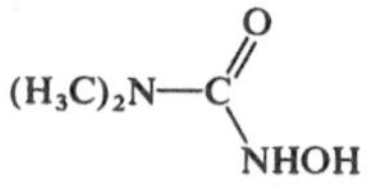

(568)

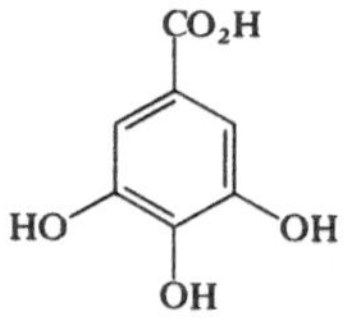

(569)

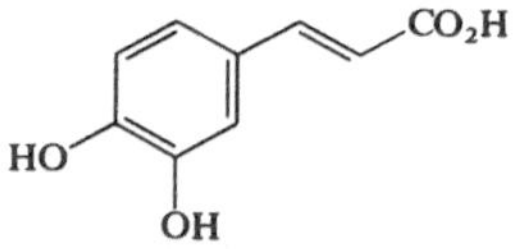

(570)

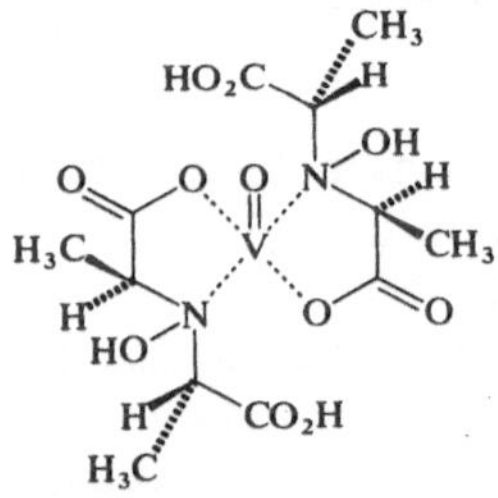

(571)

Ligand complexation of vanadium by *N*-(L-1-carboxyethyl)-*N*-hydroxy-L-alanine *in vivo* by fruit bodies of *Amanita muscaria* produces the blue complex, amavadin (571) (*69, 418*). The structure and stereochemistry of this most unusual compound has been proved by total synthesis (*418*), and the valence state of the metal in amavadin has been studied by electron spin resonance spectroscopy (*287*).

Many other reagents and chemical tests have been developed and employed by mycologists (*476, 582*). An account of the formulation of many of the more useful reagents and their effects on fungal tissue has been given by WATLING (*703*) who has also presented a valuable

Table 56. *Colour Changes by Ferric Chloride*

Species	Colour Change	Chromogen	References
Clavariadelphus pistillaris, *Ramaria* spp., *Gomphus* spp.	Colourless → dark green	Pistillarin (566)	(624)
Lyophyllum connatum	Colourless → violet	Connatin (567), and (568)	(254)
Gymnopilus spp. (Table 18)	Yellow → green	Bisnoryangonin (198)	–
Inonotus spp., *Gymnopilus* spp. (Table 18)	Yellow → green	Hispidin (199)	–
Leccinum scabrum	Colourless → green	(569) and (570)	(199)
Gloeophyllum odoratum	Yellow-brown → green	Trametin (Section 3.1)	(221)

critical appraisal of the scope and significance of chemical tests in agaricology.

The pigments of mushrooms and toadstools and the colours which they form with chemical reagents, particularly metal salts, are being exploited with growing enthusiasm and success for the dyeing of natural yarns and fabrics (640).

Acknowledgment

The authors wish to express their gratitude to Professors N. Arpin, A. Bresinsky, C.H. Eugster, J. Gripenberg and H. Musso, to Drs. H. Besl, R.L. Edwards, G. Keller and B. Oertel and to Mr. R. Schoenfeld for their invaluable comments and suggestions regarding the composition of the manuscript. Sincere thanks are also extended to colleagues cited in the bibliography who have generously provided information prior to publication.

For continuing support of their individual research efforts in the field of fungal pigments the authors thank the Deutsche Forschungsgemeinschaft (W.S.) and the Australian Research Grants Scheme (M.G.). The Australian National University and the Deutscher Akademischer Austauschdienst are gratefully acknowledged for support (to W.S. and M.G., respectively) during the preparation of the manuscript.

References

1. Aasen, A.J., and S. Liaaen-Jensen: Fungal Carotenoids II. The Structure of the Carotenoid Acid Neurosporaxanthin. Acta Chem. Scand. **19**, 1843 (1965).

2. Abrahamsson, S., and M. Innes: Molecular Structure of Xylerythrin – a Fungus Pigment. Acta Chem. Scand. **19**, 2246 (1965).

3. – – Molecular Structure of Xylerythrin, a Fungus Pigment. Acta Crystallogr. **21**, 948 (1966).

4. AGARWAL, S.C., and T.R. SESHADRI: Application of Ozonolysis to the Study of Substituted Derivatives of Vulpinic Acid. Constitution of Pinastric and Isopinastric Acids. Tetrahedron **19**, 1965 (1963).

5. – – A Reinvestigation of the Structure of Pinastric Acid and Isopinastric Acid. Indian J. Chem. **2**, 17 (1964).

6. AGHORAMURTHY, K., K.G. SARMA, and T.R. SESHADRI: The Structure of Thelephoric Acid. Tetrahedron Letters **1959**, 20.

7. AKAGI, M.: Investigation of the Mushroom Pigment from *Polyporus leucomelas* Fr. I. Isolation and Structure of a Pigment, Leucomelone. J. Pharmac. Soc. Japan **62**, 129 (1942).

8. – Synthesis of the Pigments of the Polyporic Acid Series by the Use of *N*-Nitrosoacetylarylamines. J. Pharmac. Soc. Japan **62**, 195 (1942).

9. – Investigation of the Mushroom Pigment from *Polyporus leucomelas* Fr. II. Synthesis of Leucomelone. J. Pharmac. Soc. Japan **62**, 202 (1942).

10. AKAGI, M., and K. HIROSE: Arylquinones I. Arylation of Quinone by *N*-Nitrosoacetylarylamines. J. Pharmac. Soc. Japan **62**, 191 (1942).

11. ÅKERMARK, B.: Studies on the Chemistry of Lichens 14. The Structure of Calycin. Acta Chem. Scand. **15**, 1695 (1961).

12. ALASOADURA, S.O., and S.A. VISSER: Pigment Study of *Sphaerobolus stellatus*. Mycopathol. Mycol. Appl. **47**, 295 (1972).

13. – – The Pigments of *Sphaerobolus stellatus* (Tode) Pers. Mycopathol. Mycol. Appl. **47**, 301 (1972).

14. ALLPORT, D.C., and J.D. BU'LOCK: A New Type of Melanin, and the Biogenesis of a Perylene Derivative. Proc. Chem. Soc. (London) **1957**, 264.

15. – – The Pigmentation and Cell-wall Material of *Daldinia* Sp. J. Chem. Soc. (London) **1958**, 4090.

16. – – Biosynthetic Pathways in *Daldinia concentrica*. J. Chem. Soc. (London) **1960**, 654.

17. ANCHEL, M., A.K. BOSE, K.S. KHANCHANDANI, and P.T. FUNKE: Origin of the Methylenedioxy Carbon in Phlebiarubrone: Formate and Methionine as Precursors. Phytochem. **9**, 2335 (1970).

18. ANCHEL, M., A. HERVEY, F. KAVANAGH, J. POLATNICK, and W.J. ROBBINS: Antibiotics from Basidiomycetes III. *Coprinus similis* and *Lentinus degener*. Proc. Nat. Acad. Sci. (USA) **34**, 498 (1948).

19. ANDERSON, J.M., and J. MURRAY: Isolation of 4:9-Dihydroxyperylene-3:10-quinone from a Fungus. Chem. and Ind. **1956**, 376.

20. ANDREWES, A.G., and S. LIAAEN-JENSEN: Fungal Carotenoids VII. Synthesis of β,γ- and γ,γ-Carotene with Terminal Methylene Groups. Acta Chem. Scand. **25**, 1922 (1971).

21. – – Animal Carotenoids 8. Synthesis of β,γ-Carotene and γ,γ-Carotene. Acta Chem. Scand. **27**, 1401 (1973).

22. ANKE, H., I. CASSER, R. HERRMANN, and W. STEGLICH: Neue Terphenylchinone aus Mycelkulturen von *Punctularia atropurpurascens* (Basidiomycetes). Z. Naturforsch. **39c**, 695 (1984).

23. ANKE, H., M. HOLZAPFEL, and W. STEGLICH: Unpublished work.

24. ANKE, T.: Unpublished work.

25. ANKE, T., H. BESL, R. HERRMANN, A.J. KRIEGER-BRAUER, U. MOCEK, and W. STEGLICH: Manuscript in preparation.

26. ANTKOWIAK, W.Z., and W.P. GESSNER: The Structures of Orellanine and Orelline. Tetrahedron Letters **1979**, 1931.

27. – – Photodecomposition of Orellanine and Orellinine, the Fungal Toxins of *Cortinarius orellanus* Fries and *Cortinarius speciosissimus*. Experientia **41**, 769 (1985).

28. ARCHARD, M.A., M. GILL, and R.J. STRAUCH: Anthraquinones from the Genus *Cortinarius*. Phytochem. **24**, 2755 (1985).

29. – – – Unpublished work.

30. ARIGONI, D.: Some Studies in the Biosynthesis of Terpenes and Related Compounds. Pure and Appl. Chem. **17**, 331 (1968).

31. ARIGONI, D., and B. ERB: Personal communication from Professor Arigoni.

32. ARNOLD, R., and W. STEGLICH: Unpublished work.

33. ARNOLD, S., and W. STEGLICH: Unpublished work.

34. ARPIN, N.: Recherches Chimiotaxinomiques sur les Champignons. Sur la Présence de Carotènes chez *Clitocybe venustissima* (Fries) Sacc. C. R. hebd. séances Acad. Sci., Ser. D **262**, 347 (1966).

35. – Les Caroténoides des Discomycètes: Essai Chimiotaxinomique. Thèse Doc. ès Sc., University of Lyon, 1968.

36. – Recherches Chimiotaxinomiques sur les Champignons XI. Nature et Distribution des Caroténoides chez les Discomycètes Operculés (*Sarcoscyphaceae* exclues); Conséquences Taxinomiques. Bull. soc. mycol. France **84**, 427 (1968).

37. ARPIN, N., and M.-P. BOUCHEZ: Recherches Chimiotaxinomiques sur les Champignons X. Étude Comparative de la Pigmentation de Deux Espèces du Genre *Melastiza* Boud. et de l'Espèce *Aleuria aurantia* (Pers. ex Fr.) Fuckel (= *Peziza aurantia* Pers. ex Fr.). Bull. soc. mycol. France **84**, 369 (1968).

38. ARPIN, N., and J. FAVRE-BONVIN: Les Pigments du Polypore *Fomes fomentarius* (L. ex Fr.) Kickx. Bull. soc. mycol. France **93**, 433 (1977).

39. ARPIN, N., J. FAVRE-BONVIN, and W. STEGLICH: Le Fomentariol: Nouvelle Benzotropolone Isolée de *Fomes fomentarius*. Phytochem. **13**, 1949 (1974).

40. ARPIN, N., and J.-L. FIASSON: The Pigments of Basidiomycetes: Their Chemotaxonomic Interest. In: Evolution in the Higher Basidiomycetes (R.H. PETERSEN, ed.), p. 63–98. Knoxville: University of Tennessee Press. 1971.

41. ARPIN, N., J.-L. FIASSON, M.P. BOUCHEZ-DANGYE-CAYE, G.W. FRANCIS, and S. LIAAEN-JENSEN: A New $C_{40}H_{56}$ Carotene with a Terminal Methylene Group. Phytochem. **10**, 1595 (1971).

42. ARPIN, N., J.-L. FIASSON, and P. LEBRETON: Méthodes Modernes d'Analyse Structurale des Caroténoides. Produits et problèmes pharmaceutiques **24**, 630 (1969).

43. – – – Méthodes Modernes d'Analyse Structurale des Caroténoides (suite et fin). Produits et problèmes pharmaceutiques **25**, 21 (1970).

44. ARPIN, N., H. KJØSEN, G.W. FRANCIS, and S. LIAAEN-JENSEN: The Structure of Aleuriaxanthin. Phytochem. **12**, 2751 (1973).

45. ARPIN, N., and R. KÜHNER: Les Grandes Lignes de la Classification des Boletales. Bull. Soc. Linnéenne de Lyon **1977**, 83 (1977).

46. ARPIN, N., P. LEBRETON, and J.-L. FIASSON: Recherches Chimiotaxinomiques sur les Champignons II. Les Caroténoides de *Peniophora aurantiaca* (Bres.) (Basidiomycètes). Bull. soc. mycol. France **82**, 450 (1966).

47. ARPIN, N., and S. LIAAEN-JENSEN: Recherches Chimiotaxinomiques sur les Champignons. Fungal Carotenoids IV. Les Caroténoides de *Phillipsia carminea* (Pat.) Le Gal; Isolement et Identification d'une Xanthophylle Naturelle Nouvelle. Bull. soc. chim. biol. (Paris) **49**, 527 (1967).

48. – – Recherches Chimiotaxinomiques sur les Champignons. Sur la Présence de l'Ester Méthylique de la Torularhodine chez *Cookeina sulcipes* (Berk.) Kuntze (Ascomycètes). C.R. hebd. séances Acad. Sci., Ser. D **265**, 1083 (1967).

49. – – Recherches Chimiotaxinomiques sur les Champignons. Fungal Carotenoids III. Nouveaux Caroténoides, Notamment Sous Forme d'Esters Tertiaires, Isolés de *Plectania coccinea* (Scop. ex Fr.) Fuck. Phytochem. **6**, 995 (1967).

50. ARPIN, N., S. THIVEND, and J. FAVRE-BONVIN: Substances Photoabsorbantes de

Stereum hirsutum (Willd. ex Fr.) Fr.; Isolement, Purification et Caractéristiques de P 310 (Mycosporine). Bull. soc. mycol. France **93**, 39 (1977).

51. ASAHINA, Y., and S. SHIBATA: Untersuchungen über Flechtenstoffe XCIV. Über das Vorkommen der Thelephorsäure in den Flechten. Ber. dtsch. chem. Ges. **72 b**, 1531 (1939).

52. ASANO, M., and Y. KAMEDA: Über die Konstitution des Calycins und dessen Synthese (IV. Mitteil. über Flechten-Farbstoffe der Pulvinsäure-Reihe). Ber. dtsch. chem. Ges. **68 b**, 1568 (1935).

53. ATHERTON, J., B.W. BYCROFT, J.C. ROBERTS, P. ROFFEY, and M.E. WILCOX: Studies in Mycological Chemistry XXIII. The Structure of Flavomannin, a Metabolite of *Penicillium wortmanni* Klöck. J. Chem. Soc. (London) C **1968**, 2560.

54. AYER, W.A., and L.M. BROWNE: Terpenoid Metabolites of Mushrooms and Related Basidiomycetes. Tetrahedron **37**, 2199 (1981).

55. AYER, W.A., Y. HOYANO, I. VAN ALTENA, and Y. HIRATSUKA: Metabolites of Plant Disease-causing Fungi. *Gremmeniella abietina*. Rev. Latinoam. Quim. **13**, 84 (1982).

56. AYER, W.A., and D.R. TAYLOR: Metabolites of Bird's Nest Fungi 5. The Isolation of 1-Hydroxy-6-methyl-8-hydroxymethylxanthone, a New Xanthone, from *Cyathus intermedius*. Synthesis *via* Photoenolisation. Canad. J. Chem. **54**, 1703 (1976).

57. BACHMANN, E.: Spektroskopische Untersuchungen von Pilzfarbstoffen. Wissenschaftliche Beilage zu dem Programme des Gymnasiums und Realgymnasiums zu Plauen i.V. Plauen i.V.: M. Wieprecht. Ostern 1886.

58. – Botanisch-chemische Untersuchungen über Pilzfarbstoffe. Ber. dtsch. bot. Ges. **4**, 68 (1886).

59. BACHMANN, O., B. KEMPER, and H. MUSSO: The Green Pigment from the Fungus *Roesleria hypogea*. Liebigs Ann. Chem. **1986**, 305.

60. BACKHAUS, J.: Versuche zur Synthese von 3-Amino-4-methyl-6-hydroxy-2,5-dioxo-2,5-dihydropyridin und 4-Amino-3-methyl-6-hydroxy-2,5-dioxo-2,5-dihydropyridin. Diplomarbeit, Technical University, Berlin, 1975.

61. – Ein Beitrag zur Chemie der Azachinone und einiger 2,4-substitutierter Pyridinderivate. Dissertation, University of Bonn, 1979.

62. BALENOVIĆ, K., D. CERAR, Z. PUČAR, and V. ŠKARIĆ: The Chemistry of Higher Fungi III. Contribution to the Chemistry of the Genus *Russula*. Archiv. Kem. Jugoslav. **27**, 15 (1955).

63. BAMBERGER, M., and A. LANDSIEDL: Zur Kenntnis des *Polyporus rutilans* (P.) Fr. I. Monatsh. Chem. **30**, 673 (1909).

64. BARONI, T.J.: Chemical Spot-Test Reactions – Boletes. Mycologia **70**, 1064 (1978).

65. BARTH, H., G. BURGER, H. DÖPP, M. KOBAYASHI, and H. MUSSO: Fliegenpilzfarbstoffe VII. Konstitution und Synthese des Muscaflavins. Liebigs Ann. Chem. **1981**, 2164.

66. BARTH, H., M. KOBAYASHI, and H. MUSSO: Über die Synthese des Muscaflavins. Helv. Chim. Acta **62**, 1231 (1979).

67. BARTLE, K.D., R.L. EDWARDS, D.W. JONES, and I. MIR: Constituents of the Higher Fungi VII. The Photodimerisation of Hispidin Analogues; a Proton Magnetic Resonance Study. J. Chem. Soc. (London) C **1967**, 413.

68. BATAILLE, F.: Les Réactions Macrochimiques chez les Champignons. Bull. soc. mycol. France **63** (Suppl.) (1948).

69. BAYER, E., and H. KNEIFEL: Isolation of Amavadin, A Vanadium Compound Occurring in *Amanita muscaria*. Z. Naturforsch. **27 b**, 207 (1972).

70. BEAUMONT, P.C., and R.L. EDWARDS: Constituents of the Higher Fungi IX. Bovinone, 2,5-Dihydroxy-3-geranylgeranyl-1,4-benzoquinone from *Boletus (Suillus) bovinus* (Linn. ex Fr.) Kuntze. J. Chem. Soc. (London) C **1969**, 2398.

71. – – Constituents of the Higher Fungi X. The Chromenols and Chromanols of Bo-

vinone and a Synthesis of Hydroxy(methylthio)-quinones. J. Chem. Soc. (London) C **1971**, 1000.

72. – – Constituents of the Higher Fungi XI. Boviquinone-3, (2,5-Dihydroxy-3-farnesyl-1,4-benzoquinone), Diboviquinone-3,4, Methylenediboviquinone-3,3, and Xerocomic Acid from *Gomphidius rutilus* Fr. and Diboviquinone-4,4 from *Boletus* (*Suillus*) *bovinus* (Linn. ex Fr.) Kuntze. J. Chem. Soc. (London) C **1971**, 2582.

73. BEAUMONT, P.C., R.L. EDWARDS, and G.C. ELSWORTHY: Constituents of the Higher Fungi VIII. The Blueing of *Boletus* Species. Variegatic Acid, a Hydroxytetronic Acid from *Boletus* Species and a Reassessment of the Structure of Boletol. J. Chem. Soc. (London) C **1968**, 2968.

74. BENDZ, G.: 6-Methyl-1,4-naphthaquinone Produced by *Marasmius graminum*. Acta Chem. Scand. **5**, 489 (1951).

75. BENEDICT, R.G.: Chemotaxonomic Relationships among the Basidiomycetes. Advances in Applied Microbiology **13**, 1 (1970).

76. BENEDICT, R.G., and L.R. BRADY: Antimicrobial Activity of Mushroom Metabolites. J. Pharm. Sci. **61**, 1820 (1972).

77. BENEDICT, R.G., V.E. TYLER, and R. WATLING: Blueing in *Conocybe*, *Psilocybe*, and a *Stropharia* Species and the Detection of Psilocybin. Lloydia **30**, 150 (1967).

78. BENEŠOVÁ, V., V. HEROUT, and F. ŠORM: Plant Substances III. Substances from *Lactarius deliciosus* L. Collect. Czech. Chem. Comm. **19**, 1351 (1954); Chem. Listy **48**, 882 (1954).

79. – – – On the Nature of Azulenogenic Compounds from *Lactarius deliciosus* L. Preliminary Note. Collect. Czech. Chem. Comm. **20**, 510 (1955); Chem. Listy **49**, 779 (1955).

80. BENNETT, G.J., and N. URI: Hydroxypolyporic Acids and Related Compounds as Antioxidants for Linoleic Acid and Methyl Linoleate. J. Chem. Soc. (London) **1962**, 2753.

81. BENTLEY, R., and D. CHEN: Helicobasidin: a Fungal Benzoquinone of Isoprenoid Origin. Phytochem. **8**, 2171 (1969).

82. BERGENDORFF, O., and O. STERNER: Personal communication from Dr. Sterner.

83. BERTELLI, D.J., and J.H. CRABTREE: Naturally Occurring Fulvene Hydrocarbons. Tetrahedron **24**, 2079 (1968).

84. BERTRAND, G.: Sur le Bleuissement de Certains Champignons. C. R. hebd. séances Acad. Sci. **133**, 1233 (1901).

85. – Sur l'Extraction du Bolétol. C. R. hebd. séances Acad. Sci. **134**, 124 (1902).

86. BESL, H., and A. BRESINSKY: Notizen über Vorkommen und systematische Bewertung von Pigmenten in Höheren Pilzen 2. Z. Pilzkd. **43**, 311 (1977).

87. – Unpublished work.

88. BESL, H., A. BRESINSKY, R. HERRMANN, and W. STEGLICH: Chamonixin und Involutin, zwei chemosystematisch interessante Cyclopentandione aus *Gyrodon lividus* (Boletales). Z. Naturforsch. **35c**, 824 (1980).

89. BESL, H., A. BRESINSKY, and A. KÄMMERER: Chemosystematik der Coniophoraceae. Z. Mykol. **52**, 277 (1986).

90. BESL, H., A. BRESINSKY, C. KILPERT, and W. STEGLICH: Unpublished work.

91. BESL, H., A. BRESINSKY, L. KOPANSKI, and W. STEGLICH: Pilzpigmente XXXV. 3-O-Methylvariegatsäure und verwandte Pulvinsäurederivate aus Kulturen von *Hygrophoropsis aurantiaca* (Boletales). Z. Naturforsch. **33c**, 820 (1978).

92. BESL, H., A. BRESINSKY, and I. KRONAWITTER: Notizen über Vorkommen und systematische Bewertung von Pigmenten in Höheren Pilzen 1. Z. Pilzkd. **41**, 81 (1975).

93. BESL, H., A. BRESINSKY, B. MEIXNER, U. MOCEK, and W. STEGLICH: Verpacrocin, ein Polyenpigment aus Mycelkulturen von *Verpa digitaliformis* (Pers.) Fr. (Ascomycetes). Z. Naturforsch. **38c**, 492 (1983).

94. Besl, H., A. Bresinsky, B. Oertel, and W. Steglich: Manuscript in preparation.
95. Besl, H., A. Bresinsky, W. Steglich, and K. Zipfel: Pilzpigmente XVII. Über Gyrocyanin, das blauende Prinzip des Kornblumenröhrlings (*Gyroporus cyanescens*), und eine oxidative Ringverengung des Atromentins. Chem. Ber. **106**, 3223 (1973).
96. Besl, H., R. Halbauer, and W. Steglich: Neue Anthrachinonfarbstoffe aus *Cortinarius armillatus* und *C. miniatopus* (Agaricales). Z. Naturforsch. **33c**, 294 (1978).
97. Besl, H., H.-J. Hecht, P. Luger, V. Pasupathy, and W. Steglich: Pilzpigmente XXIII. Tridentochinon, ein [13](3,6)Benzofuranophan aus *Suillus tridentinus* (Boletales). Chem. Ber. **108**, 3675 (1975).
98. Besl, H., G. Höfle, B. Jendrny, E. Jägers, and W. Steglich: Pilzpigmente XXXI. Farnesylphenole aus *Albatrellus*-Arten (Basidiomycetes). Chem. Ber. **110**, 3770 (1977).
99. Besl, H., C. Kilpert, B. Steffan and W. Steglich: Bovinsäure, ein ungewöhnliches Pulvinsäure-Derivat aus Kulturen von *Suillus bovinus* (Boletales). Submitted for publication.
100. Besl, H., J.-D. Klamann, and W. Steglich: Unpublished work.
101. Besl, H., I. Michler, R. Preuss, and W. Steglich: Pilzpigmente XXII. Grevillin D, der Hauptfarbstoff von *Suillus granulatus*, *S. luteus* and *S. placidus* (Boletales). Z. Naturforsch. **29c**, 784 (1974).
102. Birkinshaw, J.H., and R. Gourlay: The Structure of Dermocybin. Biochem. J. **80**, 387 (1961).
103. Blackburn, G.M., D.E.U. Ekong, A.H. Neilson, and Lord Todd: Xylindein. Chimia **19**, 208 (1969).
104. Blackburn, G.M., A.H. Neilson, and Lord Todd: The Structure of Xylindein. Proc. Chem. Soc. (London) **1962**, 327.
105. Blackwell, M., and A. Busard: The Use of Pigments as a Taxonomic Character to Distinguish Species of the Trichiaceae (Myxomycetes). Mycotaxon **7**, 61 (1978).
106. Bocks, S.M.: The Metabolism of Psilocin and Psilocybin by Fungal Enzymes. Biochem. J. **106**, 12P (1968).
107. Bodo, B., R.G. Tih, D. Davoust, and H. Jaquemin: Hypoxylone, a Naphthyl-naphthoquinone Pigment from the Fungus *Hypoxylon sclerophaeum*. Phytochem. **22**, 2579 (1983).
108. Boehm, R.: Beiträge zur Kenntnis der Hutpilze in chemischer und toxikologischer Beziehung. Arch. exp. Pathol. Pharmakol. **19**, 60 (1885).
109. Boekelheide, K., D.G. Graham, P.D. Mize, C.W. Anderson, and P.W. Jeffs: Synthesis of γ-L-Glutaminyl-[3,5-^{3}H]-4-hydroxybenzene and the Study of Reactions Catalysed by the Tyrosinase of *Agaricus bisporus*. J. Biol. Chem. **254**, 12185 (1979).
110. Bohlmann, F., T. Burkhardt, and C. Zdero: Naturally Occurring Acetylenes. New York: Academic Press. 1973.
111. Bollinger, P.: Über die Konstitution und Konfiguration der Lagopodine A, B und C. Dissertation, ETH, Zürich, 1965.
112. Bonnet, J.L.: Application de la Chromatographie sur Papier à l'Étude de Divers Champignons [Basidiomycètes-Hyménomycètes]. Bull. soc. mycol. France **75**, 215 (1959).
113. Bose, A.K., K.S. Khanchandani, P.T. Funke, and M. Anchel: Biosynthesis of Phlebiarubrone in *Phlebia strigosozonata*. Chem. Commun. **1969**, 1347.
114. Bottom, C.B., and D.J. Siehr: Hydroxylagopodin B, a Sesquiterpenoid Quinone From a Mutant Strain of *Coprinus macrorhizus* var. *microsporus*. Phytochem. **14**, 1433 (1975).
115. Bourquelot, E., and G. Bertrand: Le Bleuissement et le Noircissement des Champignons. C. R. séances soc. biol. **47**, 582 (1895).

116. BRADY, L.R., and R.G. BENEDICT: Occurrence of Bisnoryangonin in *Pholiota squarroso-adiposa.* J. Pharm. Sci. **61**, 318 (1972).

117. BRÄM, A., and C.H. EUGSTER: Synthese der Purpurin-8-carbonsäure; Beitrag zum Boletol-Problem. Helv. Chim. Acta **52**, 165 (1969).

118. BRANDRUD, T.E.: En Undersøkelse av Pigmentene i *Cortinarius*, Underslekt *Phlegmacium* i Norden. Hovedfagsoppgave, University of Oslo, 1980.

119. – *Cortinarius* subgen. *Cortinarius* (Agaricales) in Nordic Countries, Taxonomy, Ecology and Chorology. Nord. J. Bot. **3**, 577 (1983).

120. – Personal communication.

121. BRESINSKY, A.: Über die Natur einiger Farbstoffe des Hausschwammes (*Serpula lacrimans*). Z. Naturforsch. **28c**, 627 (1973).

122. – Zur Frage der taxonomischen Relevanz chemischer Merkmale bei Höheren Pilzen. Trav. mycol. dédiés à R. Kühner, Bull. Soc. Linnéenne de Lyon **1974**, 61.

123. – Chemotaxonomie der Pilze. In: Beiträge zur Biologie der niederen Pflanzen: Systematik, Stammesgeschichte, Ökologie (W. FREY, H. HURKA, and F. OBERWINKLER, eds.), p. 25–42. Stuttgart: G. Fischer. 1977.

124. – Zur Kenntnis der Hygrocybenpigmente. Z. Mykol. **52**, 321 (1986) [with I. Kronawitter].

125. – Unpublished work.

126. BRESINSKY, A., and R. BACHMANN: Bildung von Pulvinsäurederivaten durch *Hygrophoropsis aurantiaca* (Paxillaceae-Boletales) in vitro. Z. Naturforsch. **26b**, 1086 (1971).

127. BRESINSKY, A., and H. BESL: Notizen über Vorkommen und systematische Bewertung von Pigmenten in Höheren Pilzen 3. Untersuchungen an Boletales aus Amerika. Z. Mykol. **45**, 247 (1979).

128. – – Zum verwandtschaftlichen Anschluß von *Omphalotus*. Beih. Sydowia **8**, 98 (1979).

129. – – Giftpilze. Ein Handbuch für Apotheker, Ärzte und Biologen. Stuttgart: Wissenschaftliche Verlagsgesellschaft. 1985, and references therein.

130. BRESINSKY, A., H. BESL, and W. STEGLICH: Gyroporin und Atromentinsäure aus *Leccinum-aurantiacum*-Kulturen. Phytochem. **13**, 271 (1974).

131. BRESINSKY, A., and P. ORENDI: Chromatographische Analyse von Farbmerkmalen der Boletales und anderer Macromyzeten auf Dünnschichten. Z. Pilzkd. **36**, 135 (1970).

132. BRESINSKY, A., and A. RENNSCHMID: Pigmentmerkmale, Organisationsstufen und systematische Gruppen bei Höheren Pilzen. Ber. dtsch. bot. Ges. **84**, 313 (1971).

133. BREWER, D., W.A. JERRAM, and A. TAYLOR: The Production of Cochliodinol and a Related Metabolite by *Chaetomium* Species. Canad. J. Microbiol. **14**, 861 (1968).

134. BRIGGS, L.H., R.C. CAMBIE, I.C. DEAN, S.H. DROMGOOLE, B.J. FERGUS, W.B. INGRAM, K.G. LEWIS, C.W. SMALL, R. THOMAS, and D.A. WALKER: Chemistry of Fungi 10. Metabolites of Some Fungal Species. New Zealand J. Sci. **18**, 565 (1975).

135. BRIGGS, L.H., R.C. CAMBIE, I.C. DEAN, R. HODGES, W.B. INGRAM, and P.S. RUTLEDGE: Chemistry of Fungi XI. Corticins A, B, and C, Benzobisbenzofurans from *Corticium caeruleum*. Austral. J. Chem. **29**, 179 (1976).

136. BRITTON, G.: Later Reactions of Carotenoid Biosynthesis. Pure and Appl. Chem. **47**, 223 (1976).

137. BUCHECKER, R., N. ARPIN, and S. LIAAEN-JENSEN: Absolute Configuration of Aleuriaxanthin. Phytochem. **15**, 1013 (1976).

138. BUCK, R.W.: Psychedelic Effects of *Pholiota spectabilis*. New Engl. J. Med. **276**, 391 (1967).

139. BÜCHI, G., J.D. WHITE, and G.N. WOGAN: The Structures of Mitorubrin and Mitorubrinol. J. Amer. Chem. Soc. **87**, 3484 (1965).

140. BU'LOCK, J.D.: Constituents of the Higher Fungi IV. A Quinone from *Polyporus fumosus*. J. Chem. Soc. (London) **1955**, 575.

141. – Fungal Metabolites with Structural Function. In: Essays in Biosynthesis and Microbial Development, p. 1–18. New York: J. Wiley and Sons. 1967.

142. BU'LOCK, J.D., and J. DARBYSHIRE: Lagopodin Metabolites and Artefacts in Cultures of *Coprinus*. Phytochem. **15**, 2004 (1976).

143. BU'LOCK, J.D., P.R. LEEMING, and H.G. SMITH: Pyrones II. Hispidin, a New Pigment and Precursor of a Fungus 'Lignin'. J. Chem. Soc. (London) **1962**, 2085.

144. BU'LOCK, J.D., and H.G. SMITH: Pyrones I. Methyl Ethers of Tautomeric Hydroxypyrones and the Structure of Yangonin. J. Chem. Soc. (London) **1960**, 502.

145. – – A Fungus Pigment of Novel Type, and the Nature of Fungus 'Lignin'. Experientia **17**, 553 (1961).

146. BURTON, J.F., and B.F. CAIN: Antileukaemic Activity of Polyporic Acid. Nature **184**, 1326 (1959).

146a. BUTENANDT, A., J. KECK, and G. NEUBERT: Über Ommochrome VIII. Modell-Versuche zur Konstitution der Ommochrome: über Oxidationsprodukte der 3-Hydroxy-anthranilsäure. Liebigs Ann. Chem. **602**, 61 (1957).

147. BUTRUILLE, D., and X.A. DOMINGUEZ: Un Nouveau Produit Naturel: Dimethoxy-1,4 Nitro-2 Trichloro-3,5,6 Benzene. Tetrahedron Letters **1972**, 211.

148. CAIN, B.F.: Potential Anti-tumour Agents I. Polyporic Acid Series. J. Chem. Soc. (London) **1961**, 936.

149. – Potential Anti-tumour Agents II. Polyporic Acid Series. J. Chem. Soc. (London) **1963**, 356.

150. – Potential Anti-tumour Agents IV. Polyporic Acid Series. J. Chem. Soc. (London) C **1966**, 1041.

151. CAMBIE, R.C.: Chemistry of Fungi VI. A Phenylglyoxylic Acid from *Poria sinuosa* Fr. New Zealand J. Sci. **13**, 306 (1970).

152. CAMBIE, R.C., and R.W. LE QUESNE: Chemistry of Fungi III. Constituents of *Coriolus sanguineus* Fr. J. Chem. Soc. (London) C **1966**, 72.

153. CAMPBELL, A.C., M.S. MAIDMENT, J.H. PICK, and D.F.M. STEVENSON: Synthesis of (*E*)- and (*Z*)-Pulvinones. J. Chem. Soc. (London) Perkin Trans. I **1985**, 1567.

154. CAREY, S.T., and M.S.R. NAIR: Metabolites of Pyrenomycetes III. Production of (+) Skyrin by *Hypomyces trichothecoides*. Lloydia **38**, 357 (1975).

155. CASSER, I.: Isolierung und Strukturaufklärung von Farbstoffen aus holzbewohnenden Pilzen. Diplomarbeit, University of Bonn, 1983.

156. – Beiträge zur Chemie der Myxomyceten. Dissertation, University of Bonn, 1986.

157. CASSER, I., G. ECKHARDT, F. KNOCH, B. STEFFAN, and W. STEGLICH: Manuscript in preparation.

158. CASSER, I., B. STEFFAN, and W. STEGLICH: Zur Chemie der Plasmodienfarbstoffe des Schleimpilzes *Fuligo septica* (Myxomycetes). Angew. Chem., in press.

159. CAVILL, G.W.K., P.S. CLEZY, and J.R. TETAZ: The Chemistry of Mould Metabolites II. A Partial Structure for Polystictin. J. Chem. Soc. (London) **1957**, 2646.

160. – – – The Structure of Cinnabarin (Polystictin). Proc. Chem. Soc. (London) **1957**, 346.

161. CAVILL, G.W.K., P.S. CLEZY, J.R. TETAZ, and R.L. WERNER: The Chemistry of Mould Metabolites III. The Structure of Cinnabarin (Polystictin). Tetrahedron **5**, 275 (1959).

162. CAVILL, G.W.K., P.S. CLEZY, and F.B. WHITFIELD: The Chemistry of Mould Metabolites IV. Reductive Acetylation and Reoxidation of Some Phenoxazin-3-ones. Tetrahedron **12**, 139 (1961).

163. CAVILL, G.W.K., B.J. RALPH, J.R. TETAZ, and R.L. WERNER: The Chemistry of Mould Metabolites I. Isolation and Characterisation of a Red Pigment from *Coriolus sanguineus* (Fr.). J. Chem. Soc. (London) **1953**, 525.

164. CAVILL, G.W.K., and J.R. TETAZ: The Nucleus of Polystictin. Chem. and Ind. **1956**, 986.

165. CHANDRA, P., G. READ, and L.C. VINING: Studies on the Biosynthesis of Volucrisporin II. Metabolism of Some Phenylpropanoid Compounds by *Volucrispora aurantiaca* Haskins. Canad. J. Biochem. **44**, 403 (1966).

166. CHEN, C.-L., H.-M. CHANG, and T.K. KIRK: Betulachrysoquinone Hemiketal: A *p*-Benzoquinone Hemiketal Macrocyclic Compound Produced by *Phanerochaete chrysosporium*. Phytochem. **16**, 1983 (1977).

167. CHEN, W.-S., Y.-T. CHEN, X.-Y. WAN, E. FRIEDRICHS, H. PUFF, and E. BREITMAIER: Die Struktur des Hypocrellins und seines Photooxidationsproduktes Peroxyhypocrellin. Liebigs Ann. Chem. **1981**, 1880.

168. CHILTON, W.S., C.P. HSU, and W.T. ZDYBAK: Stizolobic and Stizolobinic Acids in *Amanita pantherina*. Phytochem. **13**, 1179 (1974).

169. CIBULA, W.G.: The Pigments of *Hygrophorus* Section *Hygrocybe* and their Significance in Taxonomy and Phylogeny. Ph. D. Thesis, University of Massachusetts, 1976.

170. CLOSSE, A., and D. HAUSER: Isolierung und Konstitutionsermittlung von Chrysodin. Helv. Chim. Acta **56**, 2694 (1973).

171. CMELIK, S.: Prilog Poznavanju Kemijskog Sastava Gljiva *Sarcoscypha coccinea* i *Polystigma rubrum*. Arch. Kem. Jugoslav. **19**, 63 (1947).

172. COHEN-ADDAD, C., and J. RIONDEL: Structure de la Géogénine, Nouvelle Quinone Lactone Isolée de *Hohenbuehelia geogenius*. Acta Crystallogr. **B 37**, 1309 (1981).

173. CROWDEN, R.K., and B.J. RALPH: Further Metabolic Products of *Polyporus tumulosus* Cooke. Austral. J. Chem. **14**, 475 (1961).

174. CZAPEK, F.: Farbstoffe bei Höheren Pilzen. In: Biochemie der Pflanzen, 2nd ed., Vol. III, p. 374–379. Jena: G. Fischer. 1921.

175. CZEZUGA, B.: Badania nad karotenoidami u grzybów II. Rodzaj *Amanita*. Acta Mycologica **12**, 265 (1976).

176. – Badania nad karotenoidami u grzybów III. Owocniki niektórych gatunków z rodzaju *Suillus*. Acta Mycologica **13**, 257 (1977).

177. CZYGAN, F.-C., and M. GRÜNSFELDER: Carotinoide der Porter-Lincoln-Reihe in *Anthurus archeri* (Clathraceae; Basidiomycetales; Mycophyta). Z. Naturforsch. **30 c**, 297 (1975).

178. DALLACKER, F., and K. DITGENS: Derivate des Methylendioxybenzols 41. Synthese des Phlebiarubrons und der Polyporsäure. Z. Naturforsch. **30 c**, 1 (1975).

179. DANGY-CAYE, M.-P., and N. ARPIN: Présence de Styryl-6α pyrones, Notamment de *Bis*-Noryangonine et d'Hispidine, chez *Gymnopilus penetrans* (Fr. ex Fr.) Murr. Trav. mycol. dédiés à R. Kühner, Bull. Soc. Linnéenne de Lyon **1974**, 109.

180. DAVID, A., B. DEQUATRE, J.-L. FIASSON, and J. BERNILLON: Specification dans le Genre *Gloeophyllum* Karst. (Polyporaceae): Utilisation des Pigments, Recherche d'Enzymes, Interfertilites. Bull. Soc. Linnéenne de Lyon **1977**, 304.

181. DAVIES, B.H., and R.F. TAYLOR: Carotenoid Biosynthesis – The Early Steps. Pure and Appl. Chem. **47**, 211 (1976).

182. DEHMLOW, E.V., and H.-J. SCHULZ: Synthesis of Orellanine, the Lethal Poison of a Toadstool. Tetrahedron Letters **26**, 4903 (1985).

183. DENNIS, R.W.G.: British Ascomycetes, 2nd ed. Vaduz: J. Cramer. 1978.

184. DEPOVERE, P., and P. MOENS: Constituents and Pigments of the Fly Agaric *Amanita* [*Amanita muscaria* (Fries) Hooker]. J. Pharm. Belg. **39**, 238 (1984).

185. DIVEKAR, P.V., G. READ, and L.C. VINING: Volucrisporin: A Novel Fungal Pigment. Chem. and Ind. **1959**, 731.

186. DIVEKAR, P.V., G. READ, L.C. VINING, and R.H. HASKINS: Volucrisporin. Isolation, Structure, and Synthesis of the Methyl Ether. Canad. J. Chem. **37**, 1970 (1959).

187. DOBLER, M.: p-Bromobenzoyl-leukopleurotin ($C_{23}H_{27}BrO_6$). Cryst. Struct. Commun. **4**, 253 (1975).

188. DÖPP, H., W. GROB, and H. MUSSO: Über die Farbstoffe des Fliegenpilzes (*Amanita muscaria*). Naturwiss. **58**, 566 (1971).

189. DÖPP, H., S. MAURER, A.N. SASAKI, and H. MUSSO: Fliegenpilzfarbstoffe VIII. Die Konstitution der Musca-aurine. Liebigs Ann. Chem. **1982**, 254.

190. DÖPP, H., and H. MUSSO: Fliegenpilzfarbstoffe II. Isolierung und Chromophore der Farbstoffe aus *Amanita muscaria*. Chem. Ber. **106**, 3473 (1973).

191. – – Die Konstitution des Muscaflavins aus *Amanita muscaria* und über Betalaminsäure. Naturwiss. **60**, 477 (1973).

192. – – Eine chromatographische Analysenmethode für Betalainfarbstoffe in Pilzen und höheren Pflanzen. Z. Naturforsch. **29c**, 640 (1974).

193. DOMINGUEZ, X.A., R. FRANCO, J. CASTILLO, S.J. VERDE, B. CANTO, and A. GILBERTO: Veracruzalone as a New Tropone from *Polyporus tomentosus*. Rev. Latinoam. Quim. **13**, 116 (1982).

194. DONNELLY, D.M.X., and J. O'REILLY: 6-Methyldehydro-α-lapachone from *Fomes annosus*. Phytochem. **19**, 277 (1980).

195. DORNBERGER, K., W. IHN, W. SCHADE, D. TRESSELT, A. ZURECK, and L. RADICS: Antibiotics from Basidiomycetes. Evidence for the Occurrence of the 4-Hydroxybenzenediazonium Ion in the Extracts of *Agaricus xanthodermus* Genevier (Agaricales). Tetrahedron Letters **27**, 559 (1986).

196. DREYER, D.L., I. ARAI, C.D. BACHMAN, W.R. ANDERSON, JR., R.G. SMITH, and G.D. DAVES, JR.: Toxins Causing Noninflammatory Paralytic Neuronopathy. Isolation and Structure Elucidation. J. Amer. Chem. Soc. **97**, 4985 (1975).

197. DUMONT, R., and H. PFANDER: Synthese von (*S*)-Plectaniaxanthin. Helv. Chim. Acta **67**, 1283 (1984).

198. EDWARDS, R.L.: Constituents of the Higher Fungi XVII. Methyl Variegatate from the Fungus *Hygrophoropsis aurantiaca* (Wulfen ex Fr.). J. Chem. Res. (S) **1977**, 276.

199. EDWARDS, R.L., and G.C. ELSWORTHY: Constituents of the Higher Fungi V. The Phenolic Constituents of *Boletus* (*Leccinum*) *scaber* (Bull. ex Fr.) Gray. J. Chem. Soc. (London) C **1967**, 410.

200. – – Variegatic Acid, a New Tetronic Acid Responsible for the Blueing Reaction in the Fungus *Suillus* (*Boletus*) *variegatus* (Swartz ex Fr.). Chem. Commun. **1967**, 373.

201. EDWARDS, R.L., G.C. ELSWORTHY, and N. KALE: Constituents of the Higher Fungi IV. Involutin, a Diphenylcyclopenteneone from *Paxillus involutus* (Oeder ex Fries). J. Chem. Soc. (London) C **1967**, 405.

202. EDWARDS, R.L., and M. GILL: Constituents of the Higher Fungi XII. Identification of Involutin as (–)-*cis*-5-(3,4-Dihydroxyphenyl)-3,4-dihydroxy-2-(4-hydroxyphenyl)-cyclopent-2-enone and Synthesis of (±)-*cis*-Involutin Trimethyl Ether from Isoxerocomic Acid Derivatives. J. Chem. Soc. (London) Perkin Trans. I **1973**, 1529.

203. – – Constituents of the Higher Fungi XIII. 2-Aryl-3-methoxymaleic Anhydrides from Pulvinic Acid Derivatives. A Convenient Method for Determination of Structure of Fungal and Lichen Pulvinic Acid Derivatives. J. Chem. Soc. (London) Perkin Trans. I **1973**, 1538.

204. – – Constituents of the Higher Fungi XIV. 3',4',4-Trihydroxypulvinone, Thelephoric Acid, and Novel Pyrandione and Furanone Pigments from *Suillus grevillei* (Klotsch) Sing. [*Boletus elegans* (Schum. per Fries)]. J. Chem. Soc. (London) Perkin Trans. I **1973**, 1921.

205. – – Constituents of the Higher Fungi XV. 3-(3,4-Dihydroxyphenyl)-2,7,8-trihydroxydibenzofuran-1,4-dione, a Precursor of Thelephoric Acid from the Fungus

Suillus grevillei (Klotsch) Sing. [*Boletus elegans* (Schum. per Fries)]. J. Chem. Soc. (London) Perkin Trans. I **1975**, 351.

206. EDWARDS, R.L., and N. KALE: The Synthesis of Aurantiacin. J. Chem. Soc. (London) **1964**, 4084.

207. – – The Structure of Xylindein. Tetrahedron **21**, 2095 (1965).

208. EDWARDS, R.L., J. KEIGHLEY, and D.G. LEWIS: The Infra-red Spectra of 2,5-Dihydroxy-3,6-diphenyl-1,4-benzoquinones. J. Appl. Chem. (London) **10**, 246 (1960).

209. EDWARDS, R.L., D.G. LEWIS, and D.V. WILSON: Constituents of the Higher Fungi I. Hispidin, a New 4-Hydroxy-6-styryl-2-pyrone from *Polyporus hispidus* (Bull.) Fr. J. Chem. Soc. (London) **1961**, 4995.

210. EDWARDS, R.L., and H.J. LOCKETT: Constituents of the Higher Fungi XVI. Bulgarhodin and Bulgarein, Novel Benzofluoranthenequinones from the Fungus *Bulgaria inquinans* (Fries). J. Chem. Soc. (London) Perkin Trans. I **1976**, 2149.

211. EDWARDS, R.L., and I. MIR: Constituents of the Higher Fungi VI. Some Analogues of Hispidin. J. Chem. Soc. (London) C **1967**, 411.

212. EDWARDS, R.L., and D.V. WILSON: Constituents of the Higher Fungi II. The Synthesis of Hispidin. J. Chem. Soc. (London) **1961**, 5003.

213. EGLE, K.: Über das Pigment von *Anthurus aseroëformis* Mac Alpine. Planta **38**, 233 (1950).

214. ENDO, M., and H. NAOKI: Antimicrobial and Antispasmodic Tetrahydroanthracenes from *Cassia singueana*. Tetrahedron **36**, 2449 (1980).

215. ENGEL, H., and I. FRIEDERICHSEN: Die Carotinoide von *Mutinus caninus* Huds. Arch. Mikrobiol. **31**, 28 (1958).

216. ENGLERT, G.: NMR of Carotenoids – New Experimental Techniques. Pure and Appl. Chem. **57**, 801 (1985).

217. ENZELL, C.R., G.W. FRANCIS, and S. LIAAEN-JENSEN: Mass Spectrometric Studies of Carotenoids 2. A Survey of Fragmentation Reactions. Acta Chem. Scand. **23**, 727 (1969).

218. EPSTEIN, E., and P.G. MILES: Identification of Indirubin as a Pigment Produced by Mutant Cultures of the Fungus *Schizophyllum commune*. Bot Mag. (Tokyo) **79**, 566 (1966).

219. ERDTMAN, H.: Corticrocin, a Pigment from the Mycelium of a Mycorrhiza Fungus II. Acta Chem. Scand **2**, 209 (1948).

220. ESCHENMOSER, W., P. UEBELHART, and C.H. EUGSTER: Synthesen der enantiomeren Aleuriaxanthine. Nachweis eines vorherrschenden (Z)-Aleuriaxanthins in *Aleuria*. Helv. Chim. Acta **66**, 82 (1983).

221. ESSER, F.: Trametin, das Pigment von *Osmoporus odoratus* und *Gloeophyllum sepiarium*. Dissertation, Technical University, Berlin, 1973.

222. EUGSTER, C.H.: Pilzfarbstoffe, ein Überblick aus chemischer Sicht mit besonderer Berücksichtigung der Russulae. Z. Pilzkd. **39**, 45 (1973).

223. – Pigments from *Russula* Species. In: Abstracts of 2nd International Mycological Congress, p. 178. Tampa: University of South Florida. 1977.

224. EUGSTER, C.H., E.F. FRAUENFELDER, and H. KOCH: *Russula*-Farbstoffe: Erkennung der roten Hauptkomponenten als dimere Pteridinglykoside; Trennung von Pterinen durch isoelektrische Fokussierung in einem pH-Saccharose-Gradienten. Helv. Chim. Acta **53**, 131 (1970).

225. EUGSTER, C.H., and P.X. ITEN: Russupteridine. In: Chemistry and Biology of Pteridines (W. PFLEIDERER, ed.), p. 881–917. Berlin: W. de Gruyter. 1976.

226. EULER, K.L., V.E. TYLER, JR., L.R. BRADY, and M.H. MALONE: Isolation and Identification of Atromentin, the Anticoagulant Principle of *Hydnellum diabolus*. Lloydia **28**, 203 (1965).

227. FALANGHE, H., and P.A. BOBBIO: Identification of Indigo Produced in Submerged

Culture of *Agaricus campestris*, Mutant Culture. Arch. Biochem. Biophys. **96**, 430 (1962).

228. FAVRE-BONVIN, J., N. ARPIN, and J. GRIPENBERG: Sur l'Anhydrodehydrofomentariol, Nouveau Pigment de *Fomes fomentarius*. Phytochem. **16**, 1852 (1977).

229. FAVRE-BONVIN, J., M. KAOUADJI, and N. ARPIN: Sur l'Anhydrofomentariol, Nouveau Pigment de *Fomes fomentarius*. Phytochem. **16**, 495 (1977).

230. FIASSON, J.-L.: Les Caroténoides des Basidiomycètes: Survoi Chimiotaxinomique. Thèse Doc. 3ème Cycle, University of Lyon, 1968.

230a. – Recherches Chimiotaxinomiques sur les Champignons. Les Caroténoides de *Phyllotopsis nidulans* (Pers. ex Fr.) Sing. (Agaricales). C.R. hebd. séances Acad. Sci., Ser. D **268**, 786 (1969).

231. – Les Caroténoides de *Cantharellus ianthinoxanthus* (R. Maire) Kühner et sa Position Taxinomique. C. R. hebd. séances Acad. Sci., Ser. D **276**, 3219 (1973).

232. – Distribution of Styrylpyrones in the Basidiocarps of Various Hymenochaetaceae. Biochemical Systematics and Ecology **10**, 289 (1982).

233. – Contribution Synthétique à la Taxinomie Phylétique des Hyménochétacées Porées d'Europe, Spécialement du Genre *Phellinus* (Champignons, Aphyllophorales). Thèse Doc. ès Sc., University of Lyon, 1983.

234. FIASSON, J.-L., and N. ARPIN: Recherches Chimiotaxinomiques sur les Champignons V. Sur les Caroténoides Mineurs de *Cantharellus tubaeformis* Fr. Bull. soc. chim. biol. (Paris) **49**, 537 (1967).

235. FIASSON, J.-L., and J. BERNILLON: Identification Chimique de Styryl-pyrones chez Quatre Hyménochétacées (Champignons, Aphyllophorales). Canad. J. Bot. **55**, 2984 (1977).

236. FIASSON, J.-L., and M.-P. BOUCHEZ: Recherches Chimiotaxinomiques sur les Champignons. Les Carotènes de *Omphalia chrysophylla* Fr. C. R. hebd. séances Acad. Sci., Ser. D **266**, 1379 (1968).

237. – – Presence de l'Ester Methylique de la Neurosporaxanthine chez l'Ascomycete *Nectria cinnabarina*. Phytochem. **9**, 1133 (1970).

238. FIASSON, J.-L., K. GLUCHOFF-FIASSON, and W. STEGLICH: Über die Farb- und Fluoreszenzstoffe des Grünblättrigen Schwefelkopfes (*Hypholoma fasciculare*, Agaricales). Chem. Ber. **110**, 1047 (1977).

239. FIASSON, J.-L., P. LEBRETON, and N. ARPIN: Les Caroténoides des Champignons. Structure, Répartition, Utilisation en Systématique. Bull. soc. naturalistes et archeologues de l'Ain **1968**, 47.

240. FIASSON, J.-L., and R.H. PETERSEN: Carotenes in the Fungus *Clathrus ruber* (Gasteromycetes). Mycologia **65**, 201 (1973).

241. FIASSON, J.-L., R.H. PETERSEN, M.-P. BOUCHEZ, and N. ARPIN: Contribution Biochimique à la Connaissance Taxinomique de Certains Champignons Cantharelloides et Clavarioides. Rev. mycol. (Paris) **34**, 357 (1969).

242. FISCHER, N., and A.S. DREIDING: Biosynthesis of Betalaines. On the Cleavage of the Aromatic Ring During the Enzymatic Transformation of Dopa into Betalamic Acid. Helv. Chim. Acta **55**, 649 (1972), and references therein.

243. FLOSS, H.G.: The Shikimate Pathway. Recent Advances in Phytochemistry **12**, 59 (1979), and references therein.

244. FRANCK, B.: The Biosynthesis of the Ergochromes. In: The Biosynthesis of Mycotoxins, A Study in Secondary Metabolism (P.S. STEYN, ed.), p. 157–191. New York: Academic Press. 1980.

245. FRANCK, B., and H. FLASCH: Die Ergochrome (Physiologie, Isolierung, Struktur und Biosynthese). Fortschr. Chem. organ. Naturstoffe **30**, 151 (1973).

246. FRANCK, B., and T. RESCHKE: Clavoxanthin and Clavorubin, zwei neue Mutterkorn-Farbstoffe. Angew. Chem. **71**, 407 (1959).

247. – – Mutterkorn-Farbstoffe II. Isolierung der Hydroxy-anthrachinon-carbonsäuren Endocrocin und Clavorubin aus Roggenmutterkorn. Chem. Ber. **93**, 347 (1960).

248. FRANCK, R.L., G.R. CLARK, and J.N. COKER: The Synthesis of Vulpinic Acid from Polyporic Acid. J. Amer. Chem. Soc. **72**, 1824 (1950).

249. FRIEDERICHSEN, I., and H. ENGEL: Der Farbstoff von *Cordyceps militaris* L. Arch. Mikrobiol. **30**, 393 (1958).

250. FUGMANN, B.: Neue niedermolekulare Naturstoffe aus Höheren Pilzen (Basidiomyceten). Isolierung, Strukturaufklärung und Synthese. Dissertation, University of Bonn, 1985.

251. FUGMANN, B., B. EBERT, and W. STEGLICH: Manuscript in preparation.

252. FUGMANN, B., E. JÄGERS, B. OERTEL, and W. STEGLICH: Farbstoffe aus *Cortinarius rufoolivaceus*. Manuscript in preparation.

253. FUGMANN, B., B. STEFFAN, and W. STEGLICH: Necatorone, an Alkaloidal Pigment from the Gilled Toadstool *Lactarius necator* (Agaricales). Tetrahedron Letters **25**, 3575 (1984).

254. FUGMANN, B., and W. STEGLICH: Ungewöhnliche Inhaltsstoffe des Blätterpilzes *Lyophyllum connatum* (Agaricales). Angew. Chem. **96**, 71 (1984); Angew. Chem. Int. Ed. Engl. **23**, 72 (1984).

255. – – Unpublished work.

256. FURTNER, W.: Versuche zur Entwicklung einer Boletolsynthese. Diplomarbeit, Technical University, München, 1967.

257. – Untersuchungen über die blauenden Pigmente der Boletaceen und das 'Boletol' von F. Kögl und W.B. Deijs. Dissertation, Technical University, München, 1969.

258. FURUSAKI, A., N. HASHIBA, T. MATSUMOTO, A. HIRANO, Y. IWAI, and S. ŌMURA: X-Ray Crystal Structure of Staurosporine: A New Alkaloid from a *Streptomyces* Strain. Chem. Commun. **1978**, 800.

259. GABRIEL, M.: Recherches sur les Pigments des Agaricales I. Comparaison des Pigments de *Gymnopilus spectabilis* (Fr.) et *hybridus* (Fr.). Ann. Univ. Sc. Nat. Lyon **10**, 65 (1958).

260. – Recherches sur les Pigments des Agaricales II. Comparaison des Pigments de *Hypholoma fasciculare* (Fr. ex Huds.), *capnoides* (Fr.) et *sublateritium* (Fr. ex Schaeff.). Ann. Univ. Sc. Nat. Lyon **10**, 159 (1958).

261. – Recherches sur les Pigments des Agaricales III. Pigments des Cortinaires des Groupes *Cinnamomei* et *Sanguinei*. Bull. soc. mycol. France **76**, 208 (1960).

262. – Recherches sur les Pigments des Agaricales IV. Deuxième Contribution à la Connaissance de la Pigmentation des Cortinaires des Groupes *Sanguinei* et *Cinnamomei*. Ann. Univ. Sc. Nat. Lyon **11–12**, 67 (1960).

263. – Recherches sur les Pigments des Agaricales V. Troisième Contribution à l'Étude des Pigments des Cortinaires des Groupes *Sanguinei* et *Cinnamomei* (Elegantes Fries). Bull. soc. mycol. France **77**, 262 (1961).

264. – Recherches sur les Pigments des Agaricales VI. Pigments des Cortinaires du Groupe *Olivascentes*: *C. venetus* Fr., *melanotus* Kalchbr., *cotoneus* Fr. sensu Quélet. Bull. soc. mycol. France **78**, 359 (1962).

265. – Contribution à la Chimiotaxinomie des Agaricales. Pigments des Bolets et des Cortinaires. Thèse Doc. ès Sc., University of Lyon, 1965.

266. GABRIEL, M., and D. LAMOURE: *Cortinarius* (*Dermocybe*) *uliginosus* Berk. var. *luteus* N. var. Bull. soc. mycol. France **81**, 258 (1965).

267. GARRETT, R.D., and G. SULLIVAN: Separation of Some Terphenylquinone and Tetronic Acid Derivatives by Thin-layer Chromatography. J. Chromatogr. **63**, 457 (1971).

268. GARRIDO, N.: Unpublished work.

269. GARTZ, J.: Extraction and Chromatography of the Blue Pigment of a *Psilocybe* Species. Pharmazie **40**, 274 (1985).
270. GASCO, A., A. SERAFINO, V. MORTARINI, and E. MENZIANI: An Antibacterial and Antifungal Compound from *Calvatia lilacina*. Tetrahedron Letters **1974**, 3431.
271. GATENBECK, S.: Incorporation of Labelled Acetate in Emodin in *Penicillium islandicum*. Acta Chem.Scand. **12**, 1211 (1958).
272. GAYLORD, M.C., R.G. BENEDICT, G.H. HATFIELD, and L.R. BRADY: Isolation of Diphenyl-Substituted Tetronic Acids from Cultures of *Paxillus atrotomentosus*. J. Pharm. Sci. **59**, 1420 (1970).
273. GAYLORD, M.C., and L.R. BRADY: Comparison of Pigments in Carpophores and Saprophytic Cultures of *Paxillus panuoides* and *Paxillus atrotomentosus*. J. Pharm. Sci. **60**, 1503 (1971).
274. GAYLORD, M.C., J.R. DE BOER, and L.R. BRADY: Qualitation and Quantitation of Some Terphenylquinones. J. Pharm. Sci. **56**, 1069 (1967).
274a. GERBER, N.N.: Phenoxazinones by Oxidative Dimerisation of Aminophenols. Canad. J. Chem. **46**, 790 (1968).
275. GIANNETTI, B.M., W. STEGLICH, W. QUACK, T. ANKE, and F. OBERWINKLER: Antibiotika aus Basidiomyceten VI. Merulinsäuren A, B und C, neue Antibiotika aus *Merulius tremellosus* Fr. und *Phlebia radiata* Fr. Z. Naturforsch. **33c**, 807 (1978).
276. GILL, M.: Studies on Some Higher Fungus Pigments. Ph. D. Thesis, University of Bradford, 1973.
277. – Polyolefinic 18-Methyl-19-oxoicosenoic Acid Pigments from the Fungus *Piptoporus australiensis* (Wakefield) Cunningham. J. Chem. Soc. (London) Perkin Trans. I **1982**, 1449.
278. – 3-[(7Z)-Hexadecenyl]-4-methylfuran-2,5-dione from *Piptoporus australiensis*. Phytochem. **21**, 1786 (1982).
279. GILL, M., A. GIMÉNEZ, and R.W. McKENZIE: Pigments of Fungi 8. Bianthraquinones from *Dermocybe austroveneta* (Cleland) Moser. Submitted for publication.
280. GILL, M., A. GIMÉNEZ, and R.J. STRAUCH: Pigments of Fungi 7. 6-Nitro-*iso*-vanillic Acid, an Unusual Chromogen from the Genus *Cortinarius*. Phytochem., in press.
281. GILL, M., N. HACKWORTH, A.F. SMRDEL, and R.J. STRAUCH: Unpublished work.
282. GILL, M., and D.A. LALLY: A Naphthalenoid Pulvinic Acid Derivative from the Fungus *Pisolithus tinctorius*. Phytochem. **24**, 1351 (1985).
283. – – Unpublished work.
283a. GILL, M., and A.F. SMRDEL: Pigments of Fungi 6. Deoxyaustrocortirubin and Deoxyaustrocortilutein, Tetrahydroanthraquinones from the Genus *Cortinarius*. Phytochem., in press.
284. GILL, M., and R.J. STRAUCH: Constituents of *Agaricus xanthodermus* Genevier: The First Naturally Endogenous Azo Compound and Toxic Phenolic Metabolites. Z. Naturforsch. **39c**, 1027 (1984).
285. – – New Tetrahydroanthraquinones from the Genus *Cortinarius*. Tetrahedron Letters **26**, 2593 (1985).
286. GILL, M., and R. WATLING: The Relationships of *Pisolithus* (Sclerodermataceae) to Other Fleshy Fungi with Particular Reference to the Occurrence and Taxonomic Significance of Hydroxylated Pulvinic Acids. Plant Systematics and Evolution **154**, 225–236 (1986).
287. GILLARD, R.D., and R.J. LANCASHIRE: Electron Spin Resonance of Vanadium in *Amanita muscaria*. Phytochem. **23**, 179 (1984).
288. GLUCHOFF, K.: Étude Chimiotaxinomique des Pigments des Russules. Thèse Doc. 3ème Cycle, University of Lyon, 1969.
289. – Sur la Forte Teneur en Riboflavine de Plusieurs Représentants du Genre *Lyophyl-*

lum Karsten *sensu* Kühn. – Romagn. (Agaricales, Basidiomycètes). C. R. hebd. séances Acad. Sci., Ser. D **279**, 473 (1974).

290. – Analyse Pigmentaire de Russules Récoltées en Zone Alpine par R. Kühner. Bull. soc. mycol. France **91**, 391 (1975).

291. GLUCHOFF, K., N. ARPIN, M.-P. DANGY-CAYE, P. LEBRETON, W. STEGLICH, E. TÖPFER, H. POURRAT, F. REGERAT, and D. DERUAZ: Recherches Chimiotaxinomiques sur les Champignons. Sur le 7,7′ Bi-physcion, Bianthraquinone Obtenue à Partir de *Tricholoma equestre* L. per Fr. (Basidiomycète, Agaricale). C. R. hebd. séances Acad. Sci., Ser. D **274**, 1739 (1972).

292. GLUCHOFF, K., and P. LEBRETON: Recherches Chimiotaxinomiques sur les Champignons. Premiers Résultats sur les Propriétés et la Structure du Pigment Majeur Rouge des *Russules* (Basidiomycètes). C. R. hebd. séances Acad. Sci., Ser. D **270**, 213 (1970).

293. GLUCHOFF, K., and W. STEGLICH: Les Pigments de *Tricholoma sulphureum* Fr. ex Bull.; Identification de Nouveaux Dérivés de la Flavomannine. Trav. mycol. dédiés à R. Kühner, Bull. Soc. Linnéenne de Lyon **1974**, 163.

294. GLUCHOFF-FIASSON, K.: Contribution à la Chimiotaxinomie des Hyménomycètes: Pigments des *Tricholomataceae* Roze et des *Strophariaceae* Sing. & Smith. Thèse Doc. ès Sc., University of Lyon, 1979.

295. GLUCHOFF-FIASSON, K., and J. BERNILLON: Les Pigments de *Pholiota flammans* (Fr.) Kummer (Basidiomycète, Agaricale); Identification de Quatre Dérivés de l'Hispidine. C. R. hebd. séances Acad. Sci., Ser. D **284**, 385 (1977).

296. GLUCHOFF-FIASSON, K., G. HÖFLE, and W. STEGLICH: Unpublished work.

297. GLUCHOFF-FIASSON, K., and R. KÜHNER: La Délimitation et la Classification des *Strophariaceae* Sing. et Smith (Agaricales) à la Lumière de Nouvelles Recherches sur la Structure des Pigments. C. R. hebd. séances Acad. Sci., Ser. D **284**, 1667 (1977).

298. GOODWIN, T.W.: Fungal Carotenoids. Bot. Rev. **18**, 291 (1952).

299. – Studies in Carotenogenesis 8. The Carotenoids Present in the Basidiomycete *Dacrymyces stillatus*. Biochem. J. **53**, 538 (1953).

300. – Carotenoids in Fungi and Non-photosynthetic Bacteria. Progr. Ind. Microbiol. **11**, 31 (1972).

301. – Carotenoids. In: The Filamentous Fungi (J.E. SMITH and D.R. BERRY, eds.) Vol. II, p. 423–444. London: E. Arnold. 1976.

302. GRANDJEAN, J., and R. HULS: Structure de la Pleurotine: Une Benzoquinone Extraite de *Pleurotus griseus*. Tetrahedron Letters **1974**, 1893.

303. GREENHALGH, G.N., and A.J.S. WHALLEY: Stromal Pigments of Some Species of *Hypoxylon*. Trans. Br. mycol. Soc. **55**, 89 (1970).

304. GRIGSBY, R.D., W.D. JAMIESON, A.G. MCINNES, W.S.G. MAASS, and A. TAYLOR: The Mass Spectra of Derivatives of Polyporic Acid. Canad. J. Chem. **52**, 4117 (1974).

305. GRIPENBERG, J.: Fungus Pigments I. Cinnabarin, a Colouring Matter from *Trametes cinnabarina* Jacq. Acta Chem. Scand. **5**, 590 (1951).

306. – Fungus Pigments II. Cortisalin, a New Polyethenoid Pigment. Acta Chem. Scand. **6**, 580 (1952).

307. – Fungus Pigments IV. Aurantiacin, the Pigment of *Hydnum aurantiacum* Batsch. Acta Chem. Scand **10**, 1111 (1956).

308. – The Structure of Cinnabarin. Proc. Chem. Soc. (London) **1957**, 233.

309. – Fungus Pigments VIII. The Structure of Cinnabarin and Cinnabarinic Acid. Acta Chem. Scand. **12**, 603 (1958).

310. – Fungus Pigments IX. Some Further Constituents of *Hydnum aurantiacum* Batsch. Acta Chem. Scand. **12**, 1411 (1958).

268   M. GILL and W. STEGLICH:

311. GRIPENBERG, J.: Fungus Pigments X. The Ultra-violet Absorption of Some Substituted 2,5-Di-phenylbenzoquinones and their Leucoacetates. Acta Chem. Scand. **12**, 1762 (1958).
312. – Fungus Pigments XI. 2-Amino-3-hydroxymethylphenol from Cinnabarin. Acta Chem. Scand. **13**, 1305 (1959).
313. – Fungus Pigments XII. The Structure and Synthesis of Thelephoric Acid. Tetrahedron **10**, 135 (1960).
314. – Fungus Pigments XIII. Tramesanguin, the Pigment of *Trametes cinnabarina* var. *sanguinea* (L.) Pilat. Acta Chem. Scand. **17**, 703 (1963).
315. – Fungus Pigments XVI. The Pigments of *Peniophora sanguinea* Bres. Acta Chem. Scand. **19**, 2242 (1965).
316. – Fungus Pigments XVII. The Synthesis of Phlebiarubrone. Tetrahedron Letters **1966**, 697.
317. – Fungus Pigments XXI. Peniophorinin, a Further Pigment Produced by *Peniophora sanguinea* Bres. Acta Chem. Scand. **24**, 3449 (1970).
318. – Fungus Pigments XXII. Peniosanguin and Its Methyl Ether. Acta Chem. Scand. **25**, 2999 (1971).
319. – Fungus Pigments XXIII. Hydnuferrugin: A Novel Type of a 2,5-Diphenylbenzoquinone-Derived Pigment. Tetrahedron Letters **1974**, 619.
320. – Fungus Pigments XXV. Penioflavin. Acta Chem. Scand. **32 B**, 75 (1978).
321. – Fungus Pigments XXIX. The Pigments of *Hydnellum ferrugineum* (Fr.) Karsten and *H. zonatum* (Batsch) Karsten. Acta Chem. Scand. **35 B**, 513 (1981).
321a. – Personal communication.
322. GRIPENBERG, J., L. HILTUNEN, and L. NIINISTÖ: Dehydrofomentariol, $C_{17}H_{14}O_7$. Cryst. Struct. Commun. **5**, 571 (1976).
323. GRIPENBERG, J., L. HILTUNEN, L. NIINISTÖ, T. PAKKANEN, and T. PAKKANEN: Fungus Pigments XXVI. A Revised Structure of Peniophorinin. Acta Chem. Scand. **33 B**, 1 (1979).
324. GRIPENBERG, J., L. HILTUNEN, T. PAKKANEN, and T. PAKKANEN: Fungus Pigments XXVII. Xylerythrinin. Acta Chem. Scand. **33 B**, 6 (1979).
325. – – – – Fungus Pigments XXVIII. The Structure of Peniophorin. Acta Chem. Scand. **34 B**, 575 (1980).
326. GRIPENBERG, J., E. HONKANEN, and O. PATOHARJU: Nucleus of Cinnabarin. Chem. and Ind. **1956**, 1505.
327. – – – Fungus Pigments V. Degradations of Cinnabarin. Acta Chem. Scand. **11**, 1485 (1957).
328. GRIPENBERG, J., and J. MARTIKKALA: Fungus Pigments XIV. On the Oxidation of Phenoxazin-3-ones. Acta Chem. Scand. **19**, 1051 (1965).
329. – – Fungus Pigments XIX. Xylerythrin and its 5-O-Methyl Derivative. Acta Chem. Scand. **23**, 2583 (1969).
330. – – Fungus Pigments XX. On the Structure of Peniophorin, One of the Pigments Produced by *Peniophora sanguinea* Bres. Acta Chem. Scand. **24**, 3444 (1970).
331. GROVE, J.F.: New Metabolic Products of *Aspergillus flavus* II. Asperflavin, Anhydroasperflavin, and 5,7-Dihydroxy-4-methylphthalide. J. Chem. Soc. (London) Perkin Trans. I **1972**, 2406.
332. GRUBER, I.: Anthrachinonfarbstoffe und Fluoreszenzerscheinungen in den Gattungen *Dermocybe* und *Cortinarius* und Versuche ihrer Auswertung für deren Systematik. Dissertation, University of Innsbruck, 1969.
333. – Fluoreszierende Stoffe der Cortinarius-Untergattung *Leprocybe*. Z. Pilzkd. **35**, 249 (1969).
334. – Anthrachinonfarbstoffe in der Gattung *Dermocybe* und Versuch ihrer Auswertung für die Systematik. Z. Pilzkd. **36**, 95 (1970).

335. – Papierchromatographische Pigmentanalyse von südamerikanischen Dermocyben und Cortinarien. Beih. z. Nova Hedwigia **52**, 524 (1975).

336. GRUND, D.W., and K.A. HARRISON: Macrochemical Reactions of Species of *Boletus* and *Tylopilus*. Canad. J. Bot. **52**, 1239 (1974).

337. GRZYMALA, S.: l'Isolement de l'Orellanine Poison du *Cortinarius orellanus* Fr. et l'Étude de ses Effet Anatomopathologique. Bull. soc. mycol. France **78**, 394 (1962).

338. GSTRAUNTHALER, G.J.A.: The Effect of Cerulenin on Fatty Acid and Anthraquinone Biosynthesis in Vegetative Mycelia of *Cortinarius orichalceus* Fr. Biochim. Biophys. Acta **750**, 424 (1983).

339. GUNAWAN, S.: Untersuchungen an Inhaltsstoffen von Ascomyceten und Sequiterpenen aus Basidiomyceten. Dissertation, University of Bonn, 1982.

340. HALLENSTVET, M., R. BUCHECKER, G. BORCH, and S. LIAAEN-JENSEN: Absolute Configuration of β,γ-Carotene and Biosynthetic Implications. Phytochem. **16**, 583 (1977).

341. HALSALL, T.G., R. HODGES, and G.C. SAYER: The Chemistry of the Triterpenes and Related Compounds XXXVI. Some Constituents of *Trametes odorata* (Wulf.) Fr. J. Chem. Soc. (London) **1959**, 2036.

342. HANNA, C., and T.J. BULAT: Pigment Study of *Dacrymyces ellisii*. Mycologia **45**, 143 (1953).

343. HARASHIMA, K.: Carotenoids of a Red Toadstool, *Phallus rugulosus*. Agric. Biol. Chem. **42**, 1961 (1978).

344. HARLAY, V.: Sur une Réaction Colorée de la Cuticule du *Lactarius turpis* Weinm. Bull. soc. mycol. France **12**, 156 (1896).

345. HARMON, A.D., K.H. WEISGRABER, and U. WEISS: Preformed Azulene Pigments of *Lactarius indigo* (Schw.) Fries (Russulaceae, Basidiomycetes). Experientia **36**, 54 (1980).

346. HARRISON, K.A., and D.W. GRUND: Macrochemical Reactions of Species in the Family Boletaceae. Canad. J. Bot. **53**, 1417 (1975).

347. HASLAM, E.: The Shikimate Pathway. London: Butterworths. 1974.

348. HATFIELD, G.M., and L.R. BRADY: Isolation of *bis*-Noryangonin from *Gymnopilus decurrens*. Lloydia **31**, 225 (1968).

349. – – Occurrence of Bis-Noryangonin in *Gymnopilus spectabilis*. J. Pharm. Sci. **58**, 1298 (1969).

350. – – Occurrence of bis-Noryangonin and Hispidin in *Gymnopilus* Species. Lloydia **34**, 260 (1971).

351. – – Biosynthesis of Hispidin in Cultures of *Polyporus schweinitzii*. Lloydia **36**, 59 (1973).

352. HATFIELD, G.M., and D.E. SLAGLE: Isolation of Skyrin from *Hypomyces lactifluorum*. Lloydia **36**, 354 (1973).

353. HATSUDA, Y., and S. KUYAMA: Metabolic Products of *Aspergillus versicolor* IV. The Antibiotic Properties of Versicolorin and Some Hydroxyxanthenones. Chem. Abs. **53**, 16125 (1959).

354. HAXO, F.: Carotenoids of the Mushroom *Cantharellus cinnabarinus*. Bot. Gaz. **112**, 228 (1950).

355. – Some Biochemical Aspects of Fungal Carotenoids. Fortschr. Chem. organ. Naturstoffe **12**, 169 (1955).

356. HAYASHI, K., K. TOKURA, and K. OKABE: Synthesis of Fomecins A and B, Antibiotics Produced by the Basidiomycete *Fomes juniperinus*. Chem. Pharm. Bull. (Japan) **28**, 1971 (1980).

357. HECHT, H.-J., R. REINHARDT, and W. STEGLICH: Structure Analysis of the Lichen Pigment Vulpinic Acid. Submitted for publication.

358. HEGNAUER, H., L.E. NYHLÉN, and D.M. RAST: Ultrastructure of Native and Synthetic *Agaricus bisporus* Melanins – Implications as to the Compartmentation of Melanogenesis in Fungi. Exptl. Mycol. **9**, 221 (1985).

359. HEILBRONNER, E., and R.W. SCHMID: Zur Kenntnis der Sesquiterpene und Azulene 113. Azulenaldehyde und Azulenketone: Die Struktur des Lactaroviolins. Helv. Chim. Acta **37**, 2018 (1954).
360. HEIM, F.: Sur les Pigments Lutéiniques des Champignons. Bull. soc. mycol. France **9**, 92 (1893).
361. HEIM, P.: Sur la Localisation des Pigments Carotiniens chez les Phalloidées. C. R. hebd. séances Acad. Sci. **222**, 1354 (1946).
362. – Sur les Pigments Carotiniens des Champignons. C. R. hebd. séances Acad. Sci. **223**, 1170 (1946).
363. HEIM, R.: Les Pigments des Champignons dans leurs Rapports avec la Systématique. Bull. soc. chim. biol. (Paris) **29**, 48 (1942).
364. HERRMANN, R.: Untersuchungen zur Konstitution, Synthese und Biosynthese von Pilzfarbstoffen. Dissertation, University of Bonn, 1980.
365. HERRMANN, R., M. HOLZAPFEL, C. KILPERT, W. STEGLICH, H.BESL, A. BRESINSKY, and G. GEIGENMÜLLER: Pilzfarbstoffe 52. Über farblose Vorstufen des Atromentins und neue Terphenylchinone aus den Blätterpilzen *Paxillus atrotomentosus* und *P. panuoides* (Basidiomycetes). Liebigs Ann. Chem., in press.
366. HERRMANN, R., C. KILPERT, W. STEGLICH, H. BESL, and A. BRESINSKY: Unpublished work.
367. HILBIG, S., T. ANDRIES, W. STEGLICH, and T. ANKE: Zur Chemie und antibiotischen Aktivität des Carbolegerlings (*Agaricus xanthoderma*). Angew. Chem. **97**, 1063 (1985); Angew. Chem. Int. Ed. Engl. **24**, 1063 (1985).
368. HILGER, C.S., B. FUGMANN, and W. STEGLICH: Synthesis of Necatorone. Tetrahedron Letters **26**, 5975 (1985).
369. HILTUNEN, L., L. NIINISTO, T. PAKKANEN, and T. PAKKANEN: Penioflavin Diacetate, $C_{25}H_{20}O_8$. Cryst. Struct. Commun. **7**, 643 (1978).
370. HÖFLE, G.: ^{13}C-NMR-Spektroskopie chinoider Verbindungen II. Substituierte 1,4-Naphthochinone und Anthrachinone. Tetrahedron **33**, 1963 (1977).
371. HOFBAUER, C.: Chemotaxonomische Untersuchungen in der Untergattung *Phlegmacium*. Dissertation, University of Innsbruck, 1983.
372. HØILAND, K.: *Cortinarius* subgenus *Leprocybe* in Norway. Norwegian J. Bot. **27**, 101 (1980).
373. – Kanel-slørhattene (*Cortinarius*, underslaegten *Dermocybe*) i Norden. Svampe **4**, 63 (1981).
374. – *Cortinarius* Subgenus *Dermocybe*. Opera Botanica **71**, 1 (1983).
375. – Personal communication.
376. HUGHES, I., and R.A. RAPHAEL: Synthesis of Arcyriaflavin B. Tetrahedron Letters **24**, 1441 (1983).
377. HUOT, R., and P. BRASSARD: Skyrin from *Hypomyces lactifluorum*. Phytochem. **11**, 2879 (1972).
378. IMPELLIZZERI, G., and M. PIATTELLI: Biosynthesis of Indicaxanthin in *Opuntia ficus-indica* Fruits. Phytochem. **11**, 2499 (1972).
379. ISLER, O., W. GUEX, R. RÜEGG, G. RYSER, G. SAUCY, U. SCHWIETER, M. WALTER, and A. WINTERSTEIN: Synthesen in der Carotenoid-Reihe 16. Carotinoide vom Typus des Torularhodins. Helv. Chim. Acta **42**, 864 (1959).
380. ITEN, P.X., S. ARIHARA, and C.H. EUGSTER: *Russula*-Farbstoffe: Zur Struktur von Russupteridin-s_{III}. Helv. Chim. Acta **56**, 302 (1973).
381. ITEN, P.X., H. MÄRKI-DANZIG, and C.H. EUGSTER: Russupteridines: Yellow Lumazines from *Russula* Species. In: Developments in Biochemistry Vol. IV, Chemistry and Biology of Pteridines (R.L. KISLIUK and G.M. BROWN, eds.), p. 105–109. New York: Elsevier. 1979.

382. ITEN, P.X., H. MÄRKI-DANZIG, H. KOCH, and C.H. EUGSTER: Isolierung und Struktur von Pteridinen (Lumazinen) aus *Russula* sp. (Täublinge: Basidiomycetes). Helv. Chim. Acta **67**, 550 (1984).

383. JADOT, J., J. CASIMIR, and M. RENARD: Isolation and Characterisation of the L-(+)-γ-(*p*-Hydroxyanilide) of Glutamic acid from *Agaricus hortensis*. Biochim. Biophys. Acta **43**, 322 (1960).

384. JÄGERS, E(lisabeth): Untersuchungen über Farbstoffe von Pilzen der Gattungen *Cortinarius, Anthracophyllum* und *Boletopsis*. Dissertation, University of Bonn, 1980.

385. JÄGERS, E(lisabeth), E. HILLEN-MASKE, H. SCHMIDT, W. STEGLICH, and E. HORAK: Acetylierte Terphenylchinone aus *Anthracophyllum*-Arten (Agaricales). Z. Naturforsch., in press.

386. JÄGERS, E(lisabeth), E. HILLEN-MASKE, and W. STEGLICH: Inhaltsstoffe von *Boletopsis leucomelaena* (Basidiomycetes): Klärung der chemischen Natur von 'Leucomelon' und 'Protoleucomelon'. Z. Naturforsch., in press.

387. JÄGERS, E(lisabeth), C. KILPERT, B. OERTEL, A. SCHERER, B. STEFFAN, and W. STEGLICH: Farbstoffe aus *Cortinarius elegantior*. Submitted for publication.

388. JÄGERS, E(lisabeth), W. STEGLICH, and M. MOSER: Unpublished work.

389. JÄGERS, E(rhard): Untersuchungen zur Konstitution und Synthese polyprenylierter Chinone und Phenole. Dissertation, University of Bonn, 1981.

390. JÄGERS, E(rhard), V. PASUPATHY, A. HOVENBITZER, and W. STEGLICH: Suillin, ein charakteristischer Inhaltsstoff von Röhrlingen der Gattung *Suillus* (Boletales). Z. Naturforsch. **41b**, 645 (1986).

391. JÄGERS, E(rhard), B. STEFFAN, R. VON ARDENNE, and W. STEGLICH: Stoffwechselprodukte des 1.2.4-Trihydroxybenzols aus Fruchtkörpern von *Gomphidius maculatus* und *G. glutinosus* (Boletales). Z. Naturforsch. **36c**, 488 (1981).

392. JÄGERS, E(rhard), and W. STEGLICH: Polyamid-katalysierte Dimerisierung von 2,5-Dihydroxybenzochinonen zu 4-Ylidentetronsäuren, ein Modell für die Biosynthese des Bovilactons-4,4. Angew. Chem. **93**, 1105 (1981); Angew. Chem. Int. Ed. Engl. **20**, 1016 (1981).

393. JARRAH, M.Y., and V. THALLER: 300 MHz ^{1}H N.m.r. Spectra of Indolo[2,1-*b*]quinazoline-6,12-dione, Tryptanthrine, and its 2- and 8-Chloro-derivatives. J. Chem. Res. (S) **1980**, 186.

394. JERRAM, W.A., A.G. MCINNES, W.S.G. MAASS, D.G. SMITH, A. TAYLOR, and J.A. WALTER: The Chemistry of Cochliodinol, a Metabolite of *Chaetomium* Spp. Canad. J. Chem. **53**, 727 (1974); Erratum: Canad. J. Chem. **54**, 2031 (1975).

395. JIRAWONGSE, V., E. RAMSTAD, and J. WOLINSKY: Isolation of Polyporic Acid from *Lopharia papyraceae*. J. Pharm. Sci. **51**, 1108 (1962).

396. JÜLICH, W.: Die Nichtblätterpilze, Gallertpilze und Bauchpilze. Aphyllophorales, Heterobasidiomycetes, Gastromycetes. In: Kleine Kryptogamenflora (H. GAMS, ed.) Vol. IIb/1. Stuttgart: G. Fischer. 1984.

397. KÄMMERER, A., H. BESL, and A. BRESINSKY: *Omphalotaceae* fam. nov. und *Paxillaceae*, ein chemotaxonomischer Vergleich zweier Pilzfamilien der *Boletales*. Plant Systematics and Evolution **150**, 101 (1985).

398. KARL, U., and W. STEGLICH: Unpublished work.

399. KARRER, P., H. RUCKSTUHL, and E. ZBINDEN: Über Lactaroviolin, einen Farbstoff aus *Lactarius deliciosus*. Helv. Chim. Acta **28**, 1176 (1945).

400. KELLER, G.: Chemotaxonomische Pigmentationsuntersuchungen in der Gattung *Dermocybe* (Fr.) Wünsche. Dissertation, University of Innsbruck, 1979.

401. – Pigmentationsuntersuchungen bei europäischen Arten aus der Gattung *Dermocybe* (Fr.) Wünsche. Sydowia **35**, 110 (1982).

402. KELLER, G., M. MOSER, E. HORAK, and W. STEGLICH: Chemotaxonomic Investigation of Species of *Dermocybe* (Fr.) Wünsche (*Agaricales*) from New Zealand, Papua New Guinea and South America. Mycotaxon, in press.

403. Keller, G., and J.F. Ammirati: Chemotaxonomic Significance of Anthraquinone Derivatives in North American Species of *Dermocybe*, Section *Sanguineae*. Mycotaxon 18, 357 (1983).
404. Keller, G., and W. Steglich: 4-Aminophyscion, A New Anthraquinone Derivative from *Dermocybe* (Agaricales). Phytochem., in press.
405. – – Unpublished work.
406. Keller-Dilitz, H.: Chemotaxonomische Pigmentationsuntersuchungen in der Gattung *Cortinarius* Fr. Untergattung *Leprocybe* Mos. Dissertation, University of Innsbruck, 1984.
407. Keller-Dilitz, H., M. Moser, and J.F. Ammirati: Orellanine and Other Fluorescent Compounds in the Genus *Cortinarius*, Section *Orellani*. Mycologia 77, 667 (1985).
408. Kelly, T.R., J.K. Saha, and R.R. Whittle: Bostrycin: Structure Correction and Synthesis. J. Org. Chem. 50, 3679 (1985).
409. Khanna, J.M., M.H. Malone, K.L. Euler, and L.R. Brady: Atromentin. Anticoagulant from *Hydnellum diabolus*. J. Pharm. Sci. 54, 1016 (1965).
410. Kidd, C.B.M., B. Caddy, J. Robertson, I.R. Tebbett, and R. Watling: Thin Layer Chromatography as an Aid for Identification of *Dermocybe* Species of *Cortinarius*. Trans. Br. mycol. Soc. 85, 213 (1985).
411. Kilecy-Ksoll, R., and W. Steglich: Unpublished work.
412. Kirk, T.K., L.F. Lorenz, and M.J. Larsen: Partial Characterisation of a Phenolic Pigment from Sporocarps of *Phellinus igniarius*. Phytochem. 14, 281 (1975).
413. Kjøsen, H., and S. Liaaen-Jensen: Fungal Carotenoids 9. Total Synthesis of Aleuriaxanthin. Acta Chem. Scand. 27, 2495 (1973).
414. Klaar, M., and W. Steglich: Pilzpigmente XXVII. Isolierung von Hispidin und 3,14'-Bihispidinyl aus *Phellinus pomaceus* (Poriales). Chem. Ber. 110, 1058 (1977).
415. – – Pilzpigmente XXVIII. Hymenochinon, der rote Farbstoff von *Hymenochaete mougeotii* (Poriales). Chem. Ber. 110, 1063 (1977).
416. Klamann, J.-D., B. Fugmann, S. Hilger, and W. Steglich: Manuscript in preparation.
417. Knackmuss, H.-J.: Zur Chemie und Biochemie der Azachinone. Angew. Chem. 85, 163 (1973); Angew. Chem. Int. Ed. Engl. 12, 139 (1973).
418. Kneifel, H., and E. Bayer: Stereochemistry and Total Synthesis of Amavadin, the Naturally Occurring Vanadium Compound of *Amanita muscaria*. J. Amer. Chem. Soc. 108, 3075 (1986).
419. Knight, D.W., and G. Pattenden: Synthesis of Pulvinones, Metabolites of *Aspergillus terreus* and *Suillus grevillei*. Chem. Commun. 1975, 876.
420. – – Synthesis of the Pulvinic Acid Pigments of Lichen and Fungi. Chem. Commun. 1976, 660.
421. – – Total Synthesis of Pulvinones, 4-Benzylidene-2-phenyltetronic Acid Pigments of Fungi. J. Chem. Soc. (London) Perkin Trans. I 1979, 70.
422. – – Synthesis of Permethylated Derivatives of Pinastric Acid and Gomphidic Acid, Pulvinic Acid Pigments of Lichen and Fungi. J. Chem. Soc. (London) Perkin Trans. I 1979, 84.
423. Kögl, F.: Untersuchungen über Pilzfarbstoffe V. Die Konstitution der Polyporsäure. Liebigs Ann. Chem. 447, 78 (1926).
424. – Pilz- und Bakterienfarbstoffe. In: Handbuch der Pflanzenanalyse (G. Klein, ed.) Vol. III, p. 1410–1445. Wien: J. Springer. 1932.
425. Kögl, F., H. Becker, A. Detzel, and G. de Voss: Untersuchungen über Pilzfarbstoffe VI. Die Konstitution des Atromentins. Liebigs Ann. Chem. 465, 211 (1928).
426. Kögl, F., H. Becker, G. de Voss, and E. Wirth: Untersuchungen über Pilzfarbstoffe VII. Die Synthese des Atromentins. Zur Kenntnis der Atromentinsäure. Liebigs Ann. Chem. 465, 243 (1928).

427. KÖGL, F., and W.B. DEIJS: Untersuchungen über Pilzfarbstoffe XI. Über Boletol, den Farbstoff der blau anlaufenden Boleten. Liebigs Ann. Chem. **515**, 10 (1935).

428. – – Untersuchungen über Pilzfarbstoffe XII. Die Synthese von Boletol und Isoboletol. Liebigs Ann. Chem. **515**, 23 (1935).

429. KÖGL, F., and H. ERXLEBEN: Untersuchungen über Pilzfarbstoffe VIII. Über den roten Farbstoff des Fliegenpilzes. Liebigs Ann. Chem. **479**, 11 (1930).

430. KÖGL, F., H. ERXLEBEN, and L. JÄNECKE: Untersuchungen über Pilzfarbstoffe IX. Die Konstitution der Thelephorsäure. Liebigs Ann. Chem. **482**, 105 (1930).

431. KÖGL, F., and J.J. POSTOWSKY: Untersuchungen über Pilzfarbstoffe I. Über das Atromentin. Liebigs Ann. Chem. **440**, 19 (1924).

432. – – Untersuchungen über Pilzfarbstoffe II. Über die Farbstoffe des blutroten Hautkopfes (*Dermocybe sanguinea* Wulf.). Liebigs Ann. Chem. **444**, 1 (1925).

433. – – Untersuchungen über Pilzfarbstoffe III. Über das Atromentin (II). Liebigs Ann. Chem. **445**, 159 (1925).

434. KÖNIG, H.: Personal communication. We thank Professor König, BASF AG, for this information.

435. KOHL, F.G.: Untersuchungen über das Carotin und seine physiologische Bedeutung in der Pflanze. Leipzig: Bornträger (1902).

436. KOPANSKI, L., E. KARBACH, G. SELBITSCHKA, and W. STEGLICH: Vesparion, ein Naphthochinon-Derivat aus dem Schleimpilz *Metatrichia vesparium*. Liebigs Ann. Chem., in press.

437. KOPANSKI, L., M. KLAAR, and W. STEGLICH: Pilzpigmente 40. Leprocybin, der Fluoreszenzstoff von *Cortinarius cotoneus* und verwandten Leprocyben (Agaricales). Liebigs Ann. Chem. **1982**, 1280.

438. KOPANSKI, L., G.-R. LI, H. BESL, and W. STEGLICH: Pilzpigmente 41. Naphthochinon-Farbstoffe aus den Schleimpilzen *Trichia floriformis* und *Metatrichia vesparium* (Myxomycetes). Liebigs Ann. Chem. **1982**, 1722.

439. KOPECKÝ, J., D. ŠAMAN, T. VANĚK, and L. NOVOTNÝ: Components of the Woodrotting Fungus *Sarcodontia setosa*. Collect. Czech. Chem. Comm. **49**, 1622 (1984).

440. KOTLABA, F., J. KŘEPINSKÝ, V. HEROUT, R. PROKEŠ, and A. VYSTRČIL: Constituents of the Wood-decaying Fungus *Sarcodontia setosa* (Pers.) Donk. Plant Substances XXV. Naturwiss. **52**, 591 (1965).

441. KOUL, S.K., S.C. TANEJA, S.P. IBRAHAM, K.L. DHAR, and C.K. ATAL: A C-Formylated Azulene from *Lactarius deterrimus*. Phytochem. **24**, 181 (1985).

442. KŘEPINSKÝ, J., V. HEROUT, F. ŠORM, A. VYSTRČIL, R. PROKEŠ, and G. JOMMI: Plant Substances XXIII. The Structure of Sarcodontic Acid. Collect. Czech. Chem. Comm. **30**, 2626 (1965).

443. KRONAWITTER, I.: Die Gattung *Hygrocybe* (Agaricales) unter besonderer Berücksichtigung von Pigmentausstattung und Sippengliederung. Dissertation, University of Regensburg, 1984.

444. KÜHNER, R.: Les Hyménomycètes Agaricoides (Agaricales, Tricholomales, Pluteales, Russulales): Etude Genérale et Classification. Bull. Soc. Linnéenne de Lyon, numéro special (1981).

445. – Some Mainlines of Classification in the Gill Fungi. Mycologia **76**, 1059 (1984).

446. KÜRNSTEINER, H., and M. MOSER: Isolation of a Lethal Toxin from *Cortinarius orellanus* Fr. Mycopathologia **74**, 65 (1981).

447. LAATSCH, H., and H. ANKE: Stoffwechselprodukte von Mikroorganismen 214. Viocristin, Isoviocristin und Hydroxyviocristin. – Struktur und Synthese natürlich vorkommender 1,4-Anthrachinone. Liebigs Ann. Chem. **1982**, 2189.

448. LANG, S., and W. STEGLICH: Unpublished work.

449. LE BLANC, G.D., and L.-M. BABINEAU: 2,5-Dimethoxybenzoquinone, 3,4-Dimethyl-8-hydroxyisocoumarin, and Eburicoic Acid Isolated from *Lenzites thermophila* Falck-75. Canad. J. Microbiol. **18**, 261 (1972).

450. LEDERER, E.: Sur les Caroténoides des Cryptogames. Bull. soc. chim. biol. (Paris) **20**, 611 (1938).

451. LEMBERG, R.: Nitrogenous Pigments from the Fungus *Coriolus sanguineus* (*Polystictus cinnabarinus*). Austral. J. Exp. Biol. Med. Sci. **30**, 271 (1952).

452. LETCHER, R.M.: Chemistry of Lichen Constituents VII: Mass Spectra of Some Pulvic Acid Derivatives. Org. Mass Spectrom. **1**, 805 (1968).

453. LETCHER, R.M., and S.H. EGGERS: Chemistry of Lichen Constituents IV. Tetrahedron Letters **1967**, 3541.

454. LEVENBERG, B.: Isolation and Structure of Agaritine, a γ-Glutamyl-substituted Arylhydrazine Derivative from *Agaricaceae*. J. Biol. Chem. **239**, 2267 (1964).

455. LEVINE, W.G.: Formation of Blue Oxidation Product from Psilocybin. Nature **215**, 1292 (1967).

456. LI, G.-R., and W. STEGLICH: Unpublished work.

457. LIAAEN-JENSEN, S.: On Fungal Carotenoids and the Natural Distribution of Spirilloxanthin. Phytochem. **4**, 925 (1965).

458. – Stereochemistry of Naturally Occurring Carotenoids. Fortschr. Chem. organ. Naturstoffe **39**, 123 (1980).

459. – Carotenoids of Lower Plants – Recent Progress. Pure and Appl. Chem. **57**, 649 (1985).

460. LOHRISCH, H.-J.: Synthese der Grevilline und ihre Umwandlung zu Terphenylchinonen. Dissertation, University of Bonn, 1977.

461. LOHRISCH, H.-J., L. KOPANSKI, R. HERRMANN, H. SCHMIDT, and W. STEGLICH: Pilzfarbstoffe 49. Synthese der Grevilline und verwandter 2*H*-Pyran-2,5(6*H*)-dione. Liebigs Ann. Chem. **1986**, 177.

462. LOHRISCH, H.-J., H. SCHMIDT, and W. STEGLICH: Pilzfarbstoffe 50. Synthese von Terphenylchinonen durch Methoxid-katalysierte Umlagerung von Grevillin-Derivaten. Liebigs Ann. Chem. **1986**, 195.

463. LOHRISCH, H.-J., and W. STEGLICH: Synthese von Grevillin C und verwandten 2H-Pyran-2,5(6H)-dionen. Tetrahedron Letters **1975**, 2905.

464. LOUNASMAA, M.: A New Synthesis of Thelephoric Acid. Acta Chem. Scand. **19**, 540 (1965).

465. LOUNASMAA, M., and A. KARJALAINEN: On the Mass Spectral Fragmentation of 2,5-Diphenyl-*p*-benzoquinones. Acta Chem. Scand. **27**, 3427 (1973).

466. LUKACS, L., and J. ZELLNER: Zur Chemie der Höheren Pilze XXII. Über *Ganoderma lucidum* Leiß, *Hydnum imbricatum* L. und *Cantharellus clavatus* Pers. Monatsh. Chem. **62**, 214 (1933).

467. MAASS, W.S.G.: Lichen Substances IV. Incorporation of Pulvinic-^{14}C Acids into Calycin by the Lichen *Pseudocyphellaria crocata*. Canad. J. Biochem. **48**, 1241 (1970).

468. MAASS, W.S.G., and A.C. NEISH: Lichen Substances II. Biosynthesis of Calycin and Pulvinic Dilactone by the Lichen, *Pseudocyphellaria crocata*. Canad. J. Bot. **45**, 59 (1967).

469. MAASS, W.S.G., G.H.N. TOWERS, and A.C. NEISH: Flechtenstoffe I. Untersuchungen zur Biogenese des Pulvinsäureanhydrids. Ber. dtsch. bot. Ges. **77**, 157 (1964).

470. MADHOSINGH, C.: Physiological Studies on the *Pycnoporus* Species. Canad. J. Bot. **40**, 1073 (1962).

471. MÄRKI-DANZIG, H., and C.H. EUGSTER: Synthesen von Aminolumazinen. Helv. Chim. Acta **67**, 570 (1984).

472. MARSHALL, D., and M.C. WHITING: Researches on Polyenes V. The Synthesis of Cortisalin. J. Chem. Soc. (London) **1957**, 537.

473. MARTIN, G.W., and C.J. ALEXOPOULOS: The Myxomycetes. Iowa City: University of Iowa Press. 1969.

474. MARUMOTO, R., C. KILPERT, and W. STEGLICH: Neue Pulvinsäurederivate aus *Pulveroboletus*-Arten (Boletales). Z. Naturforsch. **41c**, 363 (1986).

475. MASON, S.F., R.H. SEAL, and D.R. ROBERTS: Optical Activity in the Biaryl Series. Tetrahedron **30**, 1671 (1974).

476. MATHEIS, W.: Chemische Reagenzien in der Hand des Mykologen. Z. Pilzkd. **38**, 33 (1972).

477. MCINNES, A.G., D.G. SMITH, L.C. VINING, and L. JOHNSON: Use of ^{13}C in Biosynthetic Studies. Location of Isotope from Labelled Acetate and Formate in the Fungal Tropolone, Sepedonin, by ^{13}C Nuclear Magnetic Resonance Spectroscopy. Chem. Commun. **1971**, 325.

478. MCMORRIS, T.C., and M. ANCHEL: Phlebiarubrone, a Basidiomycete Pigment Related to Polyporic Acid. Tetrahedron Letters **1963**, 335.

479. – – Fomecin A and B, Phenolic Aldehydes from the Basidiomycete *Fomes juniperinus*. Canad. J. Chem. **42**, 1595 (1964).

480. – – The Structure of the Basidiomycete *Ortho* Quinone, Phlebiarubrone, and of its Novel Acetylation Product. Tetrahedron **23**, 3985 (1967).

481. MEAD, R.J., and W. SEGAL: Chemotaxonomy and Biosynthetic Relationships of Boletineae Pigments. Biochem. J. **121**, 26P (1971).

482. MEIXNER, A.: Chemische Farbreaktionen von Pilzen. Vaduz: J. Cramer. 1975.

483. MIERSCH, J.: Chemotaxonomie und Gedanken zur Evolution der Pilze. Biol. Rdsch. **20**, 75 (1982).

484. MILES, P.G., H. LUND, and J.R. RAPER: The Identification of Indigo as a Pigment Produced by a Mutant Culture of *Schizophyllum commune*. Arch. Biochem. Biophys. **62**, 1 (1956).

485. MILLER, M.W.: The Pfizer Handbook of Microbial Metabolites. London: McGraw Hill. 1961.

486. MINALE, L., M. PIATTELLI, and R.A. NICOLAUS: Pigments of Centrospermae IV. On the Biogenesis of Indicaxanthin and Betanin in *Opuntia ficus-indica* Mill. Phytochem. **4**, 593 (1965).

487. MINAMI, K., K. ASAWA, and M. SAWADA: The Structure of Amitenone. Tetrahedron Letters **1968**, 5067.

488. MISLOW, K., M.M. GREEN, P. LAUR, J.T. MELILLO, T. SIMMONS, and A.L. TERNAY, JR.: Absolute Configuration and Optical Rotatory Power of Sulphoxides and Sulfinate Esters. J. Amer. Chem. Soc. **87**, 1958 (1965).

489. MIZE, P.D., P.W. JEFFS, and K. BOEKELHEIDE: Structure Determination of the Active Sulfhydryl Reagent in Gill Tissue of the Mushroom *Agaricus bisporus*. J. Org. Chem. **45**, 3540 (1980).

490. MOCEK, U.: Neue Wirkstoffe aus Basidiomyceten. Dissertation, University of Bonn, 1985.

491. MOCEK, U., E. HILLEN-MASKE, A. KIRFEL, W. STEGLICH, and T. ANKE: Manuscript in preparation.

492. MONTFORT, M.L., V.E. TYLER, JR., and L.R. BRADY: Isolation of Aurantiacin from *Hydnellum caeruleum*. J. Pharm. Sci. **55**, 1300 (1966).

493. MOORE, H.W., and R.J. WIKHOLM: Reactions of Hydroxyquinones with Dimethyl Sulfoxide-Acetic Anhydride. Tetrahedron Letters **1968**, 5049.

494. MOSBACH, K.: On the Biosynthesis of Lichen Substances 2. The Pulvinic Acid Derivative Vulpinic Acid. Biochem. Biophys. Res. Commun. **17**, 363 (1964).

495. MOSBACH, K., H. GUILFORD, and M. LINDBERG: The Terphenyl Quinone Polyporic Acid: Production, Isolation and Characterisation. Tetrahedron Letters **1974**, 1645.

496. MOSER, M.: *Dermocybe* and *Cortinarius* Collections of R.W.G. Dennis from the Blue Mountains, Jamaica. Kew Bull. **22**, 87 (1968).

497. MOSER, M.: Die Gattung *Dermocybe* (Fr.) Wünsche (Die Hautköpfe). Schweiz. Z. Pilzkd. **50**, 153 (1972).

498. – Die Röhrlinge und Blätterpilze (Polyporales, Boletales, Agaricales, Russulales). In: Kleine Kryptogamenflora (H. GAMS, ed.) Vol. II b/2, 5th ed. Stuttgart: G. Fischer. 1983.

499. – The Relevance of Chemical Characters for the Taxonomy of Agaricales. Proc. Indian Acad. Sci., Plant Sci. **94**, 381 (1985).

500. MÜLLER, C.: Note on a British *Polyporus*: *P. rutilans?*. J. Bot. **10**, 22 (1872).

501. MURRAY, J.: Lichens and Fungi I. Polyporic Acid in *Stictae*. J. Chem. Soc. (London) **1952**, 1345.

502. MUSSO, H.: The Pigments of Fly Agaric, *Amanita muscaria*. Tetrahedron **35**, 2843 (1979).

503. – Über die Farbstoffe des Fliegenpilzes. Aufgaben und Ziele der Naturstoffchemie heute. Naturwiss. **69**, 326 (1982).

504. – Personal communication.

505. NAIR, M.S.R., and M. ANCHEL: Frustulosinol, An Antibiotic Metabolite of *Stereum frustulosum*: Revised Structure of Frustulosin. Phytochem. **16**, 390 (1977).

506. NAKAJIMA, S., K. KAWAI, S. YAMADA, and Y. SAWAI: Isolation of Oospolactone as Antifungal Principle of *Gloeophyllum sepiarium*. Agric. Biol. Chem. **40**, 811 (1976).

507. NAMBUDIRI, A.M.D., C.P. VANCE, and G.H.N. TOWERS: Effect of Light on Enzymes of Phenylpropanoid Metabolism and Hispidin Biosynthesis in *Polyporus hispidus*. Biochem. J. **134**, 891 (1973).

508. – – – Styrylpyrone Biosynthesis in *Polyporus hispidus* II. Enzymic Hydroxylation of *p*-Coumaric Acid and Bis-Noryangonin. Biochim. Biophys. Acta **343**, 148 (1974).

509. NESPIAK, A., A. NOCULAK, and A. SIEWINSKI: Bemerkungen über fluoreszierende Stoffe der Schleierlinge und ihre Auswertung für die Systematik. Acta Mycologia **9**, 205 (1973).

510. NETTLETON, D.E., T.W. DOYLE, B. KRISHNAN, G.K. MATSUMOTO, and J. CLARDY: Isolation and Structure of Rebeccamycin – A New Antitumor Antibiotic from *Nocardia aerocoligenes*. Tetrahedron Letters **26**, 4011 (1985).

511. NEVEU, A., R. BAUTE, G. BOURGEOIS, and G. DEFFIEUX: Recherches sur le Pigment Bleu du Champignon *Corticium caeruleum* (Schrad. ex Fr.) Fr. (Aphyllophorales) II. Relations Structurales avec Certains Dérivés de l'Acide Théléphorique; Étude par Spectrométrie de Masse. Bull. soc. pharm. Bordeaux **113**, 121 (1974).

512. NEVEU, A., R. BAUTE, and G. DEFFIEUX: Recherches sur le Pigment Bleu du Champignon *Corticium caeruleum* (Schrad. ex Fr.) Fr. (Aphyllophorales) I. Localisation et Principales Caractéristiques Physicochimiques. Bull. soc. pharm. Bordeaux **113**, 77 (1974).

513. NOBLES, M.K., and B.P. FREW: Studies in Wood-Inhabiting Hymenomycetes V. The Genus *Pycnoporus* Karst. Canad. J. Bot. **40**, 987 (1962).

514. NOPPEL, H.E., K.-H. SCHWEER, and F. VON MASSOW: Synthese von Pulvinsäure-[^{14}C], 4-Hydroxy-pulvinsäure-[^{14}C] und 4′-Hydroxy-pulvinsäure-[^{14}C]. J. Labelled Comp. and Radiopharm. **12**, 79 (1976).

515. OERTEL, B.: Untersuchungen zur Konstitution von Dihydroanthracenonen und Angaben zu ihrer Verbreitung in Pilzen. Dissertation, University of Bonn, 1984.

516. OERTEL, B., and D. LABER: Die Laugenreaktion an der Unterseite der Stielknolle bei Fruchtkörpern der Gattung *Cortinarius*, Untergattung *Phlegmacium* (Agaricales). Z. Mykol. **52**, 139 (1986).

517. PACKTER, N.M.: Studies on the Biosynthesis of Phenols in Fungi. Production of 4-Methoxytoluquinol, Epoxysuccinic Acid and a Diacetylenic Alcohol by Surface Cultures of *Lentinus degener* I.M.I. 110525. Biochem. J. **114**, 369 (1969).

518. PARMASTO, E., and I. PARMASTO: The Xanthochroic Reaction in Aphyllophorales. Mycotaxon **8**, 201 (1979).

519. PASTAC, I.A.: Les Matières Colorantes des Champignons. Rev. Mycol., Mémoire hors Série 2, **1942**, 1.

520. PATTENDEN, G.: Natural 4-Ylidenebutenolides and 4-Ylidenetetronic Acids. Fortschr. Chem. organ. Naturstoffe **35**, 133 (1978).

521. PATTENDEN, G., N. PEGG, and A.G. SMITH: A New Synthesis of Pulvinic Acids. Tetrahedron Letters **27**, 403 (1986).

522. PERRIN, P.W., and G.H.N. TOWERS: Metabolism of Aromatic Acids by *Polyporus hispidus*. Phytochem. **12**, 583 (1973).

523. – – Hispidin Biosynthesis in Cultures of *Polyporus hispidus*. Phytochem. **12**, 589 (1973).

524. PETTERSSON, G.: New Metabolites from *Lentinus degener*. Acta Chem. Scand. **20**, 45 (1966).

525. – On the Role of 6-Methylsalicylic Acid in the Biosynthesis of Fungal Benzoquinones. Acta Chem. Scand. **20**, 151 (1966).

526. PHIPSON, T.L.: On the Colouring Matter (Ruberine) and the Alkaloid (Agarythrine) Contained in *Agaricus ruber* (*A. sanguineus*). Chem. News (London) **56**, 199 (1882); Ber. dtsch. chem. Ges. **16**, 244 (1883).

527. PIATTELLI, M., and L. MINALE: Pigments of Centrospermae I. Betacyanins from *Phyllocactus hybridus* Hort. and *Opuntia ficus-indica* Mill. Phytochem. **3**, 307 (1964).

528. PIATTELLI, M., L. MINALE, and G. PROTA: Isolation, Structure and Absolute Configuration of Indicaxanthin. Tetrahedron **20**, 2325 (1964).

529. – – – Pigments of Centrospermae III. Betaxanthins from *Beta vulgaris* L. Phytochem. **4**, 121 (1965).

530. PINHEY, J.T., B.J. RALPH, J.J.H. SIMES, and M. WOOTTON: Extractives of Fungi I. The Constituents of *Trametes lilacinogilva*. Austral. J. Chem. **23**, 2141 (1970).

531. PLATTNER, P.A., and E. HEILBRONNER: Über die Konstitution des Lactaroviolins. Experientia **1**, 233 (1945).

532. PLATTNER, P.A., E. HEILBRONNER, R.W. SCHMID, R. SANDRIN, and A. FÜRST: The Structure of Lactaroviolin. Chem. and Ind. **1954**, 1202.

533. QUACK, N., T. ANKE, and F. OBERWINKLER: Unpublished work.

534. QUACK, W., H. SCHOLL, and H. BUDZIKIEWICZ: Ascocorynin, a Terphenylquinone from *Ascocoryne sarcoides*. Phytochem. **21**, 2921 (1982).

535. RALPH, B.J., and A. ROBERTSON: The Chemistry of Fungi XIV. 2:4:5-Trihydroxyphenylglyoxylic Acid from *Polyporus tumulosus* Cooke. J. Chem. Soc. (London) **1950**, 3380.

536. RAMAGE, R., G.J. GRIFFITHS, F.E. SHUTT, and J.N.A. SWEENEY: Dioxolanones as Synthetic Intermediates 2. Synthesis of Tetronic Acids and Pulvinones. J. Chem. Soc. (London) Perkin Trans. I **1984**, 1539.

537. RAMAGE, R., G.J. GRIFFITHS, and J.N.A. SWEENEY: Dioxolanones as Synthetic Intermediates 3. Biomimetic Synthesis of Pulvinic Acids. J. Chem. Soc. (London) Perkin Trans. I **1984**, 1547.

538. RAST, D., H. STÜSSI, and P. ZOBRIST: Self-inhibition of the *Agaricus bisporus* Spore by CO_2 and/or γ-Glutaminyl-4-hydroxybenzene and γ-Glutaminyl-3,4-benzoquinone: A Biochemical Analysis. Physiol. Plant **46**, 227 (1979).

539. READ, G., and L.C. VINING: Thelephoric Acid. Canad. J. Chem. **37**, 1442 (1959).

540. – – Biogenesis of Terphenylquinones. Chem. and Ind. **1959**, 1547.

541. READ, G., L.C. VINING, and R.H. HASKINS: Biogenetic Studies on Volucrisporin. Canad. J. Chem. **40**, 2357 (1962).

542. REININGER, W.: Synthese des Endocrocins, des Endocrocinanthrons und Aufklärung der Pigmente von *Dermocybe cinnabarina* (Fr.) Wünsche. Dissertation, Technical University, München, 1970.

543. REININGER, W., W. STEGLICH, and M. MOSER: Pilzpigmente XI. Velumpigmente einiger Cortinarien der Untergattung *Telamonia* (Agaricales). Z. Naturforsch. **27 b**, 1009 (1972).

544. REPKE, D.B., D.T. LESLIE, and N.G. KISH: GLC-Mass Spectral Analysis of Fungal Metabolites. J. Pharm. Sci. **67**, 485 (1978).

545. ROBBINS, W.J., F. KAVANAGH, and A. HERVEY: Antibiotics from Basidiomycetes II. *Polyporus biformis*. Proc. Nat. Acad. Sci. (USA) **33**, 176 (1947).

546. ROBERGE, G., and P. BRASSARD: Reactions of Ketene Acetals 13. Synthesis of Contiguously Trihydroxylated Naphtho- and Anthraquinones. J. Org. Chem. **46**, 4161 (1981).

547. RONALD, R.C., J.M. LANSINGER, T.S. LILLIE, and C.J. WHEELER: Total Synthesis of Frustulosin and Aurocitrin. J. Org. Chem. **47**, 2541 (1982).

548. RØNNEBERG, H., G. BORCH, R. BUCHECKER, N. ARPIN, and S. LIAAEN-JENSEN: Chirality of Plectaniaxanthin. Phytochem. **21**, 2087 (1982).

549. ROSS, A.E., D.L. NAGEL, and B. TOTH: Evidence for the Occurrence and Formation of Diazonium Ions in the *Agaricus bisporus* Mushroom and its Extracts. J. Agric. Food Chem. **30**, 521 (1982), and references therein.

550. ROUGERON, J.: Les Réactions Macrochimiques. Bull. Soc. mycol. France **91**, 141 (1975).

551. RÜEDI, P.: HPLC – A Powerful Tool in Carotenoid Research. Pure and Appl. Chem. **57**, 793 (1985).

552. RÜEGG, R., M. MONTAVON, G. RYSER, G. SAUCY, U. SCHWIETER, and O. ISLER: Synthesen in der Carotinoid-Reihe 15. Synthesen in der β-Carotinal- und β-Carotinol-Reihe. Helv. Chim. Acta **42**, 854 (1959).

553. SAITO, K., A. KOMAMINE, and S.-I. HATANAKA: Biosynthesis of Stizolobic and Stizolobinic Acids in *Amanita pantherina*. Z. Naturforsch. **33 c**, 793 (1978).

554. SAITO, K., A. KOMAMINE, and S. SENOH: Further Studies on the Biosynthesis of Stizolobinic Acid and Stizolobic Acid in the Etiolated Seedlings of *Stizolobium hassjoo*. Z. Naturforsch. **31 c**, 15 (1976).

555. SAWADA, M.: Studies on Pigments in Fungi I. On the Distribution of Thelephoric Acid in Fungi. J. Japan Forest Soc. **34**, 110 (1952).

556. – Studies on Pigments in Fungi II. On the Distribution of Thelephoric Acid in Fungi, 2. J. Japan. Forest Soc. **40**, 195 (1958).

557. SCHÄFFER, J.: Die chemischen Reagentien in der Hand des Pilzbestimmers. Deutsche Blätter für Pilzkd. **1/2**, 1 (1943).

558. SCHAFSTALL, K.: Versuche zur Synthese des Atrochrysons. Dissertation, University of Bonn, 1983.

559. SCHELLING, H.: Pleurotin: Struktur und Beiträge zur Biogenese. Dissertation, ETH, Zürich, 1969.

560. SCHLUNEGGER, U.P., A. KUCHEN, and H. CLÉMENÇON: Mycelprodukte Höherer Pilze 1. Phenoxazin-Derivate in *Calocybe gambosa*. Helv. Chim. Acta **59**, 1383 (1976).

561. SCHMIDT, H.: Untersuchungen zur Synthese von Farbstoffen des Maronenröhrlings (*Xerocomus badius*) und verwandten Verbindungen. Diplomarbeit, University of Bonn, 1984.

562. SCHMITT, J.A.: Strobilomycetaceae, Boletaceae, Paxillaceae und Gomphidiaceae im Saarland, mit einer chemotaxonomischen Studie von 27 Arten. Z. Pilzkd. **36**, 77 (1970).

563. – Chemotaxonomische, morphologische und pflanzensoziologische Studien an mitteleuropäischen *Lactarius*-Arten der Sektion *Dapetes* Fr. (Blutreizker). Z. Pilzkd. **39**, 219 (1974).

564. SCHNEIDER, J.C.: Der Farbstoff der Fruchtkörper von *Sphaerobolus stellatus* (Thode) Pers. Planta **63**, 351 (1964).

565. SCHÖNBEIN, C.F.: On Ozone and Ozonic Actions in Mushrooms. Phil. Mag. **11**, 137 (1856).

566. SCHÖNESEIFFEN, J.: Synthese des Grundgerüsts der *Hypholoma*-Farbstoffe und Bestimmung ihrer relativen Konfiguration. Diplomarbeit, University of Bonn, 1980.

567. – Synthese der Indolalkaloide aus dem Blätterpilz *Cortinarius infractus*. Dissertation, University of Bonn, 1983.

568. SCHOLL, T.: Synthese eines optisch aktiven Bausteins für Atrochryson ausgehend von (–)-Chinasäure. Diplomarbeit, University of Bonn, 1980.

569. SCHRAMM, G.: Neue Farbstoffe aus *Inonotus hispidus* und anderen holzbewohnenden Pilzen. Diplomarbeit, University of Bonn, 1976.

570. SCHUMACHER, T., and K. HØILAND: Mushroom Poisoning Caused by Species of the Genus *Cortinarius* Fries. Arch. Toxicol. **53**, 87 (1983).

571. SCHWARZ, H., V. PASUPATHY, and W. STEGLICH: Pilzpigmente XXV. Massenspektrometrische Untersuchung polyisoprenoider Ansabenzochinone. Org. Mass Spectrom. **11**, 472 (1976).

572. SCHWIETER, U., R. RÜEGG, and O. ISLER: Synthesen in der Carotinoid-Reihe 21. Synthese von 2,2'-Diketo-spirilloxanthin (P 518) und 2,2'-Diketo-bacterioruberin. Helv. Chim. Acta **49**, 992 (1966).

573. SEIDMAN, M.M., A. TOMS, and J.M. WOOD: Influence of Side-chain Substituents on the Position of Cleavage of the Benzene Ring by *Pseudomonas fluorescens*. J. Bacteriol. **97**, 1192 (1969).

574. SETO, S.: Biosynthesis of Aspulvinones, Metabolites from *Aspergillus terreus*. Int. Congress Pure Appl. Chem. (Proc.) **4**, A 21 (1977).

575. SEZAKI, M., T. SASAKI, T. NAKAZAWA, U. TAKEDA, M. IWATA, T. WATANABE, M. KOYAMA, F. KAI, T. SHOMURA, and M. KOJIMA: A New Antibiotic SF-2370 Produced by *Actinomadura*. J. Antibiotics **38**, 1437 (1985).

576. SHAW, B.L., and M.C. WHITING: Researches on Polyenes I. The Synthesis of Corticrocin. J. Chem. Soc. (London) **1954**, 3217.

577. SHIBATA, S., E. MORISHITA, M. KANEDA, Y. KIMURA, M. TAKIDO, and S. TAKAHASHI: Chemical Studies on the Oriental Plant Drugs XX. The Constituents of *Cassia tora* L. The Structure of Torachrysone. Chem. Pharm. Bull. (Japan) **17**, 454 (1969).

578. SHIBATA, S., S. NATORI, and S.-I. UDAGAWA: List of Fungal Products. Tokyo: University of Tokyo Press. 1964.

579. SHILDNECK, P.R., and R. ADAMS: The Synthesis of Polyporic Acid and Atromentin Dimethyl Ether. J. Amer. Chem. Soc. **53**, 2373 (1931).

580. SHIMANO, T., K. TAKI, and K. GOTO: Constituents of *Polystictus versicolor*. Ann. Proc. Gifu Coll. Pharm. **3**, 43 (1953).

581. SILVERTON, J.V.: The Crystal and Molecular Structure of Leuco-thelephoric Acid Hexamethyl Ether. Acta Crystalogr. **B 29**, 293 (1973).

582. SINGER, R.: The Agaricales in Modern Taxonomy, 3rd ed. Vaduz: J. Cramer. 1975.

583. – Notes on Bolete Taxonomy III. Persoonia **11**, 269 (1981).

584. SINGH, P., and M. ANCHEL: Atromentic Acid from *Clitocybe illudens*. Phytochem. **10**, 3259 (1971).

585. SMRDEL, A.F., and M. GILL: Unpublished work.

586. SOLACOLU, T.: Sur les Matières Colorantes de Quelques Myxomycetès. Le Botaniste **24**, 107 (1932).

587. ŠORM, F., V. BENEŠOVÁ, and V. HEROUT: Über Terpene LIV. Über die Struktur des Lactarazulens und des Lactaroviolins. Collect. Czech. Chem. Comm. **19**, 357 (1954); Chem. Listy **47**, 1856 (1953).

588. ŠORM, F., V. BENEŠOVÁ, J. KRUPIČKA, V. ŠNEBERK, L. DOLEJŠ, V. HEROUT, and J. SICHER: The Structure of Lactaroviolin. Chem. and Ind. **1954**, 1511.

589. – – – – – – – On Terpenes LXV. The Constitution of Lactaroviolin. Synthesis of 1-Ethyl-4-methyl-7-*iso*propylazulene and 4-Ethyl-1-methyl-7-*iso*propylazulene. Collect. Czech. Chem. Comm. **20**, 227 (1955).

590. SPECKENBACH, F.: Verbindungen vom Nonaketidtyp aus Pilzen der Gattung *Cortinarius*. Dissertation, University of Bonn, 1986.

591. STAHLSCHMIDT, C.: Über eine neue in der Natur vorkommende organische Säure. Liebigs Ann. Chem. **187**, 177 (1877).

592. – Beiträge zur Kenntnis der Polyporsäure. Liebigs Ann. Chem. **195**, 365 (1879).

593. STEFFAN, B.: Untersuchung über Pilz- und Flechtenfarbstoffe, insbesondere die Hutfarbstoffe des Maronenröhrlings (*Xerocomus badius*). Dissertation, University of Bonn, 1981.

594. STEFFAN, B., and W. STEGLICH: Die Hutfarbstoffe des Maronenröhrlings (*Xerocomus badius*). Angew. Chem. **96**, 435 (1984); Angew. Chem. Int. Ed. Engl. **23**, 445 (1984).

595. – – Unpublished work.

596. STEGLICH, W.: The Biosynthesis of Fungal Quinones. Hoppe-Seyler's Z. physiol. Chem. **353**, 124 (1972).

597. – Pilzfarbstoffe. Chemie in unserer Zeit **9**, 117 (1975).

598. – Pigments of Higher Fungi (Macromycetes). In: Pigments in Plants (F.-C. CZYGAN, ed.), 2nd ed., p. 393–412. Stuttgart: G. Fischer. 1980.

599. – Unpublished work.

600. STEGLICH, W., O. ABDALLAH, and A.A. ALI: Unpublished work.

601. STEGLICH, W., R. ARNOLD, W. LÖSEL, and W. REININGER: Biosynthesis of Anthraquinone Pigments in *Dermocybe*. Chem. Commun. **1972**, 102.

602. STEGLICH, W., and V. AUSTEL: Die Struktur des Dermocybins und Dermoglaucins. Tetrahedron Letters **1966**, 3077.

603. STEGLICH, W., H. BESL, and A. PROX: Zur Struktur der Grevilline, neuartiger Pigmente aus dem Goldröhrling, *Suillus grevillei* (*Boletaceae*). Tetrahedron Letters **1972**, 4895.

604. STEGLICH, W., H. BESL, and K. ZIPFEL: Pilzpigmente XIX. Festlegung der Struktur von Pulvinsäuren mit Hilfe der NMR-Spektroskopie. Z. Naturforsch. **29b**, 96 (1974).

605. STEGLICH, W., and F. ESSER: L-3,4-Dihydroxy-phenylalanin aus *Strobilomyces floccopus*. Phytochem. **12**, 1817 (1973).

606. STEGLICH, W., F. ESSER, and I. PILS: Pilzpigmente VI. Helveticon, ein Benzochinon-Derivat vom Bovinon-Typ aus *Chroogomphus helveticus* und *Ch. rutilus*. Z. Naturforsch. **26b**, 336 (1971).

607. STEGLICH, W., W. FURTNER, and A. PROX: Neue Pulvinsäure-Derivate aus *Xerocomus chrysenteron* (Bull. ex St. Amans) Quél. und Untersuchungen zur Frage des Vorkommens von Anthrachinonpigmenten bei Boletaceen. Z. Naturforsch. **23b**, 1044 (1968).

608. – – – Pilzpigmente 3. Xerocomsäure und Gomphidsäure, zwei chemotaxonomisch interessante Pulvinsäure-Derivate aus *Gomphidius glutinosus* (Schff.) Fr. Z. Naturforsch. **24b**, 941 (1969).

609. – – – Pilzpigmente V. Variegatorubin, ein Oxidationsprodukt der Variegatsäure aus *Suillus piperatus* (Bull. ex Fr.) O. Kuntze und anderen Boletaceen. Z. Naturforsch. **25b**, 557 (1970).

610. STEGLICH, W., H.-T. HUPPERTZ, and B. STEFFAN: Ein einfacher Zugang zum Naphtho[1,8-*bc*]pyrandion-System. Angew. Chem. **97**, 716 (1985); Angew. Chem. Int. Ed. Engl. **24**, 711 (1985).

611. STEGLICH, W., and B. JENDRNY: Geogenin, ein neuartiger Pyronfarbstoff aus *Hydnellum geogenium*. Festschrift R. Singer, Sydowia, Beiheft VIII, 378 (1979).

612. STEGLICH, W., M. KLAAR, and W. FURTNER: (+)-Mitorubrin Derivatives from *Hypoxylon fragiforme*. Phytochem. **13**, 2874 (1974).

613. STEGLICH, W., L. KOPANSKI, M. WOLF, M. MOSER, and G. TEGTMEYER: Indolalkaloide aus dem Blätterpilz *Cortinarius infractus* (Agaricales). Tetrahedron Letters **25**, 2341 (1984).

614. STEGLICH, W., and W. LÖSEL: Bestimmung der Stellung von *O*-Substituenten bei 1,8-Dihydroxy-Anthrachinon-Derivaten mit Hilfe der NMR-Spektroskopie. Tetrahedron **25**, 4391 (1969).

615. – – Pilzpigmente X. Anthrachinon-glucoside aus *Dermocybe sanguinea* (Wulf. ex Fr.) Wünsche. Chem. Ber. **105**, 2928 (1972).

616. STEGLICH, W., W. LÖSEL, and V. AUSTEL: Pilzpigmente IV. Anthrachinon-Pigmente aus *Dermocybe sanguinea* (Wulf. ex Fr.) Wünsche und *D. semisanguinea* (Fr.). Chem. Ber. **102**, 4104 (1969).

617. STEGLICH, W., and B. OERTEL: Untersuchungen zur Konstitution und Verbreitung der Farbstoffe von *Cortinarius*, Untergattung *Phlegmacium* (Agaricales). Sydowia **37**, 284 (1984).

618. STEGLICH, W., I. PILS, and A. BRESINSKY: Pilzpigmente VII. Nachweis und chemotaxonomische Bedeutung von Pulvinsäuren in *Rhizopogon* (Gasteromycetes). Z. Naturforsch. **26 b**, 376 (1971).

619. STEGLICH, W., and R. PREUSS: L-3,4-Dihydroxyphenylalanine from Carpophores of *Hygrocybe conica* and *H. ovina*. Phytochem. **14**, 1119 (1975).

620. STEGLICH, W., and W. REININGER: A Synthesis of Endocrocin, Endocrocin-9-anthrone, and Related Compounds. Chem. Commun. **1970**, 178.

621. – – Pilzpigmente IX. Anthrachinon-Pigmente aus *Dermocybe cinnabarina* (Fr.) Wünsche. Chem. Ber. **105**, 2922 (1972).

622. STEGLICH, W., B. STEFFAN, and A. HOVENBITZER: Unpublished work.

623. STEGLICH, W., B. STEFFAN, L. KOPANSKI, and G. ECKHARDT: Indolfarbstoffe aus Fruchtkörpern des Schleimpilzes *Arcyria denudata*. Angew. Chem. **92**, 463 (1980); Angew. Chem. Int. Ed. Engl. **19**, 459 (1980).

624. STEGLICH, W., B. STEFFAN, K. STROECH, and M. WOLF: Pistillarin, ein charakteristischer Inhaltsstoff der Herkuleskeule (*Clavariadelphus pistillaris*) und einiger *Ramaria*-Arten (Basidiomycetes). Z. Naturforsch. **39 c**, 10 (1984).

625. STEGLICH, W., A. THILMANN, H. BESL, and A. BRESINSKY: Pilzpigmente 29. 2.5-Diarylcyclopentan-1.3-dione aus *Chamonixia caespitosa* (Basidiomycetes). Z. Naturforsch. **32 c**, 46 (1977).

626. STEGLICH, W., and E. TÖPFER-PETERSEN: Pilzpigmente XII. Phlegmacin und Anhydrophlegmacin, neuartige Farbstoffe aus dem Anisklumpfuß, *Cortinarius odorifer* (Agaricales). Z. Naturforsch. **27 b**, 1286 (1972).

627. – – Pilzpigmente XIV. Neue Pigmente vom Flavomannin-Typ aus *Cortinarius vitellinus* (Agaricales). Z. Naturforsch. **28 c**, 255 (1973).

628. STEGLICH, W., E. TÖPFER-PETERSEN, and I. PILS: Pilzpigmente XVI. Neue Phlegmacin-Derivate aus *Cortinarius percomis* (Agaricales). Z. Naturforsch. **28 c**, 354 (1973).

629. STEGLICH, W., E. TÖPFER-PETERSEN, W. REININGER, K. GLUCHOFF, and N. ARPIN: Isolation of Flavomannin-6,6′-dimethyl Ether and One of Its Racemates from Higher Fungi. Phytochem. **11**, 3299 (1972).

630. STEGLICH, W., and L. ZECHLIN: Pilzpigmente XXXII. 3-Methylriboflavin aus *Panellus serotinus* (Agaricales). Z. Naturforsch. **32 c**, 520 (1977).

631. – – Pilzpigmente 33. Synthese des Fomentariols. Eine neue Methode zur Darstellung von Zimtalkoholen. Chem. Ber. **111**, 3939 (1978).

632. STEYN, P.S., and R. VLEGGAAR: The Structure of Dihydrodeoxy-8-*epi*-austdiol and the Absolute Configuration of the Azaphilones. J. Chem. Soc. (London) Perkin Trans. I **1976**, 204.

633. STRAUB, O.: Key to Carotenoids: List of Natural Carotenoids. Basel: Birkhäuser. 1976.

634. STROECH, K.D.: Anellierungen mit der Tandem-Michael-Reaktion – Ein einfacher Weg zu chiralen Anthron- und Naphtholsystemen. Dissertation, University of Bonn, 1985.

635. STÜSSI, H., and D.M. RAST: The Biosynthesis and Possible Function of γ-Gluta-minyl-4-hydroxybenzene in *Agaricus bisporus*. Phytochem. **20**, 2347 (1981).
636. SULLIVAN, G., L.R. BRADY, and V.E. TYLER, JR.: Occurrence and Distribution of Terphenylquinones in *Hydnellum* Species. Lloydia **30**, 84 (1967).
637. SULLIVAN, G., R.D. GARRETT, and R.F. LENEHAN: Occurrence of Atromentin and Thelephoric Acid in Cultures of *Clitocybe subilludens*. J. Pharm. Sci. **60**, 1727 (1971).
638. SULLIVAN, G., and W.L. GUESS: Atromentin: A Smooth Muscle Stimulant in *Clito-cybe subilludens*. Lloydia **32**, 72 (1969).
639. SULLIVAN, G., and E.D. HENRY: Occurrence and Distribution of Phenoxazinone Pigments in the Genus *Pycnoporus*. J. Pharm. Sci. **60**, 1097 (1971).
640. SUNDSTRÖM, C., and E. SUNDSTRÖM: Mit Pilzen färben. Eine Fundgrube für Kunst-gewerbler, Pilzsammler und Naturfreunde. Zürich: Orell Füssli. 1984.
641. SUORTTI, T., and A. VON WRIGHT: Isolation of a Mutagenic Fraction from Aqueous Extracts of the Wild Edible Mushroom *Lactarius necator* (A Preliminary Note). J. Chromatogr. **255**, 529 (1983).
642. SUORTTI, T., A. VON WRIGHT, and A. KOSKINEN: Necatorin, A Highly Mutagenic Compound from *Lactarius necator*. Phytochem. **22**, 2873 (1983).
643. SWACK, N.S., and P.G. MILES: Conditions Affecting Growth and Indigotin Produc-tion by Strain 130 of *Schizophyllum commune*. Mycologia **52**, 574 (1960).
644. SWAN, G.A.: Isolation, Structure, and Synthesis of Hermidin, a Chromogen from *Mercurialis perennis* L. J. Chem. Soc. (London) Perkin Trans I **1985**, 1757.
645. SZENT-GYORGYI, A., R.H. CHUNG, M.J. BOYAJIAN, M. TISHLER, B.H. ARISON, E.F. SCHOENEWALDT, and J.J. WITTICK: Agaridoxin, A Mushroom Metabolite. Isolation, Structure, and Synthesis. J. Org. Chem. **41**, 1603 (1976).
646. TACHIBANA, S., and T. MURAKAMI: Isolation and Identification of Schizoflavins. Methods Enzymol. **66 E**, 333 (1980).
647. TAKAHASHI, S., S. KITANAKA, M. TAKIDO, U. SANKAWA, and S. SHIBATA: Phlegma-cins and Anhydrophlegmacinquinones: Dimeric Hydroanthracenes from Seedlings of *Cassia torosa*. Phytochem. **16**, 999 (1977).
648. TAKAHASHI, S., M. TAKIDO, U. SANKAWA, and S. SHIBATA: Germichrysone, a Hy-droanthracene Derivative from Seedlings of *Cassia torosa*. Phytochem. **15**, 1295 (1976).
649. TAKESHITA, H., and M. ANCHEL: Production of Oosporein and Its Leuco Form by Basidiomycete Species. Science **147**, 152 (1965).
650. TAKIDO, M., S. TAKAHASHI, K. MASUDA, and K. YASUKAWA: Torosachrysone, a New Tetrahydroanthracene Derivative from the Seeds of *Cassia torosa*. Lloydia **40**, 191 (1977).
651. TEBBETT, I.R., and B. CADDY: Mushroom Toxins of the Genus *Cortinarius*. Experien-tia **40**, 441 (1984).
652. THOEN, D.: *Cortinarius sanguineus* (Wulf.) Fr. et *Cortinarius cinnabarinus* Fr., Deux Cortinaires Souvent Confondus. Les Naturalistes Belges **51–54**, 148 (1970).
653. THÖRNER, W.: Ueber einen in einer Agaricus-Art vorkommenden chinonartigen Körper. Ber. dtsch. chem. Ges. **11**, 533 (1878).
654. – Ueber den im *Ag. atrotomentosus* vorkommenden chinonartigen Körper. Ber. dtsch. chem. Ges. **12**, 1630 (1879).
655. THOMSON, R.H.: Naturally Occurring Quinones, 2nd ed. London: Academic Press. 1971.
656. TIECCO, M., M. TINGOLI, L. TESTAFERRI, D. CHIANELLI, and E. WENKERT: Total Synthesis of Orellanine The Lethal Toxin of *Cortinarius orellanus* Fries Mushroom. Tetrahedron **42**, 1475 (1986).
657. TÖPFER-PETERSEN, E.: Untersuchungen über die Farbstoffe von Phlegmacien. Disser-tation, Technical University, Berlin, 1973.

658. Toth, B., D. Nagel, and A. Ross: Gastric Tumorigenesis by a Single Dose of 4-(Hydroxymethyl)benzenediazonium Ion of *Agaricus bisporus*. Br. J. Cancer **46**, 417 (1982), and references therein.

659. Towers, G.H.N.: Metabolism of Cinnamic Acid and its Derivatives in Basidiomycetes. In: Perspectives in Phytochemistry (J.B. Harborne and T. Swain, eds.), p. 179–191. London: Academic Press. 1969.

660. – Secondary Metabolites Derived Through the Shikimate-Chorismate Pathway. In: The Filamentous Fungi (J.E. Smith and D.R. Berry, eds.) Vol. II, p. 460–473. London: E. Arnold. 1976.

661. Towers, G.H.N., and P.V. Subba Rao: Degradation Metabolism of Phenylalanine, Tyrosine and DOPA. Recent Advances in Phytochemistry **4**, 1 (1972).

662. Towers, G.H.N., C.P. Vance, and A.M.D. Nambudiri: Photoregulation of Phenylpropanoid and Styrylpyrone Biosynthesis in *Polyporus hispidus*. Recent Advances in Phytochemistry **8**, 81 (1974).

663. Tsuji, H., N. Bando, T. Ogawa, and K. Sasaoka: Identification of Two Metabolites of Radioactive Shikimic Acid in *Agaricus bisporus*. Agric. Biol. Chem. **45**, 541 (1981).

664. – – – – Studies on the Biosynthesis of N-(γ-L-Glutamyl)-4-hydroxyaniline in *Agaricus bisporus*. Identification of the Position in Shikimic Acid at which the Amination Occurs. Biochim. Biophys. Acta **677**, 326 (1981).

665. Turian, G.: Identification du Lycopène dans le Réceptacle Fructifère du Gastéromycète *Anthurus aseroiformis*. C. R. hebd. séances Acad. Sci. **241**, 764 (1955).

666. – Identification des Caroténoides Majeurs de Quelques Champignons Ascomycètes et Basidiomycètes. Neurosporène chez *Cantharellus infundibuliformis*. Arch. Mikrobiol. **36**, 139 (1960).

667. Turner, W.B.: Fungal Metabolites. London: Academic Press. 1971.

668. Turner, W.B., and D.C. Aldridge: Fungal Metabolites II. London: Academic Press. 1983.

669. Tyler, V.E.: Chemotaxonomy in the Basidiomycetes. In: Evolution in the Higher Basidiomycetes (R.H. Petersen, ed.), p. 29–58. Knoxville: University of Tennessee Press. 1971.

670. Ueno, A., S. Fukushima, Y. Saiki, and T. Harada: Studies on the Components of *Phaeolus schweinitzii* (Fr.) Pat. Chem. Pharm. Bull. (Japan) **12**, 376 (1964).

671. Umezawa, H., T. Takeuchi, H. Iinuma, M. Ito, M. Ishizuka, Y. Kurakata, Y. Umeda, Y. Nakanishi, T. Nakamura, A. Obayashi, and O. Tanabe: New Antibiotic, Calvatic Acid. J. Antibiotics **28**, 87 (1975).

672. Vacheron, M.J., G. Michel, R. Guilluy, and N. Arpin: Étude par Spectrometrie de Masse d'un Caroténoide Isolé d'un Discomycete, *Plectania coccinea*. Phytochem. **8**, 897 (1969).

673. Valadon, L.R.G.: Rubixanthin in *Peziza* (*Aleuria*) *aurantia*. Biochem. J. **92**, 19P (1964).

674. – Carotenoids as Additional Taxonomic Characters in Fungi: A Review. Trans Br. mycol. Soc. **67**, 1 (1976).

675. Valadon, L.R.G., and R.S. Mummery: Taxonomic Significance of Carotenoids. Nature **217**, 1066 (1968).

676. – – A New Carotenoid from *Laetiporus sulphureus*. Ann. Bot. **33**, 879 (1969).

677. – – The First Epoxycarotenoid in a Fungus: Possible Taxonomic Marker for Cantharelloid Fungi. Trans. Br. mycol. Soc. **65**, 485 (1975).

678. Valadon, L.R.G., R.S. Mummery, G.W. van Eijk, H.J. Roeymans, and G. Britton: Taxonomic Implications of the Carotenoids of *Iodophanus carneus*. Trans. Br. mycol. Soc. **74**, 187 (1980).

679. Vance, C.P., A.M.D. Nambudiri, C.-K. Wat, and G.H.N. Towers: Isolation and

Properties of Hydroxycinnamate: CoA Ligase from *Polyporus hispidus*. Phytochem. **14**, 967 (1975).

680. VANCE, C.P., E.B. TREGUNNA, A.M.D. NAMBUDIRI, and G.H.N. TOWERS: Styryl-pyrone Biosynthesis in *Polyporus hispidus* I. Action Spectrum and Photoregulation of Pigment and Enzyme Formation. Biochim. Biophys. Acta **343**, 138 (1974).

681. VAN EIJK, G.W.: A Naphtho[1,2-*b*]furan Derivative from the Fungus *Roesleria pallida*. Phytochem. **10**, 3263 (1971).

682. VAN EIJK, G.W., and H.J. ROEYMANS: Gas-Liquid Chromatography of Trimethyl-silyl Ethers of Naturally Occurring Anthraquinones. J. Chromatogr. **124**, 66 (1976).

683. VAN WISSELINGH, C.: Über die Nachweisung und das Vorkommen von Carotinoiden in der Pflanze. Flora **107**, 371 (1914).

684. VITERBO, D., A. GASCO, A. SERAFINO, and V. MORTARINI: *p*-Carboxyphenylazoxy-cyanide-Dimethylsulfoxide: An Antibacterial and Antifungal Compound from *Calvatia lilacina*. Acta Crystallogr. **B 31**, 2151 (1975).

685. VOKÁČ, K., Z. SAMEK, V. HEROUT, and F. ŠORM: On Terpenes CCV. The Structure of Two Native Orange Substances from *Lactarius deliciosus* L. Collect. Czech. Chem. Comm. **35**, 1296 (1970).

686. VOLC, J., P. SEDMERA, K. ROY, V. ŠAŠEK, and J. VOKOUN: Two Antibiotic Benzoqui-none-Hydroquinone Pairs from the Pyrenomycete *Camarops microspora* (Karst.) Shear. Collect. Czech. Chem. Comm. **42**, 2957 (1977).

687. VOLHARD, J.: Synthese und Constitution der Vulpinsäure. Liebigs Ann. Chem. **282**, 1 (1894).

688. VON ARDENNE, R., H. DÖPP, H. MUSSO, and W. STEGLICH: Über das Vorkommen von Muscaflavin bei Hygrocyben (Agaricales) und seine Dihydroazepin-Struktur. Z. Naturforsch. **29 c**, 637 (1974).

689. VON ARDENNE, R., and W. STEGLICH: 1.2.4-Trihydroxybenzol, ein charakteristischer Inhaltsstoff von *Gomphidius* (Boletales). Z. Naturforsch. **29 c**, 446 (1974).

690. VON MASSOW, F.: A Combination of Thin-layer Chromatographic Systems for Anal-ysis of the Pigments from *Peniophora sanguinea* (Fr.) Bres. J. Chromatogr. **105**, 391 (1975).

691. – Incorporation of Phenylpropanes into Xylerythrin-Type Pigments in *Peniophora sanguinea*. Phytochem. **16**, 1695 (1977).

692. VON MASSOW, F., and D. HUBER: Ein neues DC-System zur Identifizierung mittelpo-larer bis lipophiler Boletales-Pigmente. J. Chromatogr. **138**, 232 (1977).

693. VON MASSOW, F., and H. NIMZ: Untersuchungen an Farbstoff-produzierenden Holz-pilzen. Zur Biosynthese der *Peniophora sanguinea*-Pigmente. Arch. Mikrobiol. **88**, 147 (1973).

694. VON MASSOW, F., and H.E. NOPPEL: Biosynthesis of the Xylerythrin-Type Pigments in *Peniophora sanguinea*. Phytochem. **16**, 1699 (1977).

695. VON MASSOW, F., and I. SCHMID: Untersuchungen an Farbstoff-produzierenden Holzpilzen II. Über das Auftreten eines Enzyms vom Laccase-Typ während der Bildung von Terphenylchinon-Farbstoffen. Arch. Mikrobiol. **92**, 353 (1973).

696. VRKOČ, J., M. BUDĚŠÍNSKÝ, and L. DOLEJŠ: Phenolic Meroterpenoids from the Basi-diomycete *Albatrellus ovinus*. Phytochem. **16**, 1409 (1977).

697. WAGER, H.: A Fluorescent Colouring Matter from *Leptonia incana* Gill. Trans. Brit. mycol. Soc. **6**, 158 (1917–1919).

698. WAKSMAN DE TORRES, N.: Personal communication.

699. WALTERS, M.B.: *Pholiota spectabilis*, a Hallucinogenic Fungus. Mycologia **57**, 837 (1965).

700. WANZLICK, H.-W.: Neuere Methoden der präparativen organischen Chemie IV. Synthesen mit naszierenden Chinonen. Angew. Chem. **76**, 313 (1964); Angew. Chem. Int. Ed. Engl. **3**, 401 (1964).

701. WANZLICK, H.-W., and U. JAHNKE: Synthesen mit naszierenden Chinonen V. Synthese des Xylerythrins. Chem. Ber. **101**, 3753 (1968).
702. WAT, C.-K., and G.H.N. TOWERS: Metabolism of the Aromatic Amino Acids by Fungi. Recent Advances in Phytochemistry **12**, 371 (1979), and references therein.
703. WATLING, R.: Chemical Tests in Agaricology. In: Methods in Microbiology (C. BOOTH, ed.) Vol. IV, p. 567–597. London: Academic Press. 1971.
704. WATSON, P.: Investigation of Pigments from *Russula* Spp. by Thin Layer Chromatography. Trans. Br. mycol. Soc. **49**, 11 (1966).
705. WEAVER, R.F., K.V. RAJAGOPALAN, and P. HANDLER: Mechanism of Action of a Respiratory Inhibitor from the Gill Tissue of the Sporulating Common Mushroom, *Agaricus bisporus*. Arch. Biochem. Biophys. **149**, 541 (1972).
706. WEAVER, R.F., K.V. RAJAGOPALAN, P. HANDLER, and W.L. BYRNE: γ-L-Glutaminyl-3,4-benzoquinone. Structural Studies and Enzymatic Synthesis. J. Biol. Chem. **246**, 2015 (1971).
707. WEAVER, R.F., K.V. RAJAGOPALAN, P. HANDLER, P. JEFFS, W.L. BYRNE, and D. ROSENTHAL: Isolation of γ-L-Glutaminyl-4-hydroxybenzene and γ-L-Glutaminyl-3,4-benzoquinone: A Natural Sulfhydryl Reagent, from Sporulating Gill Tissue of the Mushroom *Agaricus bisporus*. Proc. Nat. Acad. Sci. (USA) **67**, 1050 (1970).
708. WEAVER, R.F., K.V. RAJAGOPALAN, P. HANDLER, D. ROSENTHAL, and P.W. JEFFS: Isolation from the Mushroom *Agaricus bisporus* and Chemical Synthesis of γ-L-Glutaminyl-4-hydroxybenzene. J. Biol. Chem. **246**, 2010 (1971).
709. WEEDON, B.C.L.: Carotenoids and Related Compounds V. Synthesis of Corticrocin. J. Chem. Soc. (London) **1954**, 4168.
709a. WEINSTEIN, B., and D.N. BRATTESANI: Heterocyclic Compounds VII. The Synthesis of Cinnabarin. J. Heterocyclic Chem. **4**, 151 (1967).
710. WEINSTOCK, J., J.E. BLANK, H.-J. OH, and B.M. SUTTON: A Regiospecific Synthesis of Substituted Vulpinic Acids. J. Org. Chem. **44**, 673 (1979).
711. WEISGRABER, K., U. WEISS, G.W.A. MILNE, and J.V. SILVERTON: Hexamethyl Ether of Leuco-Thelephoric Acid from *Corticium caeruleum*. Phytochem. **11**, 2585 (1972).
712. WHALLEY, A.J.S., and G.N. GREENHALGH: Chemical Races of *Hypoxylon rubiginosum*. Trans. Br. mycol. Soc. **57**, 161 (1971).
713. WHALLEY, A.J.S., and M.A. WHALLEY: Stromal Pigments and Taxonomy of *Hypoxylon*. Mycopathologia **61**, 99 (1977).
714. WHALLEY, W.B., G. FERGUSON, W.C. MARSH, and R.J. RESTIVO: The Chemistry of Fungi LXVII. The Absolute Configuration of (+)-Sclerotiorin and of the Azaphilones. J. Chem. Soc. (London) Perkin Trans. I **1976**, 1366.
715. WIKHOLM, R.J., and H.W. MOORE: Dimethyl Sulfoxide-Acetic Anhydride Oxidative Rearrangements of Hydroxyterphenylquinones. A Possible Biosynthetic Model. J. Amer. Chem. Soc. **94**, 6152 (1972).
716. WILLSTAEDT, H.: Über die Farbstoffe des echten Reizkers (*Lactarius deliciosus* L.) I. Ber. dtsch. chem. Ges. **68**, 333 (1935).
717. – Über die Farbstoffe des echten Reizkers (*Lactarius deliciosus* L.) II. Ber. dtsch. chem. Ges. **69**, 997 (1936).
718. – Pilzfarbstoffe III. Über die Carotinoide einiger *Cantharellus*-Arten. Svensk Kem. Tidskr. **49**, 318 (1937).
719. – Zur Konstitution des Lactaroviolins. Atti X. Congr. internaz. Chimica Roma **3**, 390 (1939).
720. – Pilzfarbstoffe V. Zur Natur des Sauerstoffs im Lactaroviolin. Svensk Kem. Tidskr. **58**, 23 (1946).
721. – Pilzfarbstoffe VI. Über zwei neue, lipoidlösliche Farbstoffe aus dem echten Reizker (*Lactarius deliciosus* L.). Svensk Kem. Tidskr. **58**, 81 (1946).
722. – Lactaroviolin, ein gegen Tuberkelbazillen in vitro wirksames Antibiotikum. Svensk Kem. Tidskr. **58**, 306 (1946).

723. Wright, J.L.C., A.G. McInnes, D.G. Smith, and L.C. Vining: The Structure of Sepedonin, a Tropolone Metabolite of *Sepedonium chrysospermum* Fries. Canad. J. Chem. **48**, 2702 (1970).
724. Yamamoto, Y., K.-I. Nishimura, and N. Kiriyama: Studies on the Metabolic Products of *Aspergillus terreus* I. Metabolites of the Strain IFO 6123. Chem. Pharm. Bull. (Japan) **24**, 1853 (1976).
725. Zechlin, L., M. Wolf, W. Steglich, and T. Anke: Antibiotika aus Basidiomyceten XII. Cristatsäure, ein modifiziertes Farnesylphenol aus Fruchtkörpern von *Albatrellus cristatus*. Liebigs Ann. Chem. **1981**, 2099.
726. Zellner, J.: Chemie der Höheren Pilze. Leipzig: W. Engelmann. 1907.
727. – Zur Chemie der Höheren Pilze XI. Über *Lactarius scrobiculatus* Scop., *Hydnum ferrugineum* Fr., *Hydnum imbricatum* L. und *Polyporus applanatus* Wallr. Monatsh. Chem. **36**, 611 (1915).
728. – Zur Chemie der Höheren Pilze XIV. Über *Lactarius rufus* Scopol., *Lactarius pallidus* Pers. und *Polyporus hispidus* Fr. Monatsh. Chem. **41**, 443 (1920).
729. Zopf, W.: Ueber Pilzfarbstoffe I. Ueber das Vorkommen eines dem Gummiguttgelb ähnlichen Stoffes im Pilzreich. Bot. Ztg. **47**, 53 (1889).
730. – Ueber Pilzfarbstoffe II. Ueber Thelephoren-Farbstoffe. Bot. Ztg. **47**, 69 (1889).
731. – Ueber Pilzfarbstoffe III. Farbstoffe von *Trametes cinnabarina* (Jacq.). Bot. Ztg. **47**, 85 (1889).
732. – Die Pilze in morphologischer, physiologischer, biologischer und systematischer Beziehung. Breslau: E. Trewendt. 1890.
733. – Beiträge zur Physiologie und Morphologie. Die Niederen Organismen **2**, 17 (1892).

(*Received August 1, 1986*)

Pigments of Fungi (Macromycetes)

Index of Species (Macromycetes)

Author Index

Page numbers printed in *italics* refer to References

Subject Index

Composition: Universitätsdruckerei H Stürtz AG, D-8700 Würzburg
Printed by novographic, Ing. W. Schmid, A-1238 Wien

L. Oreland / B. A. Callingham (Eds.)

Monoamine Oxidase Enzymes:
Review and Overview

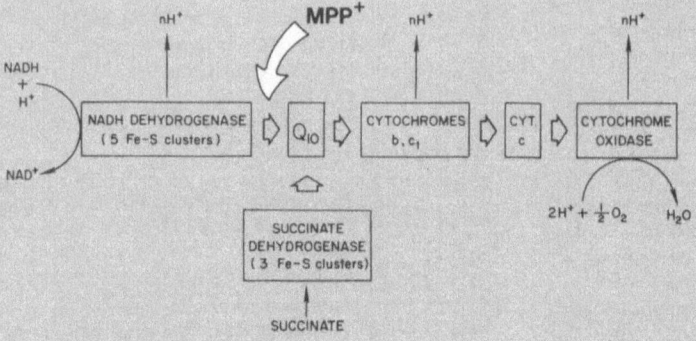

Journal of Neural Transmission / Supplementum 23

Contents: T. P. Singer: Perspectives in MAO: past, present, and future. – K. F. Tipton, Anne-Marie O'Carroll, and J. M. McCrodden: The catalytic behaviour of monoamine oxidase. – B. A. Callingham and M. A. Barrand: Some properties of semicarbazide-sensitive amine oxidases. – P. C. Waldmeier: Amine oxidases and their endogenous substrates. – A. J. Trevor, T. P. Singer, R. R. Ramsay, and N. Castagnoli, Jr.: Processing of MPTP by monoamine oxidases: implications for molecular toxicology. – U. Trendelenburg, L. Cassis, M. Grohmann, and A. Langeloh: The functional coupling of neuronal and extraneuronal transport with intracellular monoamine oxidase. – Margherita Strolin Benedetti and P. Dostert: Overview of the present state of MAO inhibitors. – J. H. Dowson: MAO inhibitors in mental disease: their current status.

Among the topics covered in the supplement is an historical overview about the knowledge of the biochemistry of Monoamine Oxidase with speculations about future prospects, a review of the present knowledge about the catalytic properties of Monoamine Oxidase, reviews of the biochemistry and pharmacology of the semicarbazide-sensitive Amine Oxidases and of the endogenous substrates both for Monoamine Oxidase and other Amine Oxidases. Reviews of the mechanism of action for the parkinsonism-inducing compound MPTP as well as of the present state of MAO-inhibitors and of MAO in relation to diseases and behaviour will also be included.

1987. 28 figures. VII, 138 pages. Soft cover DM 76,–, öS 530,– Reduced price for subscribers to the "Journal of Neural Transmission": Soft cover DM 68,40, öS 477,– ISBN 3-211-81985-1

SPRINGER-VERLAG WIEN NEW YORK

Moelkerbastei 5, A-1010 Wien · Heidelberger Platz 3, D-1000 Berlin 33 · 175 Fifth Avenue, New York, NY 10010, USA · 37-3, Hongo 3-chome, Bunkyo-ku, Tokyo 113, Japan

Fortschritte der Chemie organischer Naturstoffe

Progress in the Chemistry of Organic Natural Products

Volume 46:

1984. 7 figures. IX, 253 pages. Cloth DM 178,—, öS 1250,—.
ISBN 3-211-81804-9

Contents: O. Tanaka and R. Kasai: Saponins of Ginseng and Related Plants. — E. Fujita, M. Node: Diterpenoids of *Rabdosia* Species. — S. Johne: The Quinazoline Alkaloids.

Volume 45:

1984. 2 figures. VIII, 288 pages. Cloth DM 194,—, öS 1360,—.
ISBN 3-211-81755-7

Contents: D. A. H. Taylor: The Chemistry of the Limonoids from Meliaceae. — J. A. Elix, A. A. Whitton, and M. V. Sargent: Recent Progress in the Chemistry of Lichen Substances. — Y. Shimizu: Paralytic Shellfish Poisons.

Volume 44:

1983. 72 partly coloured figures. IX, 326 pages.
Cloth DM 208,—, öS 1460,—. ISBN 3-211-81754-9

Contents: F. J. Evans and S. E. Taylor: Pro-Inflammatory, Tumour-Promoting and Anti-Tumour Diterpenes of the Plant Families Euphorbiaceae and Thymelaeaceae. — A. Mondon and B. Epe: Bitter Principles of Cneoraceae. — S. Naylor, F. J. Hanke, L. V. Manes, and P. Crews: Chemical and Biological Aspects of Marine Monoterpenes. — J. G. Buchanan: The C-Nucleoside Antibiotics.

All Volumes and Cumulative Index 1—20 available

Price reduction for subscribers: 10%

Special reduced price (20% reduction) for the complete Series Vols. 1—51 incl. the Cumulative Index to Vols. 1—20

Springer-Verlag Wien New York

Mölkerbastei 5, A-1011 Wien
175 Fifth Avenue, New York, NY 10010, U.S.A.
Heidelberger Platz 3, D-1000 Berlin 33
37-3, Hongo 3-chome, Bunkyo-ku, Tokyo 113, Japan